Defining Mathematics Education
Presidential Yearbook Selections, 1926–2012

Seventy-fifth Yearbook

Francis (Skip) Fennell
Seventy-fifth Yearbook Editor
McDaniel College
Westminster, Maryland

William R. Speer
General Yearbook Editor
University of Nevada, Las Vegas
Las Vegas, Nevada

NATIONAL COUNCIL OF
TEACHERS OF MATHEMATICS

ISSN 0077-4103
ISBN 978-0-87353-738-4

The National Council of Teachers of Mathematics is the public voice of
mathematics education, supporting teachers to ensure equitable mathematics
learning of the highest quality for all students through vision, leadership,
professional development, and research.

Printed in the United States of America

Contents

Preface

T HE SEVENTY-FIFTH Yearbook of the National Council of Teachers of Mathematics (NCTM) is a celebration of, and reflection on, the history of this great resource for those interested in mathematics education. It was in 1926, six years after the founding of NCTM, that the first yearbook, *A General Survey of Progress in the Last Twenty-Five Years,* made its appearance. With an occasional "leave of absence" here and there in its earlier years, NCTM yearbooks have consistently focused special attention on a wide spectrum of topics of current or future interest to the professional community of mathematics educators. (For an overview of yearbook history, see the chart that begins on page 5.) This seventy-fifth volume will be the last in an illustrious series. Beginning next year, a new publication titled *Annual Perspectives in Mathematics Education (APME)* will be issued by NCTM to replace the production of annual yearbooks.

Defining Mathematics Education: Presidential Yearbook Selections, 1926–2012 begins with an introduction from Francis (Skip) Fennell, a past president of NCTM, followed by nineteen chapters, each drawn from a previous NCTM yearbook and each recommended by a past president of the Council. In his introduction, Skip describes in some detail the *process* used as this project took shape over the past two years. This Seventy-fifth Yearbook itself represents the *product* that we sincerely hope helps address some of the challenges that you face today. In rereading or, for many, in reading for the first time these chapters from the past, you'll find valuable resources and descriptions of a range of experiences that will help to refine and support your work in mathematics education. This final yearbook provides us with a vision of where we have been, helps us to reflect on where we are currently, and provides us with fodder to imagine what we can become.

Acknowledgments

One of the most challenging tasks in putting together a yearbook can normally be found in selecting from the multitude of manuscripts submitted in response to the call for entries. In the past, the selection and organization of these manuscripts eventually fell heavily on the general editor and the volume editor, but this final yearbook benefited from a different process. This time, that task became a shared exercise with past presidents of the Council. I am indebted to each participating past president for their deep expertise and wide body of experiences across the spectrum of mathematics education, elements that became the cornerstones for the success of this celebratory project. For serving as architects of the content selection for the 2013 yearbook, my special thanks go out (alphabetically!) to the following individuals for their demonstration of insight, knowledge, creativity, and dedication to the task placed before them.

Gail Burrill, 1996–98	Mary M. Lindquist, 1992–94
F. Joe Crosswhite, 1984–86	Johnny Lott, 2002–04
John Dossey, 1986–88	Jack Price, 1994–96
John Egsgard, 1976–78	Cathy Seeley, 2004–06
Francis (Skip) Fennell, 2006–08	J. Michael Shaughnessy, 2010–12
Shirley Frye, 1988–90	Max Sobel, 1980–82
Henry Kepner, 2008–10	Lee Stiff, 2000–02
Glenda Lappan, 1998–2000	Steve Willoughby, 1982–84

Of course, this yearbook would not have been possible without the outstanding contributions of the 2013 volume editor, Francis (Skip) Fennell of McDaniel College. Skip has been a keystone for us all on this project. His dedication and sense of duty and commitment to the Council and to this project—and to the profession of mathematics education—have helped in ways that are most visible in the quality of the final product.

I would be remiss if I failed to acknowledge the stellar support and guidance received from NCTM Headquarters. Several individuals worked behind the scenes to ensure the prompt and quality production of the volume. These include Joanne Hodges, senior director of publications; Myrna Jacobs, publications manager; Larry Shea, copy and production editor; Randy White, production manager; Ken Krehbiel, associate executive director for communications; Kathe Richardson, meeting planner; and Linda Cooper Foreman and Amy Roth McDuffie, Educational Materials Committee chairpersons during the production of this yearbook. These dedicated people helped along every step of the way and provided much direction and support on process and procedure. Many other unnamed members of the NCTM staff worked long and hard as well to bring this project to fruition. My sincere thanks go out to them all for their help and direction in making this Seventy-fifth Yearbook a reality.

William R. Speer
NCTM General Editor, 2011–2013 Yearbooks
Dean, College of Education, University of Nevada, Las Vegas

Introduction

N 1995, THE National Council of Teachers of Mathematics (NCTM) celebrated its seventy-fifth birthday. Now, eighteen years later, in 2013, the Council is publishing its seventy-fifth—and final—annual yearbook: *Defining Mathematics Education: Presidential Yearbook Selections, 1926–2012*.

This final yearbook is quite unique. Its chapters are actually chapters from many of the Council's previous seventy-four yearbooks (see pages 5–7 for a table of all NCTM yearbooks) and have been chosen for inclusion here by a number of NCTM's past presidents. Each past president was given the opportunity to review their collection of NCTM yearbooks and determine their three "favorite" yearbooks and chapters within these yearbooks. In this context, "favorites" were to be identified as those yearbook chapters that had a strong impact on the past presidents in some way, either in how they used the chapters or what they meant to them then or now. These yearbook and chapter selections were then organized, with consideration given to balance in their dates of publication, their mathematical and pedagogical topics, and their instructional level of emphasis. Striking this balance was a challenge, particularly because many of the "favorites" came from yearbooks of a time when it seemed like the average chapter length was fifty pages! Balance was also of concern as far as the number of selected "favorites" we could include. While some past presidents may have more than one of their favorites selected for this final yearbook, others have only one chapter that made the final cut. Such decisions were generally made on the pragmatic basis of the lengths of the available manuscripts.

One important note: The chapters presented in this yearbook are the actual chapters as they appeared originally, with no attempts to revise, rethink, or otherwise rewrite history.

The Past

For many readers, but particularly for those who have been NCTM members for three decades or even longer, our professional history as mathematics educators is linked, to an extent, to NCTM's annual yearbooks. As members "re-upped" their membership each year, it was just an expectation—a foregone conclusion—that one would also check the box that ensured receipt of the annual yearbook. Many longtime NCTM members have shelves of annual yearbooks of varying sizes, with distinctly interesting covers and, yes, titles, which we will discuss shortly. Analyzing and reviewing all the NCTM yearbooks has been a fascinating experience. I admit to not having all seventy-four yearbooks, so reviewing all the "favorites" and trying to determine the flow of this final yearbook was interesting, challenging, and, frankly, fun. I learned a lot. For example, I am guessing many readers will not know the following:

- NCTM's first yearbook was published in 1926.
- The Council did not publish a yearbook every year.
- Two yearbooks were published in 1969 and another two in 1970.
- W. D. Reeve served as yearbook editor nineteen times!
- Five yearbooks included either companion booklets or an accompanying CD.
- Some early yearbook chapters were more than seventy-five pages long.
- Only one NCTM yearbook includes *equity* and *diversity* in its title (though these subjects are topics of discussion in particular chapters in other yearbooks).

Ah, but there's so much more. During the review of the yearbooks and chapters within yearbooks, I was struck by some of the yearbook titles. Consider the following:

- *A General Survey of Progress in the Last Twenty-Five Years* (1926)
- *Approximate Computation* (1937)
- *The Slow Learner in Mathematics* (1972)
- *Estimation and Mental Computation* (1986)

The Council's First Yearbook from 1926 was selected by two NCTM past presidents as a favorite. I found the historical nature of its title to be interesting. I also found it very interesting that the 1937 and 1986 Yearbooks, while close to fifty years apart, both attempted to address the elusive challenges of number sense through their related focus on approximation and estimation and mental mathematics. I was actually shocked by the 1972 title *The Slow Learner in Mathematics*. I find this title to be grossly inappropriate and demeaning and a departure from the Council's position statements on equity and intervention. That said, a well-meaning committee found this to be acceptable forty years ago, so my concern over its inappropriateness won't change Council history. Setting its title aside, once I took a closer look at the 1972 Yearbook I found many of its chapters to be amazingly useful today—particularly at a time when challenges related to intervention regularly confront elementary and middle school mathematics teachers. My review also identified three yearbooks that made use of the word *modern*: *Mathematics in Modern Life* (1931), *The Place of Mathematics in Modern Education* (1936), and *Insights into Modern Mathematics* (1957).

What about yearbook topics? I began, and never finished, sorting all the titles by mathematical and pedagogical topics. If you wish, you can take a look at the table of yearbooks on pages 5–7 and perhaps make your own review. I did note that three yearbooks focus on algebra, while four concentrate on geometry. Measurement can be claimed as an emphasis of yearbooks published in 1976 and 2003, as well as those in 1947 (on surveying instruments) and 1948 (on the metric system). Data analysis and probability are the focus of the 1981 and 2006 yearbooks. Those struck me as the most obvious content domain considerations. I haven't forgotten number, and one could certainly consider the following titles as number-focused yearbooks: *Approximate Computation* (1937); *Instruction in Arithmetic* (1960); *Developing Computational Skills* (1978); *Estimation and Mental Computation* (1986);

The Teaching and Learning of Algorithms in School Mathematics (1998); and *Making Sense of Fractions, Ratios, and Proportions* (2002), but such a topic-driven analysis presents a challenge. Many more generally titled yearbooks include a variety of content-related chapters. Clearly it's difficult to identify a specific content or pedagogical focus for some of the yearbooks, but I think it is fair to claim that most, if not all, of them engage both mathematical content and pedagogical content knowledge.

As we consider the influence and impact of this diamond anniversary of NCTM yearbooks, take the time to reflect back on your own use of the Council's yearbooks. How did you use them? How do you use them today? What was it about making sure you had all or most of these annual Council publications? As some readers know, NCTM yearbooks have historically taken years to plan. While the process has changed over the years, the Council's Educational Materials Committee (EMC) has generally determined the focus of a yearbook several years in advance of its publication. An editor and an editorial panel were then formally appointed by the NCTM president, the call for papers developed and published, and manuscript submission deadlines determined. Manuscripts rolled in and were reviewed. Keep in mind that for most of the years of yearbook publication beginning in 1926, online submission and reviewing were unheard of. Each yearbook's editor and editorial panel were assisted by the general editor of the yearbook, a person typically appointed by the NCTM president and charged with assisting the yearbook editor and editorial panel in meeting guidelines and deadlines.

The goal of all this work, of course, was to have each yearbook available prior to the NCTM Annual Meeting. Many members looked forward to picking up their yearbook at the Meeting. One could certainly suggest that, at least to some extent, the popularity of the NCTM yearbooks was a result of the fact that the Council had, for much of its existence, produced relatively few publications. More experienced members may remember the days of only two NCTM journals: *The Mathematics Teacher,* which, in 2006, achieved its 100th anniversary and is actually older than the Council itself; and *The Arithmetic Teacher,* which was first published in 1954 and was renamed *Teaching Children Mathematics* in 1994.

The Present

And now? What publications are available for NCTM members? In a word, lots! The current journals of the Council, besides *Teaching Children Mathematics* and *Mathematics Teacher,* are *Journal for Research in Mathematics Education* (founded in 1970), *Mathematics Teaching in the Middle School* (founded in 1994), and the recently introduced, in 2012, *Mathematics Teacher Educator.* NCTM has also developed a publication voice that now includes hundreds of books and online opportunities for teachers, including the very popular Illuminations website and its accompanying reservoir of classroom activities. While this brief presentation is not a full picture of all that NCTM provides publications-wise, it is clear that the Council has moved way beyond two monthly journals and its annual yearbook.

As we consider all that NCTM does today, we should not forget how its annual yearbook has made a professional impact on mathematics educators at every level for seventy-five years. This special anniversary yearbook is merely one way to celebrate, acknowledge, and affirm this impact. As you turn to its chapters, which consider various perspectives in the field of mathematics education, you will note that the chapters are arranged in chronological order of the actual publication date of the yearbook chapters selected. Consider that this range of selections includes yearbooks published from the First Yearbook of 1926 through the Sixty-ninth Yearbook, published in 2007. Also note that the chapters represent eighteen different yearbooks, with the selections ranging from historical topics in mathematics to chapters on algebra, fractions, computational algorithms, geometry, problem solving, and so much more. The introductory statement at the beginning of each "favorite" chapter gives the reader a personal glimpse of the chapter's impact on the past president who selected it. These brief vignettes may also stimulate ideas for thinking about how you may use a particular chapter yourself.

The Future

The inaugural volume of *Annual Perspectives in Mathematics Education* (*APME*) will replace the annual yearbook in 2014. *APME* will be indexed as an annual publication. Members will continue to be able to check a Standing Order Plan as they have for the yearbook, but for now, enjoy this final NCTM annual yearbook—it's special.

Francis (Skip) Fennell
NCTM Volume Editor, 2013 Yearbook
L. Stanley Bowlsbey Professor of Education
and Graduate and Professional Studies,
McDaniel College
NCTM President, 2006–08

Seventy-Five NCTM Yearbooks

Number	Year of publication	Title	Editor
75th	2013	*Defining Mathematics Education: Presidential Yearbook Selections, 1926–2012*	Francis (Skip) Fennell
74th	2012	*Professional Collaborations in Mathematics Teaching and Learning: Seeking Success for All*	Jennifer M. Bay-Williams
73rd	2011	*Motivation and Disposition: Pathways to Learning Mathematics*	Daniel J. Brahier
72nd	2010	*Mathematics Curriculum: Issues, Trends, and Future Directions*	Barbara J. Reys and Robert E. Reys
71st	2009	*Understanding Geometry for a Changing World*	Timothy V. Craine
70th	2008	*Algebra and Algebraic Thinking in School Mathematics*	Carole E. Greenes
69th	2007	*The Learning of Mathematics*	W. Gary Martin and Marilyn E. Strutchens
68th	2006	*Thinking and Reasoning with Data and Chance*	Gail F. Burrill
67th	2005	*Technology-Supported Mathematics Learning Environments*	William J. Masalski
66th	2004	*Perspectives on the Teaching of Mathematics*	Rheta N. Rubenstein
65th	2003	*Learning and Teaching Measurement*	Douglas H. Clements
64th	2002	*Making Sense of Fractions, Ratios, and Proportions*	Bonnie Litwiller
63rd	2001	*The Roles of Representation in School Mathematics*	Albert A. Cuoco
62nd	2000	*Learning Mathematics for a New Century*	Maurice J. Burke
61st	1999	*Developing Mathematical Reasoning in Grades K–12*	Lee V. Stiff
60th	1998	*The Teaching and Learning of Algorithms in School Mathematics*	Lorna J. Morrow
59th	1997	*Multicultural and Gender Equity in the Mathematics Classroom: The Gift of Diversity*	Janet Trentacosta
58th	1996	*Communication in Mathematics, K–12 and Beyond*	Portia C. Elliott
57th	1995	*Connecting Mathematics across the Curriculum*	Peggy A. House
56th	1994	*Professional Development for Teachers of Mathematics*	Douglas B. Aichele
55th	1993	*Assessment in the Mathematics Classroom*	Norman L. Webb
54th	1992	*Calculators in Mathematics Education*	James T. Fey
53rd	1991	*Discrete Mathematics across the Curriculum, K–12*	Margaret J. Kenney

Number	Year of publication	Title	Editor
52nd	1990	*Teaching and Learning Mathematics in the 1990s*	Thomas J. Cooney
51st	1989	*New Directions for Elementary School Mathematics*	Paul R. Trafton
50th	1988	*The Ideas of Algebra, K–12*	Arthur F. Coxford
49th	1987	*Learning and Teaching Geometry, K–12*	Mary Montgomery Lindquist
48th	1986	*Estimation and Mental Computation*	Harold L. Schoen
47th	1985	*The Secondary School Mathematics Curriculum*	Christian R. Hirsch
46th	1984	*Computers in Mathematics Education*	Viggo P. Hansen
45th	1983	*The Agenda in Action*	Gwen Shufelt
44th	1982	*Mathematics for the Middle Grades (5–9)*	Linda Silvey
43rd	1981	*Teaching Statistics and Probability*	Albert P. Shulte
42nd	1980	*Problem Solving in School Mathematics*	Stephen Krulik
41st	1979	*Applications in School Mathematics*	Sidney Sharron
40th	1978	*Developing Computational Skills*	Marilyn N. Suydam
39th	1977	*Organizing for Mathematics Instruction*	F. Joe Crosswhite
38th	1976	*Measurement in School Mathematics*	Doyal Nelson
37th	1975	*Mathematics Learning in Early Childhood*	Joseph N. Payne
36th	1973	*Geometry in the Mathematics Curriculum*	Kenneth B. Henderson
35th	1972	*The Slow Learner in Mathematics*	William C. Lowry
34th	1973	*Instructional Aids in Mathematics*	Emil J. Berger
33rd	1970	*The Teaching of Secondary School Mathematics*	Myron F. Rosskopf
32nd	1970	*A History of Mathematics Education in the United States and Canada*	Phillip S. Jones
31st	1969	*Historical Topics for the Mathematics Classroom*	Arthur E. Hallerberg
30th	1969	*More Topics in Mathematics for Elementary School Teachers*	Edwin F. Beckenbach
29th	1964	*Topics in Mathematics for Elementary School Teachers*	Lenore John
28th	1963	*Enrichment Mathematics for High School*	Julius H. Hlavaty
27th	1963	*Enrichment Mathematics for the Grades*	Julius H. Hlavaty
26th	1961	*Evaluation in Mathematics*	Donovan A. Johnson
25th	1960	*Instruction in Arithmetic*	Foster E. Grossnickle

Number	Year of publication	Title	Editor
24th	1959	*The Growth of Mathematical Ideas: Grades K–12*	Phillip S. Jones
23rd	1957	*Insights into Modern Mathematics*	F. Lynwood Wren
22nd	1954	*Emerging Practices in Mathematics Education*	John R. Clark
21st	1953	*The Learning of Mathematics: Its Theory and Practice*	Howard F. Fehr
20th	1948	*The Metric System of Weights and Measures*	W. D. Reeve
19th	1947	*Surveying Instruments: Their History and Classroom Use*	Edmond R. Kiely (author); W. D. Reeve (editor)
18th	1945	*Multi-Sensory Aids in the Teaching of Mathematics*	W. D. Reeve
17th	1942	*A Sourcebook of Mathematical Application*	W. D. Reeve
16th	1941	*Arithmetic in General Education*	W. D. Reeve
15th	1940	*The Place of Mathematics in Secondary Education*	W. D. Reeve
14th	1939	*The Training of Mathematics Teachers for Secondary Schools in England and Wales and in the United States*	Ivan Stewart Turner (author); W. D. Reeve (editor)
13th	1938	*The Nature of Proof*	Harold P. Fawcett (author); W. D. Reeve (editor)
12th	1937	*Approximate Computation*	Aaron Bakst (author); W. D. Reeve (editor)
11th	1936	*The Place of Mathematics in Modern Education*	W. D. Reeve
10th	1935	*The Teaching of Arithmetic*	W. D. Reeve
9th	1934	*Relational and Functional Thinking in Mathematics*	Herbert Russell Hamley (author); W. D. Reeve (editor)
8th	1933	*The Teaching of Mathematics in the Secondary School*	W. D. Reeve
7th	1932	*The Teaching of Algebra*	W. D. Reeve
6th	1931	*Mathematics in Modern Life*	W. D. Reeve
5th	1930	*The Teaching of Geometry*	W. D. Reeve
4th	1929	*Significant Changes and Trends in the Teaching of Mathematics Throughout the World Since 1910*	W. D. Reeve
3rd	1928	*Selected Topics in the Teaching of Mathematics*	J. R. Clark and W. D. Reeve
2nd	1927	*Curriculum Problems in Teaching Mathematics*	W. D. Reeve
1st	1926	*A General Survey of Progress in the Last Twenty-Five Years*	Raleigh Schorling

1

From *A General Survey of Progress in the Last Twenty-Five Years*, NCTM's First Yearbook (1926)

We begin this final NCTM Yearbook with an entry suggested by Johnny Lott, who served as NCTM president from 2002 to 2004. Johnny recommended **"On the Foundations of Mathematics**," which was originally published in 1926 as the second chapter in NCTM's First Yearbook. This selection was also recommended by Steve Willoughby, NCTM president from 1982 to 1984, and it was previously reprinted in 1967 in *The Mathematics Teacher*.

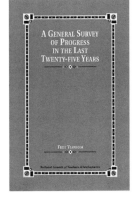

This chapter was actually the address delivered by Eliakim Hastings Moore, the president of the American Mathematical Society (AMS), at that organization's ninth annual meeting on December 29, 1902. Moore's speech was his final presentation of his AMS presidency. As many readers know, NCTM presidents also prepare and deliver an annual presentation during each of the two years of their presidency.

In his address, Moore proposes a series of reforms related to mathematics learning and teaching. Readers interested in the history of policy and reform in mathematics education will find this address, now more than 110 years old, to be of special value. In particular, take note of Moore's attention to pedagogy and to what readers today will consider as K–12 mathematics teaching and learning.

On the Foundations of Mathematics

Eliakim Hastings Moore

THE American Mathematical Society gives its retiring president the privilege of speaking on whatever he may have at heart.[1] Accordingly, this afternoon I propose to consider with you some matters of importance—indeed, perhaps of fundamental importance—in the development of mathematics in this country; and it will duly appear in what non-technical sense I am speaking 'On the Foundations of Mathematics.'

A View

Abstract Mathematics.—The notion within a given domain of defining the objects of consideration rather by a body of properties than by particular expressions or intuitions is as old as mathematics itself. And yet the central importance of the notion appeared only during the last century—in a host of researches on special theories and on the foundations of geometry and analysis. Thus has arisen the general point of view of what may be called *abstract mathematics*. One comes in touch with the literature very conveniently by the mediation of Peano's *Revue des Mathématiques*. The Italian school of Peano and the *Formulaire Mathématique*, published in connection with the *Revue*, are devoted to the codification in Peano's symbolic language of the principal mathematical theories, and to researches on abstract mathematics. General interest in abstract mathematics was aroused by Hilbert's Gauss-Weber Festschrift of 1899: 'Ueber die Grundlagen der Geometrie,' a memoir rich in results and suggestive in methods; I refer to the reviews by Sommer,[2] Poincaré,[3] Halsted,[4] Hedrick[5] and Veblen.[6]

We have as a basal science logic, and as depending upon it the special deductive sciences which involve undefined symbols and whose propositions are not all capable of

---

[1] Presidential address delivered before The American Mathematical Society at its ninth annual meeting, December 29, 1902. Reprinted here from *Science*, N. S., Vol. XVII, pp. 401–416, March 13, 1903; punctuation and other matters of style are as in the original.

[2] *Bull. Amer. Math. Soc.* (2), vol. 6 (1900), p. 287.

[3] *Bull. Sciences Mathém.*, vol. 26 (1902), p. 249.

[4] *The Open Court*, September, 1902.

[5] *Bull. Amer. Math. Soc.* (2), vol. 9 (1902), p. 158.

[6] *The Monist*, January, 1903.

proof. The symbols denote either classes of elements or relations amongst elements. In any such science one may choose in various ways the system of undefined symbols and the system of undemonstrated or primitive propositions, or postulates. Every proposition follows from the postulates by a finite number of logical steps. A careful statement of the fundamental generalities is given by Padoa in a paper[7] before the Paris Congress of Philosophy, 1900.

Having in mind a definite system of undefined symbols and a definite system of postulates, we have first of all the notion of the compatibility of these postulates; that is, that it is impossible to prove by a finite number of logical steps the simultaneous validity of a statement and its contradictory statement; in the next place, the question of the independence of the postulate or the irreducibility of the system of postulates; that is, that no postulate is provable from the remaining postulates. Padoa introduces the notion of the irreducibility of the system of undefined symbols. A system of undefined symbols is said to be reducible if for one of the symbols, X, it is possible to establish, as a logical consequence of the assumption of the validity of the postulates, a nominal or symbolic definition of the form $X=A$, where in the expression A there enter only the undefined symbols distinct from X. For the purpose of practical application, it seems to be desirable to modify the definition so as to call the system of undefined symbols reducible if there is a nominal definition $X=A$ of one of them X in terms of the others such that in any interpretation of the science the postulates retain their validity when instead of the initial interpretation of the symbol X there is placed the interpretation of the symbol A of that symbol. If the system of symbols is reducible in the sense of the original definition it is in the sense of the new definition, but not necessarily conversely, as appears for instance from the following example, occurring in the foundations of geometry.

Hilbert uses the following undefined symbols: 'point,' 'line,' 'plane,' 'incidence' of point and line, 'incidence' of point and plane, 'between,' and 'congruent.' Now it is possible to give for the symbol 'plane' a symbolic definition in terms of the other undefined symbols—for instance, a plane is a certain class of points (as Peano showed in 1892), or again, a plane is a certain class of lines; while the notion 'incidence' of point and plane receives convenient definition. It is apparent from the fact that these definitions may be given in these two ways that Hilbert's system of undefined symbols is not in Padoa's sense irreducible, at least, in so far as the symbols 'plane,' 'incidence' of point and plane are concerned—while it is equally clear that these symbols are in the abstract geometry superfluous.

In his dissertation on Euclidean geometry, Mr. Veblen, following the example of Pasch and Peano, takes as undefined symbols 'point' and 'between,' or 'point' and 'segment.' In terms of these two symbols alone he expresses a set of independent fundamental postulates of Euclidean geometry, in the first place developing the projective geometry, and then as to congruence relating himself to the point of view of Klein in his 'Erlangen

---

[7] 'Essai d'une théorie algébrique des nombres entiers, précédé d'une introduction logique à une théorie déductive quelconque.' *Bibliothèque du Congrès International de Philosophie*, vol. 3, p. 309.

Programm,' whereby the group of movements of Euclidean geometry enters as a certain subgroup of the group of collineations of projective geometry. Here arises an interesting question as to the sense in which the undefined symbol 'congruence' is superfluous in the Euclidean geometry based upon the symbols 'point,' 'between.' One sees at once that a definition of 'congruence' involves parametric points in its expression, while on the other hand a definition of the system of all 'planes,' that is, of the general concept 'plane,' involves no such parametric elements. But, again, just as there exist distinct definitions of 'congruence,' owing to a variation of the parametric points, so there exist distinct definitions of the general concept 'plane,' as was indicated a moment ago. One has the feeling that the state of affairs must be as follows: In any interpretation of, say, Hilbert's symbols, wherein the postulates of Hilbert are valid, every valid statement which does not involve the symbol 'plane' in direct connection with the general logical symbol ($=$) of symbolic definition, remains valid when we modify it in accordance with either of the definitions of 'plane' previously referred to. On the other hand, this state of affairs does not hold for the symbol 'congruence.' The proof of the former statement would seem to involve fundamental logical niceties.

The compatibility and the independence of the postulates of a system of postulates of a special deductive science have been up to this time always made to depend upon the self-consistency of some other deductive science; for instance, geometry depends thus upon analysis, or analysis upon geometry. The fundamental and still unsolved problem in this direction is that of the direct proof of the compatibility of the postulates of arithmetic, or of the real number system of analysis. (To the society this morning Dr. Huntington exhibited two sets of independent postulates for this real number system.) This is the second of the twenty-three problems listed by Hilbert in his address before the Paris Mathematical Congress of 1900.

The Italian writers on abstract mathematics for the most part make use of Peano's symbolism. One may be tempted to feel that this symbolism is not an essential part of their work. It is only right to state, however, that the symbolism is not difficult to learn, and that there is testimony to the effect that the symbolism is actually of great value to the investigator in removing from attention the concrete connotations of the ordinary terms of general and mathematical language. But of course the essential difficulties are not to be obviated by the use of any symbolism, however delicate.

Indeed the question arises whether the abstract mathematicians in making precise the metes and bounds of logic and the special deductive sciences are not losing sight of the evolutionary character of all life-processes, whether in the individual or in the race. Certainly the logicians do not consider their science as something now fixed. All science, logic and mathematics included, is a function of the epoch—all science, in its ideals as well as in its achievements. Thus with Hilbert let a special deductive or mathematical science be based upon a finite number of symbols related by a finite number of compatible postulates, every proposition of the science being deducible by a finite number of logical steps from the postulates. The content of this conception is far from absolute. It involves what presuppositions as to general logic? What is a finite number? In what sense is a postulate—for example, that any two distinct points determine a line—a single postulate? What are the permissible logical steps of deduction? Would the usual syllogistic steps of formal logic suffice? Would they suffice even with the aid of the principle of mathematical induction, in which Poincaré finds[8] the essential synthetic element of mathematical argumentation the basis of that generality without which there would be no science? In what sense is mathematical induction a single logical step of deduction?

One has then the feeling that the carrying out in an absolute sense of the program of the abstract mathematicians will be found impossible. At the same time, one recognizes the importance attaching to the effort to do precisely this thing. The requirement of rigor tends toward essential simplicity of procedure, as Hilbert has insisted in his Paris address, and the remark applies to this question of mathematical logic and its abstract expression.

Pure and Applied Mathematics.—In the ultimate analysis for any epoch, we have general logic, the mathematical sciences,[9] that is, all special formally and abstractly deductive self-consistent sciences, and the natural sciences, which are inductive and informally deductive. While this classification may be satisfactory as an ideal one, it fails to recognize the fact that in mathematical research one by no means confines himself to processes which are mathematical according to this definition; and if this is true with respect to the research of professional mathematicians, how much more is it true with respect to the study, which should throughout be conducted in the spirit of research, on the part of students of mathematics in the elementary schools and colleges and universities. I refer to the articles[10] of Poincaré on the role of intuition and logic in mathematical research and education.

It is apparent that this ideal classification can be made by the devotee of science only when he has reached a considerable degree of scientific maturity, that perhaps it would fail to appeal to non-mathematical experts, and that it does not accord with the definitions given by practical work in mathematicians. Indeed, the attitude of practical mathematicians toward this whole subject of abstract mathematics, and especially the symbolic form of abstract mathematics, is not unlike that of the practical physicist toward the whole subject of theoretic mathematics, and in turn not unlike that of the practical engineer toward the whole subject of theoretical physics and mathematics. Furthermore, every one understands that many of the most important advances of pure mathematics have arisen in connection with investigations originating in the domain of natural phenomena.

Practically then it would seem desirable for the interests of science in general that

---

[8] 'Sur la nature du raisonnement mathématique,' *Revue de Métaphysique et de Morale*, vol. 2 (1894), pp. 371–384.

[9] Of which none is at present known to exist.

[10] 'La logique et l'intuition dans la science mathématique et dans l'enseignement,' *L'Enseignement Mathématique*, vol. 1 (1899), pp. 157–162. 'Du role de l'intuition et de la logique en mathématiques,' *Compte Rendu du Deuxième Congrès International des Mathématiciens, Paris* (1900), 1902, pp. 115–130. 'Sur les rapports de l'analyse pure et de la physique mathématique,' Conference, Zurich, 1897; *Acta Mathematica*, vol. 21, p. 238.

there should be a strong body of men thoroughly possessed of the scientific method in both its inductive and its deductive forms. We are confronted with the questions: What is science? What is the scientific method? What are the relations between the mathematical and the natural scientific processes of thought? As to these questions, I refer to articles and addresses of Poincaré,[11] Boltzmann[12] and Burkhardt,[13] and to Mach's 'Science of Mechanics' and Pearson's 'Grammar of Science.'

Without elaboration of metaphysical or psychological details, it is sufficient to refer to the thought that the individual, as confronted with the world of phenomena in his effort to obtain control over this world, is gradually forced to appreciate a knowledge of the usual coexistences and sequences of phenomena, and that science arises as the body of formulas serving to epitomize or summarize conveniently these usual coexistences and sequences. These formulas are of the nature of more or less exact descriptions of phenomena; they are not of the nature of explanations. Of all the relations entering into the formulas of science, the fundamental mathematical notions of number and measure and form were among the earliest, and pure mathematics in its ordinary acceptation may be understood to be the systematic development of the properties of these notions, in accordance with conditions prescribed by physical phenomena. Arithmetic and geometry, closely united in mensuration and trigonometry, early reached a high degree of advancement. But after the development of the generalizing literal notation of algebra, and largely in response to the insistent demands of mechanics, astronomy and physics, the seventeenth century, binding together arithmetic and geometry infinitely more closely, created analytic geometry and the infinitesimal calculus, those mighty methods of research whose application to all branches of the theoretical and practical physical sciences so fundamentally characterizes the civilization of to-day.

The eighteenth century was devoted to the development of the powers of these new instruments in all directions. While this development continued during the nineteenth century, the dominant note of the nineteenth century was that of critical reorganization of the foundations of pure mathematics, so that, for instance, the majestic edifice of analysis was seen to rest upon the arithmetic of positive integers alone. This reorganization and the consequent course of development of pure mathematics were independent of the question of the application of mathematics to the sister sciences. There has thus arisen a chasm between pure mathematics and applied mathematics. There have not been lacking, however, influences making toward the bridging of this chasm; one thinks especially of the whole influence of Klein in Germany and of the École Polytechnic in

[11] In addition to those already cited: 'On the Foundations of Geometry,' *The Monist*, vol. 9, October, 1898, pp. 1–43. 'Sur les principes de la mécanique,' *Bibliothèque du Congrès Intemational de Philosophie*, vol. 3, pp. 457–494.

[12] 'Ueber die Methoden der theoretischen Physik,' *Dyck's Katalog mathematischer und mathematisch-physikalischer Modelle, Apparate und Instrumente*, pp. 89–98, Munich, 1892.

[13] 'Mathematisches und Naturwissenschaftliches Denken,' *Jahresbericht der Deutschen Math.-Ver.*, vol. 11 (1902), pp. 49–57.

15

France. As a basis of union of the pure mathematicians and the applied mathematicians, Klein has throughout emphasized the importance of a clear understanding of the relations between those two parts of mathematics which are conveniently called 'mathematics of precision' and 'mathematics of approximation,' and I refer especially to his latest work of this character, 'Anwendung der Differential und Integral-Rechnung auf Geometrie: Eine Revision der Principien' (Göttingen, summer semester, 1901, Teubner, 1902). This course of lectures is designed to present particular applications of the general notions of Klein, and furthermore, it is in continuation of the discussion between Pringsheim and Klein and others, as to the desirable character of lectures on mathematics in the universities of Germany.

Elementary Mathematics.—This separation between pure mathematics and applied mathematics is grievous even in the domain of elementary mathematics. In witness, in the first place: The workers in physics, chemistry and engineering need more practical mathematics; and numerous textbooks, in particular, on calculus, have recently been written from the point of view of these allied subjects. I refer to the works by Nernst and Schoenflies,[14] Lorentz,[15] Perry,[16] and Mellor,[17] and to a book on the very elements of mathematics now in preparation by Oliver Lodge.

In the second place, I dare say you are all familiar with the surprisingly vigorous and effective agitation with respect to the teaching of elementary mathematics which is at present in progress in England, largely under the direction of John Perry, professor of mechanics and mathematics of the Royal College of Science, London, and chairman of the Board of Examiners of the Board of Education in the subjects of engineering, including practical plane and solid geometry, applied mechanics, practical mathematics, in addition to more technical subjects, and in this capacity in charge of the education of some hundred thousand apprentices in English night schools. The section on Education of the British Association had its first session at the Glasgow meeting, 1901, and the session was devoted to the consideration, in connection with the section on Mathematics and Physics, of the question of the pedagogy of mathematics, and Perry opened the discussion by a paper on 'The Teaching of Mathematics.' A strong committee under the chairmanship of Professor Forsyth, of Cambridge, was appointed 'to report upon improvements that might be effected in the teaching of mathematics, in the first instance, in the teaching of elementary mathematics, and upon such means as they think likely to effect such improvements.' The paper of Perry, with the discussion of the subject at Glasgow, and additions including the report

---

[14] Nernst und Schoenflies, 'Einführung in die mathematische Behandlung der Naturwissenschaften' (Munich and Leipsic, 1895); the basis of Young and Linebarger's 'Elements of Differential and Integral Calculus,' New York, 1900.

[15] Lorentz, 'Lehrbuch der Differential- und Integralrechnung,' Leipsic, 1900.

[16] Perry, 'Calculus for Engineers' (second edition, London, E. Arnold, 1897); German translation by Fricke (Teubner, 1902). Cf. also the citations given later on.

[17] Mellor, 'Higher Mathematics for Students of Chemistry and Physics, with special reference to Practical Work,' Longmans, Green & Co., 1902, pp. xxi + 543.

of the committee as presented to the British Association at its Belfast meeting, September, 1902, are collected in a small volume, 'Discussion on the Teaching of Mathematics,' edited by Professor Perry (Macmillan, second edition, 1902).[18]

One should consult the books of Perry, 'Practical Mathematics,'[19] 'Applied Mechanics,'[20] 'Calculus for Engineers'[21] and 'England's Neglect of Science,'[22] and his address[23] on 'The Education of Engineers'—and furthermore the files from 1899 on of the English journals, *Nature, School World, Journal of Education* and *Mathematical Gazette.*

One important purpose of the English agitation is to relieve the English secondary school teachers from the burden of a too precise examination system, imposed by the great examining bodies; in particular, to relieve them from the need of retaining Euclid as the sole authority in geometry, at any rate with respect to the sequence of propositions. Similar efforts made in England about thirty years ago were unsuccessful. Apparently the forces operating since that time have just now broken forth into successful activity; for the report of the British Association committee was distinctly favorable, in a conservative sense, to the idea of reform, and already noteworthy initial changes have been made in the regulations for the secondary examinations by the examination syndicates of the universities of Oxford, Cambridge, and London.

The reader will find the literature of this English movement very interesting and suggestive. For instance, in a letter to *Nature* (vol. 65, p. 484, March 27, 1902) Perry mildly apologizes for having to do with the movement whose immediate results are likely to be merely slight reforms, instead of thoroughgoing reforms called for in his pronouncements and justified by his marked success during over twenty years as a teacher of practical mathematics. He asserts that the orthodox logical sequence in mathematics is not the only possible one; that, on the contrary, a more logical sequence than the orthodox one (because one more possible of comprehension by students) is based upon the notions underlying the infinitesimal calculus taken as axioms; for instance, that a map may be drawn to scale; the notions underlying the many uses of squared paper; that decimals may be dealt with as ordinary numbers. He asserts as essential that the boy should be *familiar* (by way of experiment, illustration, measurement, and by every possible means) with the ideas to which he applies his logic; and moreover that he should be thoroughly *interested* in the subject studied; and he closes with this peroration:

> "'Great God! I'd rather be
> A pagan, suckled in a creed outworn.'

---

[18] Cf. also 'Report on the Teaching of Elementary Mathematics issued by the Mathematical Association,' G. Bell & Sons, London, 1902.

[19] Published for the Board of Education by Eyre and Spottiswoode, London, 1899.

[20] D. Van Nostrand Co., New York, 1898.

[21] Second edition, London, E. Arnold, 1897.

[22] T. Fisher Unwin, London, 1900.

[23] In opening the discussion of the sections on Engineering and on Education at the Belfast, 1902, meeting of the British Association; published in *Science*, November 14, 1902.

I would rather be utterly ignorant of all the wonderful literature and science of the last twenty-four centuries, even of the wonderful achievements of the last fifty years, than not to have the sense that our whole system of so-called education is as degrading to literature and philosophy as it is to English boys and men."

As a pure mathematician, I hold as the most important suggestion of the English movement the suggestion of Perry's, just cited, that by emphasizing steadily the practical sides of mathematics, that is, arithmetic computations, mechanical drawing and graphical methods generally, in continuous relation with problems of physics and chemistry and engineering, it would be possible to give very young students a great body of the essential notions of trigonometry, analytic geometry, and the calculus. This is accomplished, on the one hand, by the increase of attention and comprehension obtained by connecting the abstract mathematics with subjects which are naturally of interest to the boy, so that, for instance, all the results obtained by theoretic process are capable of check by laboratory process, and, on the other hand, a diminution of emphasis on the systematic and formal sides of the instruction in mathematics. Undoubtedly many mathematicians will feel that this decrease of emphasis will result in much, if not irreparable, injury to the interests of mathematics. But I am inclined to think that the mathematician with the catholic attitude of an adherent of science, in general (and at any rate with respect to the problems of the pedagogy of elementary mathematics there would seem to be no other rational attitude) will see that the boy will be learning to make practical use in his scientific investigations—to be sure, in a naive and elementary way—of the finest mathematical tools which the centuries have forged; that under skillful guidance he will learn to be interested not merely in the achievements of the tools, but in the theory of the tools themselves, and that thus he will ultimately have a feeling towards his mathematics extremely different from that which is now met with only too frequently—a feeling that mathematics is indeed itself a fundamental reality of the domain of thought, and not merely a matter of symbols and arbitrary rules and conventions.

The American Mathematical Society.—The American Mathematical Society has, naturally, interested itself chiefly in promoting the interests of research in mathematics. It has, however, recognized that those interests are closely bound up with the interests of education in mathematics. I refer in particular to the valuable work done by the committee appointed, with the authorization of the Council, by the Chicago section of the society, to represent mathematics in connection with Dr. Nightingale's committee of 1899 of the National Educational Association in the formulation of standard curricula for high schools and academies, and to the fact that two committees are now at work, one appointed in December, 1901, by the Chicago Section, to formulate the desirable conditions for the granting, by institutions of the Mississippi valley, of the degree of Master of Arts for work in mathematics, and the other appointed by the society at its last summer meeting to cooperate with similar committees of the National Educational Association and of the Society for the Promotion of Engineering Education, in formulating standard definitions of requirements in mathematical subjects for admission to colleges and technological schools; and furthermore I refer to the fact that (although not formally) the society has made a valuable contribution to the interests of secondary education in that the College Entrance Examination Board has as its secretary the principal founder of the society. I have accordingly felt at liberty to bring to the attention of the society these matters of pedagogy of elementary mathematics, and I do so with the firm conviction that it would be possible for the society, by giving still more attention to these matters, to further most effectively the highest interests of mathematics in this country.

A Vision

An Invitation.—The pure mathematicians are invited to determine how mathematics is regarded by the world at large, including their colleges of other science departments and the students of elementary mathematics, and to ask themselves whether by modification of method and attitude they may not win for it the very high position in general esteem and appreciative interest which it assuredly deserves.

This general invitation and the preceding summary view invoke this vision of the future of elementary mathematics in this country.

The Pedagogy of Elementary Mathematics.—We survey the pedagogy of elementary mathematics in the primary schools, in the secondary schools and in the junior colleges (the lower collegiate years). It is, however, understood that there is a movement for the enlargement of the strong secondary schools, by the addition of the two years of junior college work and by the absorption of the last two or three grades of the primary schools, into institutions more of the type of the German gymnasia and the French lycée[24]; in favor of this movement there are strong arguments, and among them this, that in such institutions, especially if closely related to strong colleges or universities, the mathematical reforms may the more easily be carried out.

The fundamental problem is that of *the unification of pure and applied mathematics.* If we recognize the branching implied by the very terms 'pure,' 'applied,' we have to do with a special case of *the correlation of different subjects* of the curriculum, a central problem in the domain of pedagogy from the time of Herbart on. In this case, however, the fundamental solution is to be found rather by way of indirection—by arranging the curriculum so that throughout the domain of elementary mathematics the branching be not recognized.

The Primary Schools.—Would it not be possible for the children in the grades to be trained in power of observation and experiment and reflection and deduction so that always their mathematics should be directly connected with matters of thoroughly concrete character? The response is immediate that this is being done to-day in the kindergartens and in the better elementary schools. I understand that serious difficulties arise with children of from nine to twelve years of age, who are no longer contented with the simple,

---

[24] As to the mathematics of these institutions, one may consult the book on 'The Teaching of Mathematics in the Higher School of Prussia' (New York, Longmans, Green & Co., 1900) by Professor Young, and the article (*Bulletin Amer. Math. Soc.* (2), vol. 6, p. 225) by Professor Pierpont.

concrete methods of earlier years and who, nevertheless, are unable to appreciate the more abstract methods of the later years. These difficulties, some say, are to be met by allowing the mathematics to enter only implicitly in connection with the other subjects of the curriculum. But rather the material and methods of the mathematics should be enriched and vitalized. In particular, the grade teachers must make wiser use of the foundations furnished by the kindergarten. The drawing and the paper folding must lead on directly to systematic study of intuitional geometry,[25] including the construction of models and the elements of mechanical drawing, with simple exercises in geometrical reasoning. The geometry must be closely connected with the numerical and literal arithmetic. The cross-grooved tables of the kindergarten furnish an especially important type of connection, viz., a conventional graphical depiction of any phenomenon in which one magnitude depends upon another. These tables and the similar cross-section blackboards and paper must enter largely into all the mathematics of the grades. The children are to be taught to represent, according to the usual conventions, various familiar and interesting phenomena and to study the properties of the phenomena in the pictures: to know, for example, what concrete meaning attaches to the fact that a graph curve at a certain point is going down or going up or is horizontal. Thus the problems of percentage—interest, etc.—have their depiction in straight or broken line graphs.

The Secondary Schools.—Pending the reform of the primary schools, the secondary schools must advance independently. In these schools at present, according to one type of arrangement, we find algebra in the first year, plane geometry in the second, physics in the third, and the more difficult parts of algebra and solid geometry, with review of all the mathematics in the fourth.

Engineers[26] tell us that in the schools algebra is taught in one water-tight compartment, geometry in another, and physics in another, and that the student learns to appreciate (if ever) only very late the absolutely close connection between these different subjects, and then, if he credits the fraternity of teachers with knowing the closeness of this relation, he blames them most heartily for their unaccountably stupid way of teaching him. If we contrast this state of affairs with the state of affairs in the solid four years'

---

[25] Here I refer to the very suggestive paper of Benchara Branford, entitled 'Measurement and Simple Surveying. An Experiment in the Teaching of Elementary Geometry' to a small class of beginners of about ten years of age (*Journal of Education*, London, the first part appearing in the number for August, 1899.)

[26] Why is it that one of the sanest and best-informed scientific men living, a man not himself an engineer, can charge mathematicians with killing off every engineering school on which they can lay hands? Why do engineers so strongly urge that the mathematical courses in engineering schools be given by practical engineers? And why can a reviewer of 'Some Recent Books of Mechanics' write with truth: "The students' previous training in algebra, geometry, trigonometry, analytic geometry and calculus as it is generally taught has been necessarily quite formal. These mighty algorithms of formal mathematics must be learned so that they can be applied with readiness and precision. But with mechanics comes the application of these algorithms, and formal, do-by-rote methods, though often possible, yield no results of permanent value. How to elicit and cultivate thought is now of primary importance"? (E. B. Wilson, *Bulletin Amer. Math. Soc.*, October, 1902.) But is it conceivable that in any part of the education of the student the problem of eliciting and cultivating thought should not be of primary importance?

course in Latin, I think we are forced to the conclusion that the organization of instruction in Latin is much more perfect than that of the instruction in mathematics.

The following question arises: *Would it not be possible to organize the algebra, geometry and physics of the secondary school into a thoroughly coherent four years' course,* comparable in strength and closeness of structure with the four years' course in Latin? (Here under physics I include astronomy, and the more mathematical and physical parts of physiography.) It would seem desirable that, just as the systematic development of theoretical mathematics is deferred to a later period, likewise much of theoretical physics might well be deferred. Let the physics also be made thoroughly practical. At any rate, so far as the instruction of boys is concerned, the course should certainly have its character largely determined by the conditions which would be imposed by engineers. What kind of two or three years' course in mathematics and physics would a thoroughly trained engineer give to boys in the secondary school? Let this body of material postulated by the engineer serve as the basis of the four years' course. Let the instruction in the course, however, be given by men who have received expert training in mathematics and physics as well as in engineering and let the instruction be so organized that with the development of the boy, in appreciation of the practical relations, shall come simultaneously his development in the direction of theoretical physics and theoretical mathematics.

Perry is quite right in insisting that it is scientifically legitimate in the pedagogy of elementary mathematics to take a large body of basal principles instead of a small body and to build the edifice upon the larger body for the earlier years, reserving for the later years the philosophic criticism of the basis itself and the reduction of the basal system.

To consider the subject of geometry in all briefness: with the understanding that proper emphasis is laid upon all the concrete sides of the subject, and that furthermore from the beginning exercises in informal deduction[27] are introduced increasingly frequently, when it comes to the beginning of the more formal deductive geometry why should not the students be directed each for himself to set forth a body of geometric fundamental principles, on which he would proceed to erect his geometric edifice? This method would be thoroughly practical and at the same time thoroughly scientific. The various students would have different systems of axioms, and the discussions thus arising naturally would make clearer in the minds of all precisely what are the functions of the axioms in the theory of geometry. The students would omit very many of the axioms, which to them would go without saying. The teacher would do well not to undertake to make the system of axioms thoroughly complete in the abstract sense. "Sufficient unto the day is the precision thereof." The student would very probably wish to take for granted all the ordinary properties of measurement and of motion, and would be ready at once to

---

[27] In an article shortly to appear in the *Educational Review*, on 'The Psychological and the Logical in the Teaching of Geometry,' Professor John Dewey, calling attention to the evolutionary character of the education of an individual, insists that there should be no abrupt transition from the introductory, intuitional geometry to the systematic, demonstrative geometry.

accept the geometrical implications of coordinate geometry. He could then be brought with extreme ease to the consideration of fundamental notions of the calculus as treated concretely, and he would find those notions delightfully real and powerful, whether in the domain of mathematics or of physics or of chemistry.

To be sure, as Study has well insisted, for a thorough comprehension of even the elementary parts of Euclidean geometry the non-Euclidean geometries are absolutely essential. But the teacher is teaching the subject for the benefit of the students, and it must be admitted that beginners in the study of demonstrative geometry can not appreciate the very delicate considerations involved in the thoroughly abstract science. Indeed, one may conjecture that, had it not been for the brilliant success of Euclid in his effort to organize into a formally deductive system the geometric treasures of his times, the advent of the reign of science in the modern sense might not have been so long deferred. Shall we then hold that in the schools the teaching of demonstrative geometry should be reformed in such a way as to take account of all the wonderful discoveries which have been made—many even recently— in the domain of abstract geometry? And should similar reforms be made in the treatment of arithmetic and algebra? To make reforms of this kind, would it not be to repeat more gloriously the error of those followers of Euclid who fixed his 'Elements' as a textbook for elementary instruction in geometry for over two thousand years? Every one agrees that professional mathematicians should certainly take account of these great developments in the technical foundations of mathematics, and that ample provision should be made for instruction in these matters; and on reflection, every one agrees further that this provision should be reserved for the later collegiate and university years.

The Laboratory Method.—This program of reform calls for the development of a thoroughgoing laboratory system of instruction in mathematics and physics, a principal purpose being as far as possible to develop on the part of every student the true spirit of research, and an appreciation, practical as well as theoretic, of the fundamental methods of science.

In connection with what has already been said, the general suggestions I now add will, I hope, be found of use when one enters upon the questions of detail involved in the organization of the course.

As the world of phenomena receives attention by the individual, the phenomena are described both graphically and in terms of number and measure; the number and measure relations of the phenomena enter fundamentally into the graphical depiction, and furthermore the graphical depiction of the phenomena serves powerfully to illuminate the relations of number and measure. This is the fundamental scientific point of view. Here under the term graphical depiction I include representation by models.

To provide for the needs of laboratory instruction, there should be regularly assigned to the subject two periods, counting as one period in the curriculum.

As to the possibility of effecting this unification of mathematics and physics in the secondary schools, objection will be made by some teachers that it is impossible to do well more than one thing at a time. This pedagogic principle of concentration is undoubtedly sound. One must, however, learn how to apply it wisely. For instance, in the physical laboratory it is undesirable to introduce experiments which teach the use of the calipers or of the vernier or of the slide rule. Instead of such uninteresting experiments of limited purpose, the students should be directed to extremely interesting problems which involve the use of these instruments, and thus be led to learn to use the instruments as a matter of course, and not as a matter of difficulty. Just so the smaller elements of mathematical routine can be made to attach themselves to laboratory problems, arousing and retaining the interest of the students. Again, everything exists in its relations to other things, and in teaching the one thing the teacher must illuminate these relations.

Every result of importance should be obtained by at least two distinct methods, and every result of especial importance by two essentially distinct methods. This is possible in mathematics and the physical sciences, and thus the student is made thoroughly independent of all authority.

All results should be checked, if only qualitatively or if only 'to the first significant figure.' In setting problems in practical mathematics (arithmetical computation or geometrical construction) the teacher should indicate the amount or percentage of error permitted in the final result. If this amount of percentage is chosen conveniently in the different examples, the student will be led to the general notion of closer and closer approximation to a perfectly definite result, and thus in a practical way to the fundamental notions of the theory of limits and of irrational numbers. Thus, for instance, uniformity of convergence can be taught beautifully in connection with the concrete notion of area under a monotonic curve between two ordinates, by a figure due to Newton, while the interest will be still greater if in the diagram area stands for work done by an engine.

The teacher should lead up to an important theorem gradually in such a way that the precise meaning of the statement in question, and further, the practical, i. e., computational or graphical or experimental—truth of the theorem is fully appreciated; and, furthermore, the importance of the theorem is understood, and, indeed, the desire for the formal proof of the proposition is awakened, before the formal proof itself is developed. Indeed, in most cases, much of the proof should be secured by the research work of the students themselves.

Some hold that absolutely individual instruction is the ideal, and a laboratory method has sometimes been used for the purpose of attaining this ideal. The laboratory method has as one of its elements of great value the flexibility which permits students to be handled as individuals or in groups. The instructor utilizes all the experience and insight of the whole body of students. He arranges it so that the students consider that they are studying the subject itself, and not the words, either printed or oral, of any authority on the subject. And in this study they should be in the closest cooperation with one another and with their instructor, who is in a desirable sense one of them and their leader. Instructors may fear that the brighter students will suffer if encouraged to spend time in cooperation with those not so bright. But experience shows that just as every teacher learns by teaching, so even the brightest students will find themselves much the gainers for this cooperation with their colleagues.

In agreement with Perry, it would seem possible that the student might be brought into vital relation with the fundamental elements of trigonometry, analytic geometry and the calculus, on condition that the whole treatment in its origin is and in its development remains closely associated with thoroughly concrete phenomena. With the momentum of such practical education in the methods of research in the secondary school, the college students would be ready to proceed rapidly and deeply in any direction in which their personal interests might lead them. In particular, for instance, one might expect to find effective interest on the part of college students in the most formal abstract mathematics.

For all students who are intending to take a full secondary school course in preparation for colleges or technological schools, I am convinced that the laboratory method of instruction in mathematics and physics, which has been briefly suggested, is the best method of instruction—for students in general, and for students expecting to specialize in pure mathematics, in pure physics, in mathematical physics or astronomy, or in any branch of engineering.

Evolution, not Revolution.—In contemplating this reform of secondary school instruction we must be careful to remember that it is to be accomplished as an evolution from the present system, and not as a revolution of that system. Even under the present organization of the curriculum the teachers will find that much improvement can be made by closer cooperation one with another; by the introduction, so far as possible, of the laboratory two-period plan; and in any event by the introduction of laboratory methods; laboratory record books, cross-section paper, computational and graphical methods in general, including the use of colored inks and chalks; the cooperation of students; and by laying emphasis upon the comprehension of propositions rather than upon the exhibition of comprehension.

The Junior Colleges.—Just as secondary schools should begin to reform without waiting for the improvement of the primary schools, so the elementary collegiate courses should be modified at once without waiting for the reform of the secondary schools. And naturally, in the initial period of reform, the education in each higher domain will involve many elements which later on will be transferred to a lower domain.

Further, by the introduction into the junior colleges of the laboratory method of instruction it will be possible for the colleges and universities to take up a duty which for the most part has been neglected in this country. For, although we have normal schools and other training schools for those who expect to teach in the grades, little attention has as yet been given to the training of those who will become secondary school teachers. The better secondary schools of today are securing the services of college graduates who have devoted special attention to the subjects which they intend to teach, and as time goes on the positions in these schools will as a rule be filled (as in France and Germany) by those who have supplemented their college course by several years of university work. Here these college and university graduates proceed at once to their work in the secondary schools. Now in the laboratory courses of the junior college, let those students of the senior college and graduate school who are to go into the teaching career be given training in the pedagogy of mathematics according to the laboratory system; for such a student the laboratory would be a laboratory in the pedagogy of mathematics; that is, he would be a colleague-assistant of the instructor. By this arrangement, the laboratory instruction of the colleges would be strengthened at the same time that well equipped teachers would be prepared for work in the secondary schools.

The Freedom of the Secondary Schools.—The secondary schools are everywhere preparing students for colleges and technological schools, and whether the requirements of those institutions are expressed by way of examination of students or by way of the conditions for the accrediting of schools or teachers, the requirements must be met by the secondary schools. The stronger secondary school teachers too often find themselves shackled by the specific requirements imposed by local or college authorities. Teaching must become more of a profession. And this implies not only that the teacher must be better trained for his career, but also in his career he be given with greater freedom greater responsibility. To this end closer relations should be established between the teachers of the colleges and those of the secondary schools; standing provisions should be made for conferences as to improvement of the secondary school curricula and in the collegiate admission requirements; and the leading secondary school teachers should be steadily encouraged to devise and try out plans looking in any way toward improvement.

Thus the proposed four years' laboratory course in mathematics and physics will come into existence by way of evolution. In a large secondary school, the strongest teachers, finding the project desirable and feasible, will establish such a course alongside the present series of disconnected courses and as time goes on their success will in the first place stimulate their colleagues to radical improvements of method under the present organization and finally to a complete reorganization of the courses in mathematics and physics.

The American Mathematical Society.—Do you not feel with me that the American Mathematical Society, as the organic representative of the highest interests of mathematics in this country, should be directly related with the movement of reform? And, to this end, that the society, enlarging its membership by the introduction of a large body of the strongest teachers of mathematics in the secondary schools, should give continuous attention to the question of improvement of education in mathematics, in institutions of all grades? That there is need for the careful consideration of such questions by the united body of experts, there is no doubt whatever, whether or not the general suggestions which we have been considering this afternoon turn out to be desirable and practicable. In case the question of pedagogy does come to be an active one, the society might readily hold its meetings in two divisions—a division of research and a division of pedagogy.

Furthermore, there is evident need of a national organization having its center of gravity in the whole body of science instructors in the secondary schools; and those of us interested in these questions will naturally relate ourselves also to this organization. It is possible that the newly formed Central Association of Physics Teachers may be the nucleus of such an organization.

Conclusion

The successful execution of the reforms proposed would seem to be of fundamental importance to the development of mathematics in this country. I urge that individuals and organizations proceed to the consideration of the general question of reform with all the related questions of detail. Undoubtedly in many parts of the country improvements in organization and methods of instruction in mathematics have been made these last years. All persons who are, or may become, actively interested in this movement of reform should in some way unite themselves, in order that the plans and the experience, whether of success or failure, of one may be immediately made available in the guidance of his colleagues.

I may refer to the centers of activity with which I am acquainted. Miss Edith Long, in charge of the Department of Mathematics in the Lincoln (Neb.) High School, reports upon the experience of several years in the correlation of algebra, geometry and physics, in the October, 1902, number of the *Educational Review*. In the Lewis Institute of Chicago, Professor P. B. Woodworth, of the Department of Electrical Engineering, has organized courses in engineering principles and electrical engineering in which are developed the fundamentals of practical mathematics. The general question came up at the first meeting[28] (Chicago, November, 1902) of the Central Association of Physics Teachers, and it is to be expected that this association will enlarge its functions in such a way as to include teachers of mathematics and of all sciences, and that the question will be considered in its various bearings by the enlarged association. At this meeting informal reports were made from the Bradley Polytechnic Institute of Peoria, the Armour Institute of Technology of Chicago, and the University of Chicago. The question is evoking much interest in the neighborhood of Chicago.

I might explain how I came to be attracted to this question of pedagogy of elementary mathematics. I wish, however, merely to express my gratitude to many mathematical and scientific friends, in particular, to my Chicago colleagues, Mr. A. C. Lunn and Professor C. R. Mann, for their cooperation with me in the consideration of these matters, and further to express the hope that we may secure the active cooperation of many colleagues in the domains of science and of administration, so that the first carefully chosen steps of a really important advance movement may be taken in the near future.

I close by repeating the questions which have been engaging our attention this afternoon.

In the development of the individual in his relations to the world, there is no initial separation of science into constituent parts, while there is ultimately a branching into the many distinct sciences. The troublesome problem of the closer relation of pure mathematics to its applications: can it not be solved by indirection, in that through the whole course of elementary mathematics, including the introduction to the calculus, there be recognized in the organization of the curriculum no distinction between the various

---

[28] Subsequent to the meeting of organization in the spring of 1902. Mr. Chas. H. Smith of the Hyde Park High School, Chicago, is president of the Association. Reports of the meetings are given in *School Science* (Ravenswood, Chicago).

branches of pure mathematics, and likewise no distinction between pure mathematics and its principal applications? Further, from the standpoint of pure mathematics: will not the twentieth century find it possible to give to young students during their impressionable years, in thoroughly concrete and captivating form, the wonderful new notions of the seventeenth century?

By way of suggestion these questions have been answered in the affirmative, on condition that there be established a thoroughgoing laboratory system of instruction in primary schools, secondary schools and junior colleges—a laboratory system involving a synthesis and development of the best pedagogic methods at present in use in mathematics and the physical sciences.

2

From *A General Survey of Progress in the Last Twenty-Five Years,* NCTM's First Yearbook (1926)

Steve Willoughby, NCTM president from 1982 to 1984, also recommended that we include a chapter from our inaugural yearbook. His choice, **"A General Survey of the Progress of Mathematics in Our High Schools in the Last Twenty-Five Years"** by David Eugene Smith, will be of particular interest to readers who enjoy reading about the history of mathematics education and about issues of policy and curriculum from a historical perspective.

In this chapter, Smith considers an assortment of factors that influence high school mathematics, including the work of international and national committees, the rise of the junior high schools, the role of the College Examination Board, the work of schools of education, and textbooks. The chapter also has some very interesting sections discussing the progress of algebra and geometry.

In one of his concluding comments, Smith remarks, "There has been a notable advance in the testing of pupils' abilities and achievements, hampered only by the fact that many of the tests in algebra have tended to perpetuate some of the most objectionable features of the science,—a difficulty that will, of course, tend to disappear under the combined efforts of those with some mathematical vision and those who know the technique of testing." Though eighty-seven years have passed since those words were written, it appears we have still not yet managed to eradicate *all* of those objectionable features!

A General Survey of the Progress of Mathematics in Our High Schools in the Last Twenty-Five Years

David Eugene Smith

I. Early Attempts at Improving the Syllabi

A T the beginning of the present century the syllabi in mathematics in the American high schools were determined largely by the requirements for entering our colleges. As a rule examinations were set by each college for its own candidates, the requirements being dictated by the department of mathematics.

As President Butler said in an address delivered on November 6, 1925,—"Twenty-five years ago, the colleges throughout the United States were going their several ways with sublime unconcern for the policies of other colleges, for the needs of the secondary schools, or for the general public interest. They regarded themselves as wholly private institutions and each indulged in some peculiar idiosyncrasy having to do with the admission of students to its freshman class. The colleges made no attempt to agree among themselves, either as to what subjects should be prescribed for admission or what content should be given to any particular subject. The several colleges held admission examinations when it was most convenient for them to do so, and, with the rarest of exceptions, only at the college itself. No secondary school could adjust its work and its program to the requirements of several colleges without a sort of competence as a pedagogic acrobat that was rare to the point of non-existence. The situation would have been comic were it not so preposterous."

The purpose of the examination at that time seems to have been chiefly to assure the entrance of students who gave at least a fair degree of promise of becoming mathematicians. Although set by such a large number of different examining bodies, the subject matter was fairly uniform, based as it was upon a tradition that was generally known throughout the country.

In 1902 a committee of the American Mathematical Society, on "definitions of college entrance requirements," made a recommendation that elementary algebra should cover the usual topics through progressions; that higher algebra should cover permutations and combinations, the applications of mathematical induction, logarithms, theory of equations (with graphic methods), Horner's method, determinants, and complex numbers; that plane geometry should cover "the usual theorems and constructions of good standard text-books, the solution of original exercises, applications to problems of mensuration of lines and of plane figures, and to loci problems," and similarly for solid geometry. Plane trigonometry was to cover the six functions, the "proofs of principal formulas," logarithms, and the solution of triangles.

This report was evidently rather inclusive through its very lack of precision. It kept open the way for every eccentric examiner to propose almost any question he wished, and yet it served fairly well as a starting point for reform. At any rate, it was the expression of a national instead of a local opinion.

In the year 1900 the College Examination Board was organized. This was a great step in advance. It sought to unify the examinations and to prepare them with much greater care than was usually the case with local efforts. It also gave an opportunity for the schools to be consulted by and become a part of a central organization, thus being represented in the preparation of the papers. While the range of the examinations soon became that which was set by the committee of the American Mathematical Society, and was therefore rather indefinite as to limitations, the papers themselves became more standardized and represented in general a better selection of material. The traditional still played a leading role, but at last there was some hope of modernizing the syllabus and there was a feeling of assurance that this improvement would in due time be realized.

Tradition still demanded the retention of a large amount of abstract manipulation of polynomials, including long problems in the multiplication and division of integral and fractional expressions, with extended work in the finding of roots, in factoring, in lowest common multiple, and in highest common factor, and with equally useless manipulations of complex fractions and radicals. Simultaneous linear equations extended to four and more unknowns, and simultaneous quadratics of the trick variety were in evidence. As to the higher algebra it is not necessary to speak, since that concerned and still concerns a relatively small number of pupils, these being a rather selected lot who always look forward to a certain amount of work in college mathematics.

Geometry was still more stagnant, as would naturally be expected of a subject that had been many centuries longer in the making.

In neither subject did there seem to be any clear conception of the purpose of teaching mathematics in the twentieth century as distinguished from that which came into being with the rise of analysis and algebraic symbolism three hundred years ago.

It is hardly necessary to consider other syllabi. In general they were more conservative than the somewhat indefinite ones followed by the College Entrance Examination Board and suggested by the committee of the American Mathematical Society, and they showed a lower range of scholarship.

II. Influences for the Betterment of the Courses

In the period in question there were various influences making for the betterment of the syllabi and of the courses offered in the schools. These include the following:

1. The work of the International Commission on the Teaching of Mathematics, serving as it did, to let American teachers see the curricula of the best types of secondary schools in all the other leading countries of the world and to compare our progress with that which is found abroad. Naturally this comparison showed us that conservatism existed elsewhere as well as here, but it also served to show that various other countries were ahead of us in achievement and that it was well for us to ascertain the cause and to see if we had any advantages to counterbalance this apparent disadvantage. A brief statement of the work of the Commission is set forth in Section III of this survey.

2. The work of the National Committee on Mathematical Requirements. This committee was appointed by the Mathematical Association of America in 1916 and was financially assisted by the General Education Board. Its investigations were thorough and its report was fully considered by representative associations of teachers throughout the country. It made a careful study of the purposes which should determine the teaching of mathematics in our secondary schools and suggested a syllabus which should eliminate nonessentials, retain those things which should best meet the needs of pupils of the present generation, and introduce such modern material as should strengthen the work without attempting to make it unreasonably difficult. The nature of the report of this committee is briefly summarized in Section IV. Since certain portions of the report itself were the object of study and discussion all over the country for at least two years before they were finally approved, and since the complete report has been very widely circulated and discussed, its influence has been very great.

3. The revised requirements of the College Entrance Examination Board. These were prepared by a commission appointed by the Board in 1922. They were reported in 1923 and are now the basis of the examinations. They represent the combined judgment of the colleges and the secondary schools and are thus no longer subject to the criticism that the college is assuming undue rights in dictating what shall be taught in the high school. They are more precise than any others that have been heretofore set forth for the guidance of schools, and even where schools are not preparing for these examinations they have tended to set a new standard and to eliminate much of the work for which no reasonable justification could be found. The work of this commission is set forth more fully in Section V.

4. The rise of the Junior High Schools. The rise of these schools, a development of the "six-and-six plan," is one of the most significant movements in the last quarter of a century. Not hampered in its early development by any examination system,

this type of school was free to formulate its own course and to set its own standards. The result was a more uniform curriculum than at first was thought probable. It enabled our schools to introduce intuitive geometry in a satisfactory manner in Grade VII, to allow algebra to grow naturally out of the need for the formula in connection with this geometry, to show that a simple form of trigonometry grows out of algebra, and to give some slight notion of the significance of a demonstration. At the same time it continued the work in arithmetic as applied to the intuitive treatment of geometry, as related to algebra, and as bearing upon the everyday needs of our people. The result was a perfectly natural correlation of the various parts of elementary mathematics that are suitable for this school period, a process of discovery of mathematical ability, and an interesting type of work that had been lacking in the older kind of arithmetic and algebra which it displaced. Courses differed in different schools, but this was rather in unimportant details than in the large features. Although the plan has not met with universal approval, the general feeling has been very favorable to it and, with progressive and well-trained teachers, the new curriculum has been decidedly successful.

5. The work of the schools of education in the universities. As effective agencies in any noteworthy improvement in teaching, these schools may be said to be largely a product of the twentieth century. Naturally their achievements have been manifested largely in the line of mental measurements which, for our purposes include various types of tests. The results have been very encouraging, although the published tests have, as stated later, had one unfortunate feature that has to a great extent counterbalanced the good which might have been accomplished and which will eventually result when they are prepared with more care. In general, however, these schools of education have shown that mathematics can easily be adjusted to the capacities of young people, while the capacities of these pupils cannot be so readily adjusted to the old-style mathematics; that the science can be made part of the lives of children as well as those of adults; and that students may rightly expect to enjoy learning as they enjoy other phases of life. Whether this attitude of mind in the work of the school has lowered our ideals of scholarship is a mooted question, but that it must necessarily do so can hardly be asserted by anyone who carefully considers the future work of our schools. In any case our departments of education have, through their experiments and their studies, created a healthy spirit in our schools and have of late become more internationally minded in their outlook. Although they have not fostered sound mathematical scholarship as much as we might wish, they have probably done so as fully as present conditions permit, and it must be recalled that their work is still in its infancy.

6. The textbook. Among the various influences that have worked for the progress of mathematics in the last twenty-five years it is but a just tribute to the makers of textbooks to say that the progress that has been made would have been absolutely impossible without their aid. While certain books have appeared that were impossible as aids to instruction, and while commercialism occasionally enters into an effort of this kind, it is only fair to say that a large majority of textbooks on mathematics have been prepared solely with a view to assist in the bettering of instruction in our elementary and high schools. There is no other country in the world that produces as fine pieces of bookmaking as those issued from our best presses. At the same time, in no other country is the work set forth in such textbooks with as great attention to the needs and interests of children as is found here, and in none has the advance in teaching been so rapid. On the other hand, in sound scholarship many foreign books surpass ours, but not in meeting the practical needs of the people.

7. The Spirit of the Times,—a phrase which certain writers assert has no meaning, but which is convenient as representing the mass psychology of the moment. At any rate, within the past quarter of a century there has been a general recognition the world over that the traditional education of the nineteenth century is not adapted to present conditions; that there must be a well-accepted reason for teaching algebra or else the subject must be discarded, and similarly for the other mathematical disciplines. The result of this feeling has been very salutary as may be seen in the present requirements in mathematics as set by the College Entrance Examination Board.

III. The Work of the International Commission

The section on philosophy, history, and instruction of the Fourth International Congress of Mathematicians, held in Rome, April 6 to 11, 1908, submitted to the Congress a resolution to create an International Commission on the Teaching of Mathematics. The suggestion was indorsed by the Congress on April 11, and an organizing committee was appointed consisting of Professor Klein of Göttingen, Sir George Greenhill of London, and Professor Fehr of Geneva, besides others appointed immediately thereafter. Delegates were afterwards chosen from the countries which had taken part in at least two of the congresses. The result was the publication of a large number of reports showing the nature of the work done in mathematics in schools of all types throughout most of the world. In the United States reports were prepared relating to various topics, including the following:

(1) Mathematics in the elementary schools; (2) Mathematics in the public and private secondary schools; (3) Training of teachers of elementary and secondary mathematics; (4) Influences tending to improve the work of the teacher of mathematics; (5) Mathematics in the technological schools of collegiate grade; (6) Undergraduate work in mathematics in colleges of liberal arts and universities; (7) Mathematics at West Point and Annapolis;

(8) Graduate work in mathematics in universities; (9) Report of the American committee; (10) Curricula in mathematics in various countries; (11) Mathematics in the lower and middle commercial and industrial schools of various countries; (12) The training of elementary school teachers in mathematics in various countries; (13) The training of teachers of mathematics for the secondary schools in various countries. Allied to this work there were published two other bulletins as follows: (1) Bibliography of the teaching of mathematics, 1900–1912; (2) Union list of mathematical periodicals taken by the larger libraries in the United States.

These reports were published by the United States Bureau of Education in the years 1911–1918 and were widely circulated in this country. They served to show to our schools the range of our system on instruction in mathematics and the general purposes in view in the various types of school. Perhaps the chief value to our country, however, was the comparison which was thus made possible between the work done here and that done in the other leading countries of the world. This showed that we were distinctly behind other countries, as to subject matter, particularly after Grade IV, although we might properly claim to be at least equal to them in the spirit of the work done in our schools. It raised the question, however, as to whether a good spirit could compensate for poor work, and it caused a large amount of discussion in bodies of teachers throughout the country. The past ten years have shown some gratifying results of this discussion.

IV. The Work of the National Committee on Mathematical Requirements

The work of the National Committee is too well known for detailed remarks. It is set forth in its report, *The Reorganization of Mathematics in Secondary Education,* published in 1923 by the Mathematical Association of America, under whose auspices the Committee was established. This report was prepared in close cooperation with bodies of teachers throughout the country. It set forth very clearly the aims of mathematical instruction in the several years of the junior high school, the senior high school, and the older type of four-year high school. It presented the model courses for these several types of school and made suggestions for carrying out the work. It considered the question of college entrance requirements, the basal propositions of geometry, the role of the function concept, and the terms and symbols which might properly have place in the schools. It fostered various other investigations, including the present status of the theory of disciplinary values, the theory of correlation applied to school grades, a comparison of our curricula with those in use abroad, experimental courses in mathematics, standardized tests, and the training of teachers.

It is not too much to say that the advance in the teaching of mathematics in our secondary schools in the last decade has been due in large part to the work of this committee. Since the report is available in most high school and public school libraries in the country, it would only seem to lessen its value to attempt any further résumé of its contents.

V. The Influence of the College Entrance Examination Board

No other influence in the reform of the teaching of mathematics in our secondary schools of the present time has been more potent than that exerted by this board. Often thought to be very conservative and to represent the views of the college professor alone, it has shown itself in the best sense radical in its reforms, and representative of the secondary schools to fully as great a degree as of the colleges. Among its reforms only the most noteworthy can be mentioned in this brief report.

In algebra it has eliminated the extended and largely useless manipulation of polynomials in connection with the elementary operations. It distinctly says:

"It is not expected that pupils will be called upon to perform long and elaborate multiplications or division of polynomials, but that they will have complete mastery of those types that are essential in the subsequent work with ordinary fractional equations, and with such other topics as are found in elementary algebra. In other words, these operations should be looked upon chiefly as a means to an end."

Thus in two sentences it has struck out a large amount of entirely useless and uninteresting work that had cumbered up the inherited course.

It then eliminated most of the work in factoring, a subject which began to occupy an undue amount of space in the closing quarter of the nineteenth century, reaching its culmination at the opening of the present one. The requirement was reduced to only three types,—

(1) Monomial factors;

(2) The difference of two squares;

(3) Trinomials of the type $x^2 + px + q$.

When we consider the fact that the subject was used almost exclusively in those fractions and equations which were made up merely for the purpose of using it, the board's attitude was most salutary. Indeed, from the standpoint of practical use it may be doubted if the value of the factoring of a quadratic trinomial is not even now overrated.

The requirement in fractions has been simplified, and the result has been the elimination of long and useless operations that consumed time and led to no worthy end. The report says:

"The meaning of the operations with fractions should be made clear by numerical illustrations, and the results of algebraic calculations should be frequently checked by numerical substitution as a means of the attainment of accuracy in arithmetical work with fractions."

The requirement includes complex fractions of about the following degree of difficulty:

$$\frac{p + \dfrac{a}{b}}{q - \dfrac{c}{d}}, \qquad \frac{\dfrac{a+3b}{c-5d}}{\dfrac{a-3b}{c+5d}}, \qquad \frac{\dfrac{a}{b} + \dfrac{c}{d}}{\dfrac{m}{n} - \dfrac{p}{q}} \, . \text{''}$$

This is a great gain over the plan of even a dozen years preceding. Such forms as those here given actually enter into the simple formulas which the pupil will see in elementary science. They can therefore be justified.

Furthermore, ratio is treated merely as a case of simple fractions, and proportion is treated as a simple type of fractional equation, so that at once the whole subject has been simplified materially,—indeed, as a separate topic, it is substantially discarded. Such terms as "alternation" and "composition" have naturally been abandoned by this action.

The position of the formula is a great advance over what was the case a generation ago. Few would now deny that the formula is the element of algebra that will be most often used by the student when be begins his serious work in science. This should therefore be the subject of greatest emphasis in the first year's work. The commission of the Board recognizes the fact and has this to say with respect to promoting the subject:

"In the work done with formulas, the general idea of the dependence of one variable upon another should be repeatedly emphasized. The illustrations should include formulas from science, mensuration, and the affairs of everyday life. Throughout the course, there should be opportunity for a reasonable amount of numerical work and for the clarification of arithmetical processes."

The graph also has the proper kind of recognition, being introduced for the definite purpose of illustrating and making more clear the formulas needed in science and in business.

In the subject of linear equations the pupil is no longer expected to solve an array of abstract types that admit of no application, but to devote his energies to the solution of those types which have some chance of being used. For this reason the first year's work excludes cases with three or more unknowns. The report further makes the statement:

"Besides numerical linear equations in one unknown, involving numerical or algebraic fractions, the pupil will be expected to solve such literal equations as contribute to an understanding of the elementary theory of algebra. For example, he should be able to solve the equation

$$s = \frac{ar^n - a}{r - 1} \qquad \text{for } a.$$

"In the case of simultaneous linear equations, he should be able to solve such a set of equations as

$$ax + by = k,$$
$$cx + dy = l,$$

in order to establish general formulas. But the instruction should include a somewhat wider range of cases, as for example:

$$\begin{cases} ax + (a+b)y = ab, \\ ax + (a-b)y = -ab, \end{cases} \quad \text{or} \quad \begin{cases} ax + by = ab, \\ x + \left(1 + \dfrac{b}{a}\right)y = a. \end{cases}$$

"The work in equations will include cases of fractional equations of reasonable difficulty; but, in general, cases will be excluded in which long and unusual denominators appear and in which the common factor of the denominators, or the lowest common denominator, cannot be found by inspection.

"Problems in linear equations, as in ratio, proportion, and variation, will whenever practicable be so framed as to express conditions that the pupil will meet in his later studies."

In the latter part of the first year's work two notable improvements have been made. The first relates to the simplification of the work in surds, an inheritance from the past that has lost much of its former significance. The report considers exponents and radicals under the following head:

1. "The proof of the laws for positive integral exponents.

2. "The reduction of radicals, confined to transformation of the following types:

$$\sqrt{a^2 b} = a\sqrt{b}, \qquad \sqrt{\frac{a}{b}} = \frac{\sqrt{ab}}{b} \qquad \frac{\sqrt{a}}{\sqrt{b}} = \sqrt{\frac{a}{b}}$$

$$\frac{1}{\sqrt[3]{a}} = \frac{\sqrt[3]{a^2}}{a}, \qquad \frac{1}{\sqrt[3]{a^2}} = \frac{\sqrt[3]{a}}{a},$$

and to the evaluation of simple expressions involving the radical sign.

3. "The meaning and use of fractional exponents, limited to the treatment of the radicals that occur under 2) above.

4. "A process for finding the square root of a number, but no process for finding the square root of a polynomial.

   "In all work involving radicals, such theorems as $\sqrt{\dfrac{a}{b}} = \dfrac{\sqrt{a}}{\sqrt{b}}$ and $\sqrt{\dfrac{a}{b}} = \dfrac{\sqrt{a}}{\sqrt{b}}$

   may be assumed. Proofs of these theorems should be given only in so far as they make clearer the reasonableness of the theorems; and the reproduction of such proofs is explicitly excepted from the requirements here formulated."

The second noteworthy feature of the latter part of the first year is the introduction of simple numerical trigonometry. This is made possible by the elimination of a considerable amount of relatively useless material and by the selection of the minimum essentials of the subject.

Upon the range of this work the report recommends the following:

"The use of the sine, cosine, and tangent in solving right triangles.

"The use of four-place tables of natural trigonometric functions is assumed, but the teacher may find it useful to include some preliminary work with three-place tables.

"The recognition of the fact that the pupil should acquire facility in simple interpolation; in general, emphasis should be laid on carrying the computation to the limit of accuracy permitted by the table."

With respect to geometry the Commission makes three noteworthy recommendations:

1. That the number of "book theorems" required on any examination shall be materially reduced; in fact, only eighty-nine theorems are included in the syllabus for plane geometry, and of these only about a third are required for examination purposes. This allows plenty of time for the important subject of "originals," a subject which has assumed an entirely new position of importance within the last quarter of a century.

2. That a year's course involving both plane and solid geometry be allowed in place of the single course in plane geometry. This allows a pupil to secure a fair knowledge of both phases of the subject in a single year. This is rendered possible by the reduction to fifty-nine of the number of propositions in plane geometry and to twenty-four as the number required for examination, with a similar reduction in solid geometry.

3. That there be offered a certain amount of work in mensuration of a type more frequently met with in various lines of industry. The treatment of this work is modern in spirit and the work itself is outlined in the commission's report.

In brief, the report shows a tendency to break away from too much formalism, to depend much more upon originals, to combine plane and solid geometry (if desired) in a single year, and to approach European standards in the field of practical mensuration.

VI. The Progress of Arithmetic

Some idea of the progress of arithmetic, of the type used in the junior high school, in the period in question may be obtained by a comparison of the nature of the topics as set forth in some of the most prominent arithmetics of the close of the nineteenth century with that of the present time. The following synoptic presentation of the case shows the nature of the changes that have taken place in this brief period:*

---

* This synopsis and certain other portions of this section have been taken from the author's essay on *The Progress of Arithmetic in the Last Quarter of a Century*, Boston, 1923.

Then	*Now*
Arithmetic of special and unusual occupations. For example,	Arithmetic of the daily life of the people. For instance,
alligation	arithmetic of the home
equation of payments	a simple bank account
arbitrated exchange	the check book
partnership involving time	arithmetic of the store
true discount	organization of common corporations
general average	cost of production and overhead charges
tax collectors' commissions	transportation
marine insurance	the common industries
partial payments	farm problems of today
measurement of hogsheads, granaries, and cisterns	Short methods. For example,
Obsolete processes. For example, greatest common divisor and least common multiple of large numbers	in making change
	in checking bills
	in common multiplication
Work with long and unusual fractions	Fractions limited to those of everyday life
Arithmetic progression	Thrift and savings
Geometric progression	Safe types of investment
Cube root	Percentage related more closely to decimals
Present worth:	Decimals related more closely to U. S. money
at simple interest	
at compound interest	Graphs
Troy weight	Compound numbers limited to a few really useful types of work
Extensive work in compound numbers of unusual types	Our duty to the government
Gregorian and Julian calendars	Government expenses:
Proportion as a means of solving commercial problems	city, state, national: necessity for thrift
Ratio without applications	Systematic reviews of principles but with new problems
No reviews except by going over the same work	Minimum essentials emphasized
All topics of equal importance	

Not only have such changes as these in the topics of arithmetic been made, but even more noteworthy ones appear in the nature of the problems. The following list, taken from a popular textbook of a quarter of a century ago, contains a fair sampling of what can be found in most of the works of that period:

1. Find the value of $[84 - 7 \times 6 + (3 \times 5) - 3] \div 9$.

2. Divide 19/42 of 28/33 of 11/14 of 7 1/9 by 23/35 of 5/8 of 16/23 of 8/35 of 24 5/12.

3. I bought 26 yards of carpet at $1 9/10 a yard, 3 curtains at $5 3/5 each, and 6 chairs at $1 3/4 each. What is my bill? (As if we ever used these common fractions of a dollar in this way!)

4. A vessel sailed from Portland, Me., for New Orleans with a cargo of 1528.375 tons of ice. On the way 94.58 tons of it melted. How much ice reached New Orleans? (The weight of the cargo of ice is given to within 2 lb., which is rather close when we consider that it probably varied 1000 lb. while being stowed away.)

5. Reduce to ounces 5T. 10cwt. 24lb. 8oz.

6. From 5 lb. 7 oz. take 3 lb. 10 oz. 5 drams, 1 scruple, 15 grains.

7. Find the compound interest on $4921.50 for 4 yr. 9 mo. 24 da. at 7%, using the table.

8. What is the present worth of $3180.50 payable in 2 yr. 3 mo. 21 da., when money is worth 5 1/2% ?

9. A, B, and C formed a partnership. A put in $3000 for 5 mo., and then increased it $1500 for 4 mo. more. B put in $9000 for 4 mo., and then, withdrawing half his capital, continued the remainder 3 mo. longer. C put in $5500 for 7 mo. They gained $3630. What was each partner's share of the gain?

10. If 5 horses eat as much as 6 cattle, and 8 horses and 12 cattle eat 12 tons of hay in 40 da., how much hay will be needed to keep 7 horses and 15 cattle 65 da.?

11. Three men bought a grindstone 20 inches in diameter. How much of the diameter must each grind off so as to share the stone equally, making no allowance for the eye?

12. A man bequeathed his property in such a way that his wife received $7 for every $5 received by each of his two sons and every $4 received by each of his three daughters. If his estate was worth $250,000, what was the sum bequeathed to each of the heirs?

13. Find the greatest common divisor of 462, 882, and 546.

14. A farmer wishes to put 336 bushels of wheat and 576 bushels of corn into the least number of bins possible of uniform size, without mixing the two kinds of grain. How many bushels must each bin hold?

15. Find the least common multiple of 2520 and 2772 and also of 11 1/9, 14 2/7, and 33 1/3.

16. Change 268*te* on the duodecimal scale to the decimal scale.

17. Multiply 3424 on the quinary scale by 234 on the same scale.

18. Take 3/5 of 4 mi. from 7/8 of 3 mi. 18 rd. 3 yd. 2 ft.

19. Divide 19 T. 17 cwt. 29 lb. 7 oz. by 4/5.

20. Find the weight of an ivory ball 2 in. in diameter, the weight of ivory being 1825 oz. a cubic foot. (It would be interesting to see ivory sold by the cubic foot.)

21. A man walked 23 2/3 mi. the first day of a trip, 25 3/20 mi. the second, 28 14/64 mi. the third, and 26 53/100 mi. the fourth. How far did he walk in all?

22. Find the value of 8 3/7 + 5 4/9 + 9 2/3 − 3 8/21 − 3 6/7.

23. Find the value of 23/49 × 7 3/4 × 9/10.

24. Find the value of $\dfrac{9/19 \text{ of } 13 \ 7/12}{18/38 \text{ of } 7 \ 5/16}$

25. Reduce 6 mi. 37 rd. 4 yd. 3 ft. 6 in. to inches, and 5/7 of a rod to yards, feet and inches.

26. Reduce 721327 inches to miles. (The number was not even written in periods of three figures.)

27. Reduce 7 sq. mi. 17 A. 13 sq. ch. to square chains.

28. Reduce 9230 scruples to higher denominations.

29. Reduce 7 hr. 32 min. 49 sec. to seconds.

30. Find the sum of 10 mi. 172 rd. 2 yd. 2 ft. 9 in., 12 mi. 172 rd. 4 yd. 11 in., 16 mi. 74 rd. 1 yd. 2 ft. 3 in., 19 mi. 198 rd. 4 yd. 9 in., and 39 mi. 131 rd. 5 yd. 1 ft. 7 in.

Whatever may be said of many of the problems set in our schools today, a reading of the above list shows that there has been a decided advance in the quality of the material and in adapting the exercises to the needs and interests of the pupils.

As to the methods of presenting the subject of arithmetic or, indeed, of the other branches of mathematics, this report is not directly concerned. It is desirable, however, to call attention to one change that has become more and more evident in the last two or three decades, and that is our sympathy with childhood,—not our affection, probably not always our good judgment, but certainly our sympathy with the child in school. The severe discipline of two generations ago had begun to relax at the close of the nineteenth century, and at the present time it has become very much less pronounced in the better type of school. The pupil has come to live the child life more freely instead of trying to live the adult life that the world not long ago sought to impose upon him. We have still a long way to go to reach the goal, and we run continued risk of so reducing the mental food supply as to make education a poor affair and one that requires so little effort as to have neither interest nor value. On the whole, however, the average elementary pupil gets much more joy out of his school life than his parents did out of theirs, and his general range of knowledge is rather better than theirs was at the same age. It is also probable that his powers of computation in those ordinary problems of life that he is capable of understanding do not suffer by the same comparison. In spite of the easy fashion to deny this assertion, there seems no reason to think that it is not perfectly true.

All this has been a distinct gain. It has not come from any of the "methods" that loomed up so large in the eyes of the teachers of a generation ago, nor has it come to any great extent from the results of psychological studies; it has come largely from the use of plain common sense in adapting arithmetic to this new view of education,—that of letting the child live as natural a life as possible while in school,—and in adapting the work in arithmetic to his mental powers at each stage of his growth.

Perhaps the most important change of all is seen in the purpose of teaching arithmetic. A quarter of a century ago it was felt that the subject should be hard in order to be valuable, and it sometimes looked as if it did not make so much difference to the school as to what a pupil studied so long as he hated it. The old idea that this was good for the mind and soul was not at that time fully discarded. There was also prevalent the idea that as many applications of arithmetic should be introduced as the time allowed, irrespective of whether they were within the mental horizon of the pupil or within the probable needs of his life after leaving school. This view has now been changed; the purpose of teaching arithmetic has come to be recognized as the acquisition of power to calculate within the limits of the needs of the average well-informed citizen. It has also come to be recognized that the problem is primarily designed to show a need for computation, by giving applications that add to the interest in calculation and by introducing the puzzle element of problem-solving, which may add further interest. A secondary purpose of the problem is the imparting of some knowledge of the economic conditions, that the pupil will find in daily life, this being presented to him in a simple manner that will make it seem interesting and worth while.

We should not fail, moreover, to recognize the value of the tests in arithmetic which have been devised during the past quarter of a century. These precede by some years the tests in algebra and geometry and have been much less open to legitimate criticism. They have accomplished much in the improvement of the work in arithmetic, in diagnosing pupils' difficulties, and in the measurement of their capacities. It is reasonably certain that the newer tests in the high school subjects will, when purged of certain objectionable features, especially as to their work in traditional subject matter, accomplish similarly beneficial results.

VII. The Progress of Algebra

Encouraging as has been the progress of arithmetic in our schools, the progress of algebra has been none the less noteworthy. Twenty-five years ago the subject was usually taught as if it were a purely mathematical discipline, unrelated to life except as life might enjoy the meaningless puzzle. Valuable as the teacher might feel it to be, the majority of pupils looked upon it as a fairly interesting way of getting nowhere.

If we were to seek the most significant step taken in the improvement of the teaching of algebra in the last twenty-five years, it would probably be found in the clearer vision that we have of the real purpose of the subject. To take our current educational phraseology, we have been concerned, and properly so, with establishing our "objectives." The purpose a quarter of a century ago seems to have been to make mathematicians; the purpose today is to make well-informed American citizens. A man or a woman is not well informed if he or she is ignorant of the general meaning of geography, of the simpler natural sciences, of a few masterpieces of our language, of the significance of foreign tongues, of the qualities of good art (including music), of the social and economic needs of people, of the nature of government, of one's duties as a citizen, and of the significance of religion,—most of these being taught to best advantage in school. In the same way both the man and the woman needs to know something of the significance of mathematics.

As a result of this view of the reason for teaching algebra, we have come to see that we should not expect everyone to solve two simultaneous quadratic equations, although out of an entire class there will be found a few who can do so. Nor should we expect to have all the pupils able to factor $ax^2 + bx + c$ (a useless accomplishment for most people), even though a considerable number will take pleasure in performing such a task and will thereby acquire some special skill which they may find useful in later work. The purpose of teaching algebra is found in none of these details: it consists in giving to everyone a general idea of the meaning of algebra, together with a few definite and useful applications which everyone is likely to meet. If the subject is to be valuable, the learning should be a pleasure, and it may properly be expected that this pleasure will carry the pupil into such manipulations of algebraic expressions as will fix the habit of using algebra in the cases to which it can be applied.*

This has led to a consideration of those topics of algebra that are of most worth to the average citizen, and herein the change has been very marked. If, today, the consensus of opinion were to be taken among progressive teachers it would probably result in the naming of the formula, the graph, the directed number, the linear equation with one unknown, and (by way of application) numerical trigonometry as the five important topics to be considered. Facility in algebraic manipulation, which played such an unduly important part a generation ago would be relegated to a relatively minor role at the present time.

---

* For an amplification of this subject see the author's *The Progress of Algebra in the Last Quarter of a Century*, Boston, 1925.

Painfully precise definitions and attempts at ultra scientific explanations are no longer felt to be either necessary or desirable.

One of the most popular texts of twenty-five years ago had eight pages of definitions and theory before a single example was given, and out of nearly 1600 exercises in the first 147 pages only 111 were verbal problems and only two could lay claim to relating to any apparent human need. Another text of that period gave about 1800 exercises in the first 128 pages; of these only 109 were of the verbal variety, and only one had any apparent application to any condition that would arise in daily life.

Any good modern text, however, would show the need for algebra on the first page; would begin its real problems immediately; and would give a large number of verbal and written exercises at once, with as many genuine applications as reasonably possible,— applications of a kind that pupils can understand and in which they will have a real interest. Some of our modern books have more verbal problems than both of the other two already referred to, and many times the number of genuine applications relating to daily life.

In neither of these two algebras of a quarter of a century ago (and they were among the best of their time) was there a single example showing the meaning of or the need for the directed number, whereas in a good modern text the pupil will find dozens of them, not to mention numerous illustrations, showing its value in our daily lives.

These are only a few of the evidences of progress in the purpose of school algebra in the last quarter of a century,—a progress which, without exaggeration, may be characterized as revolutionary. Probably no other subject found in the course of study in the average high school has undergone so marked a change.

The earlier type of algebra was arranged on the same plan as the earlier type of arithmetic. On the theory that we must scientifically define all terms before they can safely be used, the book began with definitions—a plan which would make it necessary to define "elephant" before visiting a menagerie. If the book gave any idea at first about the purpose of algebra, it was that it was a science in which letters were used in solving the most impractical sort of number puzzles.

The book next proceeded to introduce strange terms, such as monomial, binomial, residual (now discarded), polynomial, coefficient, and exponent,—not as they were needed, but in order to provide for their use at some time in the future.

It then took up the four operations which had been developed in arithmetic, but which have only a slight use in practical algebra, and spent some weeks of the pupil's time in mastering a technique that was of little value—at least in the beginning of the science.

Having covered this ground for integral expressions, the book then considered the question of fractions by giving work of a type that few students would ever need in subsequent mathematics. Linear equations were then introduced; after which followed a large amount of work in incommensurable numbers (involving such names as "surds" and "radicals"), and then quadratics, proportions, series, and other advanced topics. There were but few attempts to frame verbal problems, even of the fictitious type, and none to develop the real applications of the science.

At the present time every leading writer of school algebras is making the attempt, with more or less success, to arrange the topics on a more rational basis. The sanely progressive books begin with the formula and show its meaning, its practical use, and the method of deriving one formula from another. This being done, the most valuable part of pure algebra has been presented, and it is a part that, a quarter of a century ago, was practically ignored. The graph, the negative number, and the linear equation are then presented, the equation having already been encountered in connection with the formula. Numerical trigonometry appears later in the course. As to the division of polynomials by polynomials, elaborate algebraic fractions, highest common factor and the lowest common multiple, most of factoring, roots, most of the work in surds, linear equations with more than two unknowns, and simultaneous quadratics,—the relative value of all these has diminished greatly in the estimation of those who wish to salvage the parts of algebra that the pupil will really use in his later work.

In the matter of algebraic problems there has also been a notable advance. Some idea of the types in current use a quarter of a century ago can be formed by considering a few of the best of the problems contained in a popular work of that period:

A man, being asked if he had 100 head of cattle, replied that if he had twice as many as he then had and 4 more, he would have 100. How many had he?

If B were 5 yr. younger, A's age would be twice B's. The sum of their ages is 20. How old is each?

A's capital was ¾ of B's. If A's had been $500 less, it would have been ½ of B's. What is the capital of each?

Paving a square court with stones at 40¢ a square yard will cost as much as inclosing it with a fence at $1 per yard. What is the length of a side of the court?

Bought 8 horses, a number of cows, and 100 sheep for $2500. The number of cows was equal numerically to 4 times the price of a sheep, and a sheep and a horse cost $5 less than 1/5 the cost of all the cows. Find the cost of a horse, and a sheep, and the number of cows, if a cow cost $40. (As an example of English, in which a *number* is equal *numerically*, this is interesting.)

It must not be thought that such problems are without value, that there are no good reasons for giving them, or that the older books are to be condemned for having a reasonable number of this type. Some of them have stood the test of time and have maintained their own throughout the centuries because pupils could easily visualize them and were interested in their solution, which is rarely the case with any of the multitude of real problems of a technical nature in physics, in shop practice, in the biological sciences, or in the field of commerce. It is probable that we shall always find it best to draw upon certain types of puzzle problems as exercises in algebra, in arithmetic, and in various other branches, for the reason that most technical problems are too difficult to be understood by the pupil when he is studying these subjects. To postpone algebra until such time as he could understand these applications would be to put off taking up something for which the pupil is mentally ready until a time when he would deem it too childish to be of interest. Indeed, it may safely be said that we are probably not making enough use of the interest afforded by the puzzle element in any of our work in mathematics.

All this, however, is no excuse for giving nothing but unreal problems in algebra, which was the situation at the beginning of the century. That we have made a gratifying advance is seen from an examination of various leading textbooks of the present day, the genuine applications of algebra, particularly in the case of formulas and of other types of equations (as in the study of ratio), being much in evidence.

It is, however, in the introduction of numerical trigonometry as a legitimate, interesting, simple, and valuable part of algebra that the most notable step in the last quarter of a century has been taken. It has long been recognized that trigonometry has much more practical importance in the world than most of the work given in the older type of algebras. The tradition that this subject must necessarily follow demonstrative geometry has no merit except its antiquity: the subject is easier than any of the topics in the second half of the old-time algebra, it is more interesting, and it admits of attractive outdoor work. It thus opens up a new field of interest for the pupil—the field of indirect measurement, in which there are discovered the first steps in the measuring of the distance to the stars and in the understanding of some of the former secrets of the universe in which we live. It follows naturally after the study of proportion, and its inclusion in algebra has now met with the approval of all leaders in the teaching of elementary mathematics. The initial work requires no knowledge of logarithms, a subject that may properly be left to a later course in algebra, simply because the time is hardly sufficient to allow for its introduction in the early stages of elementary algebra.

The fact that this topic has been recommended by the National Committee as part of elementary algebra, that it is required by the College Entrance Examination Board, that it is generally taught in close connection with algebra in other countries, that the plan has been generally approved by American associations of teachers, and that it has been followed in various recent textbooks assures its status in our elementary courses.

Much has been written of the advance in appreciation of the function concept in recent years. This advance is, of course, particularly noticeable in algebra, and the topic of trigonometry is the one in which it is most in evidence. It is also seen in the treatment of the formula and in the entire subject of variation as a part of ratio and of fractions. It has of late come to be looked upon as a kind of unifying principle running through all parts of algebra, and as such, when not too consciously forced into the language of science, has undoubted value.

VIII. The Progress of Geometry

As a scientifically organized part of mathematics geometry is the oldest of its branches. For this reason it has had a longer period in which to perfect itself. It is therefore looked upon as less capable of reform or improvement than algebra and arithmetic.

The last quarter of a century has shown, however, that as a school subject it is capable of improvement in the same spirit if not to the same extent as these other branches.

For one thing, the recent years have clearly differentiated between intuitive and demonstrative geometry. While this has always been recognized in a small way, as in the treatment of simple mensuration in arithmetic, it was not until the Cambridge meeting of the International Mathematical Congress in 1912 that intuitive geometry was brought prominently before the educational section of that organization and began to be seriously considered by bodies of teachers throughout the world. Since then it has come, in this country, to occupy a worthy place in all our courses for the junior high schools. This place is properly in the seventh and eighth school years and to some extent even earlier. The subject naturally precedes demonstrative geometry, and our schools have come generally to recognize that it has but little sanction in the latter and more mature branch.

Demonstrative geometry twenty-five years ago consisted of at least one year of plane geometry, following the course in algebra, and at least a half year in solid geometry. In most schools there was a good deal of memorizing of demonstrations and the original exercise still played an almost negligible part, being, for many pupils, without either purpose or pleasure. A few teachers enlivened the work by applications of doubtful value, but on the whole it was generally looked upon as an intellectual grind.

The progress since that time has been steady and encouraging. Its nature may be summarized briefly as follows:

1. There has been a more definite recognition by the schools that the chief purpose of demonstrative geometry is to show the application of logic to the proof of mathematical statements. It therefore requires a maturity of mind hardly found before the tenth school year, although for purposes of information a little work in demonstration may properly be given to the abler pupils in the preceding grade.

2. Therefore the purpose of demonstrative geometry is not mensuration, this being sufficiently cared for in the work in intuitive geometry; its purpose is, in part, to demonstrate the truths already known intuitively. For this reason the work in the mensuration of the circle has little sanction in demonstrative geometry, the rules being already known from intuitive geometry and the demonstrations as given not being very satisfactory from the standpoint of logic. The subject is therefore no longer required in college entrance examinations or for high school graduation. The same is true as to the mensuration of the rectangle, the rectangular solid, the cylinder, the sphere, and the cone.

3. The number of demonstrated theorems, and especially of the corollaries, has been greatly reduced, the purpose being to retain only the basal propositions that are of most use in the demonstrations of the "originals." This has shifted the emphasis from the book proofs, which usually constituted all the geometry of a century ago and most of that of the last quarter of the nineteenth century, to the original exercises where it properly belongs. Recent textbooks have an amount of original work of a simpler character that was hardly imagined a generation back. Indeed, the older geometries may be compared to an algebra that had all its

examples fully worked out, and no exercises for the pupils. The purpose of "book propositions" in geometry is largely that of worked-out examples in algebra,—to set a model for the pupil and to furnish a basis for his original work.

4. The number of solved problems has been proportionately reduced quite as much as the number of demonstrated theorems. The simpler constructions with ruler and compasses are given in intuitive geometry and their demonstration is not of much value as compared with the demonstrations of the theorems, leading as they do to only a small number of exercises and depending chiefly upon two or three simple theorems.

5. The exercises have greatly increased in number, but they have decreased in difficulty. The increase is due, as already stated, to the shifting of emphasis from that which an author has thought out for the pupil to that which the pupil is to think out for himself. The decrease in difficulty has arisen from the fact that the ability of pupils can certainly not be said to have increased during the period in which the schools have tended to the education of everyone rather than to that of a selected body of pupils of high intellectual promise. The tendency toward some form of universal high-school education is probably for the happiness of the race and the strengthening of the state, so that we shall have to accept, for many pupils, this lower standard.

There has, however, been another reason,—the feeling that a large number of simple exercises trains the immature pupil better than a small number of difficult ones. In our efforts to conform to this belief we are still in the experimental stage. The pupil of mathematical inclinations will prefer a more difficult type, and for him it will probably be better to pass rapidly over a few of the easy exercises and to come as soon as possible to those requiring more thought.

6. The discussion and generalization of propositions now holds higher place than it did a score of years ago; at least it is rather more in evidence in our courses of study and in our textbooks.

7. There is a strong movement on foot to cover the essential parts of plane and solid geometry in a single year. This is often met by the assertion that it is impossible. This assertion, however, depends upon the meaning assigned to the expression "essential parts." It would be feasible to frame a course in plane geometry that would require three years of hard work, but it would not lead to the most profitable use of the pupil's time. If we eliminate most of the construction problems, assume all the work in inequalities, eliminate all mention of incommensurables as applied to line segments and the circle, all the theory of proportion (treated of in algebra), the work in the mensuration of solids, and the rather purposeless treatment of spherical triangles, we can readily frame a very satisfactory course for a single year. This can be done by making selections from any standard geometry.

No mention has been made of the efforts toward developing courses in general mathematics. These refer rather to the method of presentation than to the improvement in subject matter. It may, however, be said that the recent development of the junior high school affords a natural field for combining different parts of mathematics in a single course, and this has been recognized in all our modern textbooks on the subject. Numerical trigonometry, also, naturally blends with algebra, and this is recognized in the recent courses of study. Demonstrative geometry, however, offers a different problem. It can use the algebraic equation in its proofs, although it can get on about as well without it; but neither algebra nor trigonometry makes use of the demonstrations of geometry in its work. Our successful courses in general mathematics, therefore, tend to segregate demonstrative geometry, and this, psychologically, will have to be the case in the future. Either demonstrative geometry must be considered largely by itself or else it will tend to drop out of the curriculum or, at the best, to remain as a feeble memory of the world's effort to show how truth is logically established in the mathematical sciences.

IX. Conclusion

The progress of mathematics in our schools in the last quarter of a century may, then, be summarized briefly as follows:

1. Early attempts at improving the courses were greatly hampered by the force of tradition.

2. The most potent of the later influences for betterment have been the work of the International Commission on the Teaching of Mathematics; that of the National Committee on Mathematical Requirements; that of the College Entrance Examination Board, which brought together the secondary schools and the colleges; the rise of the junior high schools; the work of the schools of education; the improvements in textbooks; and the general Spirit of the Times.

3. The results of these labors are seen in the setting forth with greater clearness the aims which should guide in the teaching of each branch of mathematics. This is one of the two greatest gains. It has led to the elimination of much obsolete or relatively valueless material in arithmetic and algebra, to the introduction of new topics in each, to the merging of the first course in numerical trigonometry with the work in elementary algebra, to the elimination from geometry of matter of doubtful value, and to the general union of related parts of mathematics through such coordinating influences as that of the function concept and that of the social needs of our people.

4. The second of the gains of greatest importance has been the recognition of the rights of children to see the purposes of their studies, to find that the subjects synchronize with the development of their intellectual capacities, and to enjoy the work in mathematics as they should enjoy their work in other lines of intellectual activity.

5. There has been a notable advance in the testing of pupils' abilities and achievements, hampered only by the fact that many of the tests in algebra have tended to perpetuate some of the most objectionable features of the science,—a difficulty that will, of course, tend to disappear under the combined efforts of those with some mathematical vision and those who know the technique of testing.

6. In no field of elementary or secondary education has advancement in the last twenty-five years been more marked than in that of mathematics. If teachers feel discouraged with the reactionary attitude of certain administrators or of boards of control in state or city, they may well take courage by considering the state of high school mathematics at the beginning of the century and comparing it with the state of the subject at the present time.

3

From *The Nature of Proof,* NCTM's Thirteenth Yearbook (1938)

Two NCTM past presidents—F. Joe Crosswhite (who served from 1984 to 1986) and Jack Price (1994 to 1996)—recommended that we include selections from our Thirteenth Yearbook. This volume, published in 1938 and titled *The Nature of Proof,* was written entirely by Harold P. Fawcett. Joe Crosswhite noted that this entry in the series was of special interest to him because of Fawcett's advocacy for a new approach to teaching. Jack Price remarked that the Fawcett yearbook was his favorite of all time, particularly because of its focus on geometry. For our special Seventy-fifth Yearbook, we have excerpted here **chapter 1, "The General Nature of the Problem,"** and **chapter 3, "A Description of Procedures."**

Fawcett writes in his 1938 introduction that the purpose of his work is "to describe classroom procedures by which geometric proof may be used as a means for cultivating critical and reflective thought. . . ." These excerpts—and, in fact, the entire Thirteenth Yearbook—may be intriguing to any mathematics educator with an interest in geometry and how it has evolved historically and as part of the curriculum.

The General Nature of the Problem

Harold P. Fawcett

T HERE has probably never been a time in the history of American education when the development of critical and reflective thought was not recognized as a desirable outcome of the secondary school. Within recent years, however, this outcome has assumed increasing importance and has had a far-reaching effect on the nature of the curriculum. Teachers in all areas have felt the effect of this change in emphasis and their general acceptance of it has been accompanied by an increasing modification of classroom procedures. Teachers of mathematics, however, have felt that this new emphasis called for little change in their field since demonstrative geometry has long been justified on the ground that its chief contribution to the general education of the young people in our secondary schools is to acquaint them with the nature of deductive thought and to give them an understanding of what it really means to prove something. While verbal allegiance is paid to these large general objectives related to the nature of proof, actual classroom practice indicates that the major emphasis is placed on a body of theorems to be learned rather than on the *method* by which these theorems are established. The pupil feels that these theorems are important in themselves and in his earnest effort to "know" them he resorts to memorization. The tests most commonly used emphasize the importance of factual information, and there is little evidence to show that pupils who have studied demonstrative geometry are less gullible, more logical and more critical in their thinking than those who did not follow such a course.

It is the purpose of this study to describe classroom procedures by which geometric proof may be used as a means for cultivating critical and reflective thought and to evaluate the effect of such experiences on the thinking of the pupils.

The Origin and Background of the Problem

The history of mathematical education in the United States reveals the fact that demonstrative geometry was taught only in colleges until a comparatively recent date. Whatever the values derived from the study of this subject may have been they were reserved for the selected group of young men and women who were interested in continuing their formal education beyond that offered in the public school. However, as new and interesting subjects claimed the attention of those responsible for college curricula it gradually developed that there was no room in these curricula for demonstrative geometry, while the rapid and increasing development of the secondary school presented an opportunity for continued study of this subject.[1] It thus happened that by the middle of the nineteenth century demonstrative geometry became a definite part of the high school curriculum, but with the change in the maturity of the pupils to whom this work was offered there occurred no fundamental change in the nature of the content. In considering the teaching of geometry John Wesley Young writes, "Our texts in this subject are still patterned more or less closely after the model of Euclid, who wrote over two thousand years ago, and whose text, moreover, was not intended for the use of boys and girls, but for mature men."[2]

It is true that teachers of mathematics recognized the advisability of so modifying the subject matter as to make it more palatable to the less mature pupils in the secondary schools and a number of important committees[3] have studied this problem. As a result many theorems have disappeared from modern texts, different arrangements of those remaining have been suggested, new theorems have been introduced, and methods of teaching have changed; but in general there has been no significant change in the nature of the subject matter. The great majority of geometry students, regardless of their interests and capacities, are required to work through ninety or more theorems selected by the author of the particular text used in any given situation. Certain properties of geometric figures are assumed and the student is asked to establish other predetermined properties by logical proof.

The Values Claimed for Demonstrative Geometry

The subject matter relates to rectilinear figures, circles, proportion and areas. After the pupil has covered the allotted number of theorems and has demonstrated his ability to work out a number of original exercises, it is assumed that the values to be derived from the study of demonstrative geometry have been added to his educational equipment. What are these values? What is the unique contribution which demonstrative geometry makes to the general education of the young people in our secondary schools? The National Committee on the Reorganization of Mathematics in Secondary Education answers this question by saying that purposes of instruction in this subject are: "To exercise further

---

[1] A complete and thorough treatment of the significant changes and trends in the teaching of geometry may be found in *The History of the Teaching of Elementary Geometry* by Alvin W. Stamper, Bureau of Publications, Teachers College, Columbia University, 1909, and in *Recent Developments in the Teaching of Geometry* by J. Shibli, Pennsylvania State College, State College, Pennsylvania, 1932.

[2] John Wesley Young, *Lectures on Fundamental Concepts of Algebra and Geometry*, p. 1. The Macmillan Co., New York, 1925.

[3] Committee of Ten on Secondary School Studies, 1894; Committee on College Entrance Requirements, 1899; National Committee of Fifteen on Geometry Syllabus, 1912; National Committee on Mathematical Requirements, 1923; College Entrance Examination Board, Document 108, 1923; First Committee on Geometry, 1929; Second Committee on Geometry, 1930; Third Committee on Geometry, 1932.

the spatial imagination of the student, to make him familiar with the great basal propositions and their applications, to develop understanding and appreciation of a deductive proof and the ability to use this method of reasoning where it is applicable, and to form habits of precise and succinct statement, of the logical organization of ideas, and of logical memory."[4] This important question has also been discussed by many of the leading teachers in the field of mathematics. While it is not possible to quote all of the answers which have been made, some of the most significant are presented here.

In the Fifth Yearbook of the National Council,* Professor Reeve writes that "The purpose of geometry is to make clear to the pupil the meaning of demonstration, the meaning of mathematical precision, and the pleasure of discovering absolute truth. If demonstrative geometry is not taught in order to enable the pupil to have the satisfaction of proving something, to train him in deductive thinking, to give him the power to prove his own statements, then it is not worth teaching at all."[5]

In similar vein Professors Birkhoff and Beatley of Harvard University say that "In demonstrative geometry the emphasis is on reasoning. . . . To the extent that the subject fails to develop the power to reason and to yield an appreciation of scientific method in reasoning, its fundamental value for purposes of instruction is lessened."[6]

The Third Committee on Geometry of which Professor Beatley is chairman reports that "Teachers agree that the main outcomes of demonstrative geometry pertain to logical thinking and wish to maintain the distinction between this subject and informal geometry which emphasizes the factual aspects of geometry. . . . There is equally enthusiastic response to the proposal that instruction in demonstrative geometry call attention to logical chains of theorems, to the gaps in Euclid's logic and bring the pupil to appreciate the nature of a mathematical system, the need of undefined terms, the arbitrariness of assumptions and the possibility of other arrangements of propositions than that given in his own text."[7]

In discussing this problem Dr. H. C. Christofferson says, "Geometry achieves its highest possibilities if, in addition to its direct and practical usefulness, it can establish a pattern of reasoning; if it can develop the power to think clearly in geometric situations, and to use the same discrimination in non-geometric situations; if it can develop

---

[4] National Committee on Mathematical Requirements, *The Reorganization of Mathematics in Secondary Education* (Part I), p. 48. Houghton Mifflin Co., Boston, 1923.

* Hereinafter the *Yearbooks of the National Council of Teachers of Mathematics* will be referred to as Yearbook.

[5] W. D. Reeve, "The Teaching of Geometry," *Fifth Yearbook of the National Council of Teachers of Mathematics*, 1930, pp. 13–14.

[6] G. D. Birkhoff and Ralph Beatley, "A New Approach to Elementary Geometry," *Fifth Yearbook*, 1930, p. 86.

[7] Ralph Beatley, "The Third Report of the Committee on Geometry," *The Mathematics Teacher*, Vol. xxviii, No. 6, 1935, p. 334.

the power to generalize with caution from specific cases, and to realize the force and all-inclusiveness of deductive statements; if it can develop an appreciation of the place and function of definitions and postulates in the proof of any conclusion, geometric or non-geometric; if it can develop an attitude of mind which tends always to analyze situations, to understand their interrelationships, to question hasty conclusions, to express clearly, precisely, and accurately non-geometric as well as geometric ideas."[8]

As a final contribution to this discussion let us consider the clear and forceful statement of Professor C. B. Upton who says, "I firmly believe that the reason we teach demonstrative geometry in our high schools today is to give pupils certain ideas about the nature of proof. The great majority of teachers of geometry hold this same point of view. Some teachers may at first think our purpose in teaching geometry is to acquaint pupils with a certain body of geometric facts or theorems, or with the applications of these theorems in everyday life, but on second reflection they will probably agree that our great purpose in teaching geometry is to show pupils how facts are proved. I will go still further in clarifying our aims. . . . The purpose in teaching geometry is not only to acquaint pupils with the methods of proving geometric facts, but also to familiarize them with that rigorous kind of thinking which Professor Keyser has so aptly called 'the If-Then kind, a type of thinking which is distinguished from all others by its characteristic form: If this is so, then that is so.' Our great aim in the tenth year is to teach the nature of deductive proof and to furnish pupils with a model of all their life thinking."[9] Professor Keyser calls this kind of thinking "autonomous thinking" or "postulational thinking" and proceeds to point out that in the "Elements" of Euclid we find "The most famous example of autonomous thought in the history of science."[10]

It is noticeable that in these statements concerning the chief contribution which the study of demonstrative geometry makes to the general education of young people very little, if any, reference is made to the facts of geometry. The truth of the matter is that demonstrative geometry is no longer justified on the ground that it is necessary for the purpose of giving students control of useful geometric knowledge, since the facts of geometry which may at one time or another actually serve some useful purpose in the developing life of a boy or girl are learned or can be learned in the junior high school. In the "Third Report of the Committee on Geometry" already quoted, the committee states, "It is generally agreed that the important facts of geometry can be mastered below the tenth grade through inductions based on observation, measurement, constructions with drawing instruments, cutting and pasting, and also through simple deductions from the

---

[8] H. C. Christofferson, *Geometry Professionalized for Teachers*, p. 28. George Banta Publishing Co., Menasha, Wis., 1933.

[9] C. B. Upton, "The Use of Indirect Proof in Geometry and Life," *Fifth Yearbook*, 1930, pp. 131–132.

[10] C. J. Keyser, *Thinking about Thinking*, p. 25. E. P. Dutton and Co., New York, 1926.

foregoing inductions as well as from geometric notions intuitively held."[11] The consensus of opinion therefore seems to be that the most important values to be derived from the study of demonstrative geometry are an acquaintance "with the nature of proof" and a familiarity with "postulational thinking" as a method of thought which is available, not only in the field of mathematics, but also "in every field of thought, in the physical sciences, in the moral or social sciences, in all matters and situations where it is important for men and women to have logically organized bodies of doctrine to guide them and save them from floundering in the conduct of life."[12]

General Acceptance of These Values

These purposes are recognized as worthy and desirable, not only by teachers of mathematics but by most thoughtful men and women who are interested in the general education of young people. There is no disagreement concerning the educational value of any experience which leads children to recognize the necessity for clarity of definition, to weigh evidence, to look for the assumptions on which conclusions depend, and to understand what proof really means. John Dewey has defined reflective thinking as "active, persistent and careful consideration of any belief or supposed form of knowledge in the light of the grounds that support it and the further conclusions to which it tends."[13] And in view of the stated purposes for teaching demonstrative geometry is it not reasonable to expect that effective work in the study of this subject should lead the pupil to examine critically any conclusion he is pressed to accept "in the light of the grounds that support it and the further conclusions to which it tends"? The "reflective thinking" of our young people should be improved through experience in analyzing situations which involve "the nature of proof."

The Columbia Associates in Philosophy have stated that "The function of education, in large part, is the moulding of minds capable of taking and using the best that the world has given. Such minds must be well stored with information, free from prejudice, critical of new ideas presented, and fitted to understand the kind and quantity of proof required before they may adopt the pronouncements of the generals of the society of minds."[14] While one might question the use of the word "moulding" in connection with an educational process, is there any subject in the curriculum of the secondary school which should make a greater contribution toward the development of minds that are "fitted to understand the kind and quantity of proof required" in the acceptance of conclusions than that subject which has as its fundamental purpose the leading of the student to understand what it really means to prove something?

---

[11] Ralph Beatley, *op. cit.*, p. 334.

[12] C. J. Keyser, *op. cit.*, p. 35.

[13] John Dewey, *How We Think,* p. 6. D. C. Heath and Co., Boston, 1910.

[14] Columbia Associates in Philosophy, *An Introduction to Reflective Thinking,* p. 11. Houghton Mifflin Co., Boston, 1923.

Achievement of Values Questioned

The reasons which mathematicians offer to justify the continued teaching of demonstrative geometry to the young people in our secondary schools are not questioned so far as their educational value is concerned. There is, however, serious question as to whether or not these desirable results are actually achieved through the usual course in this subject. After a careful and thorough analysis of the results in the December "Every Pupil Plane Geometry Test" for Ohio, Dr. H. C. Christofferson writes, "The reasons given by pupils for statements often seem to disregard entirely the thought of the situation. Often it seems that it is mere habit that dictates the response, not a thought process. Pupils have often used the various theorems as reasons and with satisfaction. They seem in some cases to have used them so often without meaning that they give them as so many memorized nonsense syllables. Guesses would be right more often than the type of reasoning attempted in some cases."[15] This statement would seem to indicate that the study of demonstrative geometry has not greatly improved the ability of the students to reason accurately even within the narrow confines of the subject. In view of this is it reasonable to expect that the results of such study will be more helpful in non-geometric situations? The Third Committee on Geometry, composed of twenty-six prominent teachers in the field of mathematics, prepared a questionnaire which raised pertinent questions concerning the teaching of geometry.[16] This questionnaire was sent to each member of the committee and also to 101 of the outstanding teachers in Maine, Massachusetts, Ohio, Illinois, Minnesota, Kansas, Oklahoma and Colorado. The replies indicate that "there is almost unanimous agreement that demonstrative geometry can be so taught that it will develop the power to reason logically more readily than other school subjects, and that the degree of transfer of this logical training to situations outside geometry is a fair measure of the efficacy of the instruction. However great the partisan bias in this expression of opinion, the question 'Do teachers of geometry ordinarily teach in such a way as to secure transfer of those methods, attitudes, and appreciations which are commonly said to be most easily transferable?' elicits an almost unanimous but sorrowful 'No.'"[17]

A more vigorous statement concerning the outcomes of demonstrative geometry is made by Eric Bell in a recent volume in which he writes, "A diluted sort of Euclid . . . is one of the mainstays of American education today. It is supposed to quicken the reason, and there is no doubt that it does in the hands of a thoroughly competent and modernized teacher, who lets the children use their heads and see for themselves exactly how nonsensical some of the stuff presented as 'proof' really is. . . . Uncritical reverence for the

---

[15] H. C. Christofferson, *A State Wide Survey of the Learning and Teaching of Geometry,* p. 41. State Department of Education, Columbus, Ohio, 1930.

[16] Ralph Beatley, "The Second Report of the Committee on Geometry," *The Mathematics Teacher,* Vol. xxvi, No. 6, 1933, p. 366.

[17] Ralph Beatley, "The Third Report of the Committee on Geometry," *The Mathematics Teacher,* Vol. xxviii, No. 6, 1935, p. 336.

supposed rigidity of Euclid's geometry had much to do with the retardation of progress in close reasoning. . . . If school children fail to get some conception of geometry and close reasoning out of their course in 'geometry' they get nothing, except possibly a permanent inability to think straight and a propensity to jump at conclusions which nothing in reason or sanity warrants."[18]

Now it is probably safe to assume that the values emphasized in the testing program of any school are those values which receive emphasis in the classroom, and a study of commonly used tests in geometry is sufficient to reveal that little, if any, attempt is made to measure the degree to which the purposes claimed for demonstrative geometry are realized. Perhaps the tests which have the widest use are the College Entrance Board Examinations, the Breslich Geometry Survey Test, the Cooperative Mathematics Tests, and the mathematics section of the Sones-Harry High School Achievement Test. It is claimed that together these tests are given to more than two million young people, which means that these students are being examined on the facts and skills of geometry, as there is little, if anything, in these tests which by any stretch of the imagination could be interpreted as examining children on their understanding of "the nature of proof" and their ability to apply postulational thinking to "situations outside the field of geometry."

The chief objective of mathematical study, according to Young,[19] is "to make the pupil think" and "if mathematical teaching fails to do this, it fails altogether." Young then continues with the following significant statement: "The mere memorizing of a demonstration in geometry has about the same educational value as the memorizing of a page from the city directory. And yet it must be admitted that a very large number of our pupils do study mathematics in just this way. There can be no doubt that the fault lies with the teaching."

The assumption which mathematics teachers are making is that since demonstrative geometry offers possibilities for the development of critical thinking, this sort of thinking is *necessarily* achieved through a study of the subject. Such an assumption has not been validated and the results of past experience indicate that it should be seriously questioned. To theorize concerning values which are believed to be the unique contribution of demonstrative geometry to the general education of young people is not a difficult matter, but to plan and carry out this program in such a way that these desired outcomes are actually realized is a problem which has not been squarely faced by teachers of mathematics.

The Problem Defined

While teachers of mathematics agree in general as to the unique contribution which the study of demonstrative geometry should make to the general education of young people, there may be some disagreement as to just what they mean by "the nature of proof." How is this concept defined? What is it that a pupil has learned when he understands what a

---

[18] Eric Bell, *The Search for Truth*, pp. 124–126. Williams and Wilkins Co., Baltimore, 1934.

[19] John Wesley Young, *op. cit.*, pp. 4–5.

proof really means? While some teachers of mathematics will answer this question in one way and some in another, the importance of the answer should not be overlooked for on it will depend the sort of activity going on in the classroom. For purposes of this study it is assumed that a pupil understands the nature of deductive proof when he understands:

1. The place and significance of undefined concepts in proving any conclusion.
2. The necessity for clearly defined terms and their effect on the conclusion.
3. The necessity for assumptions or unproved propositions.
4. That no demonstration proves anything that is not implied by the assumptions.

In speaking of this topic Young states as follows: "If we consider the nature of a deductive proof, we recognize at once that there must be a hypothesis. It is clear, then, that the starting point of any mathematical science must be a set of one or more propositions which remain entirely unproved. This is essential; without it a vicious circle is unavoidable. Similarly we may see that there must be some undefined terms. In order to define a term we must define it in terms of some other term or terms, the meaning of which is assumed known. In order to be strictly logical, therefore, a set of one or more terms must be left entirely undefined."[20] It is further assumed that a pupil who understands these things will also understand that the conclusions thus established can have universal validity only if the definitions and assumptions which imply these conclusions have universal validity. The conclusions are "true" only to the extent that the fundamental bases from which they were derived are "true." Truth is relative and not absolute.

While teachers of mathematics say they want the young people in our secondary schools to understand the nature of proof, that should not be and probably is not their total concern. What these teachers really want is not only that these young people should understand the nature of proof but that their way of life should show that they understand it. Of what value is it for a pupil to understand thoroughly what a proof means if it does not clarify his thinking and make him more "critical of new ideas presented"? The real value of this sort of training to any pupil is determined by its effect on his behavior, and for purposes of this study we shall assume that if he clearly understands these aspects of the nature of proof his behavior will be marked by the following characteristics:

1. He will select the significant words and phrases in any statement that is important to him and ask that they be carefully defined.
2. He will require evidence in support of any conclusion he is pressed to accept.
3. He will analyze that evidence and distinguish fact from assumption.
4. He will recognize stated and unstated assumptions essential to the conclusion.
5. He will evaluate these assumptions, accepting some and rejecting others.
6. He will evaluate the argument, accepting or rejecting the conclusion.

---

[20] John Wesley Young, *op. cit.*, p. 3.

7. He will constantly re-examine the assumptions which are behind his beliefs and which guide his actions.

While the total educational experience of the student in the secondary school should contribute and doubtless does to some extent contribute to the development of this kind of behavior, there seem to the writer to be possibilities in demonstrative geometry which no other subject offers. While Professor Young considers the subject matter of mathematics to be of importance he writes as follows: "still more important than the subject matter of mathematics is the fact that it exemplifies most typically, clearly and simply certain modes of thought which are of utmost importance to everyone."[21] In this area the concepts considered and the ideas studied are devoid of strong emotional content. The student's native ability to think is not stifled by prejudice or bias. He becomes conscious of the fact that his conclusions are determined by the definitions and assumptions which he, himself, makes and he recognizes the far-reaching effect of these basic ideas. He sees a method of thought applied to idealized concepts and "without that ideal, thinking is without a just standard for self criticism; it is without light upon its course; it is a wanderer like a vessel at sea without a compass or star."[22] The logical rigor of geometric proof illustrates the needed "ideal," and it is the purpose of this study to show that by placing the major emphasis on those aspects of demonstrative geometry which serve to illustrate the nature of proof and not on the factual content of the subject, it is possible to improve the reflective thinking of young people and to develop minds that are "critical of new ideas presented, and fitted to understand the kind and quantity of proof required before they may adopt the pronouncements of the generals of the society of minds."

It is not sufficient, however, to study only those situations wherein the concepts are idealized and "the material to be presented is simple and wholly unobscured by the emotions."[23] If the kind of thinking which is to result from an understanding of the nature of proof is to be used in non-mathematical situations such situations must be considered during the learning process. Wheeler says that "No transfer will occur unless the material is learned in connection with the field to which transfer is desired. Isolated ideas and subjects do not integrate. Learning is not bond-forming. It is an orderly and organized process of differentiating general grasps of situations with respect to experience. The details emerge organized, as they differentiate from previous knowledge, in the face of new situations, not repeated ones."[24] Transfer is secured only by training for transfer and teachers of mathematics can no longer expect that the careful study of ninety or more geometric theorems will alone enable their students to distinguish between a sound argument and a tissue of nonsense.

[21] J. W. A. Young, *The Teaching of Mathematics*, pp. 17–18. Longmans, Green and Co., New York, 1924.

[22] C. J. Keyser, "The Human Worth of Rigorous Thinking," *The Mathematics Teacher*, Vol. xv, No. 1, 1922, pp. 1–5.

[23] Eric T. Bell, "The Meaning of Mathematics," *Eleventh Yearbook*, p. 138, 1936.

[24] R. H. Wheeler, "The New Psychology of Learning," *Tenth Yearbook*, p. 239, 1935.

William Betz, who made a comprehensive study of the problem of transfer with particular reference to geometry, presents in a summary the following findings and states that they "might well be incorporated in the creed and daily practice of every progressive teacher":

1. Training for transfer is a worth while aim of instruction; from the standpoint of life it is the most important aim.

2. Transfer is not automatic. "We reap no more than we sow."

3. Every type of "specific" training, if it is to rise above a purely mechanical level, should be used as a vehicle for generalized experience.

4. "The cultivation of thinking is the central concern of education."[25]

It is thus evident that the general problem to be attacked consists of three related problems:

1. The problem of leading the pupil to understand the nature of deductive proof through the study of geometric situations.

2. The problem of generalizing this experience so that effective transfer will result.

3. The problem of evaluating the resulting change in the behavior of the student.

A Summary of Related Studies

It has become increasingly evident in the last twenty-five years that the teaching of demonstrative geometry in secondary schools must be greatly improved if the values claimed for it are to be realized. Efforts to effect this improvement have, to a large extent, been directed to a rearrangement of geometric theorems and this rearrangement has been logically developed from an adult point of view. In discussing this situation E. Russell Stabler points out among other things that "The sequence of theorems tends to be arranged to meet the logical or traditional requirements of the subject as seen by the author, and not with a view to obtaining the maximum amount of cooperation from the pupils in developing and appreciating the logical structure."[26] Furthermore, little attention has been given to changes in the nature of the content which are necessary if the habits of thought, which it is hoped will be developed through a study of this subject, are to transfer to non-geometric situations.

The results of an interesting and suggestive experiment related to this problem are reported by Elsie Parker.[27] Assuming that under favorable conditions transfer of training

---

[25] William Betz, "The Transfer of Training with Particular Reference to Geometry," *Fifth Yearbook*, pp. 149–198, 1930.

[26] E. Russell Stabler, "Teaching an Appreciation of Mathematics: The Need of Reorganization in Geometry," *The Mathematics Teacher*, Vol. xxvii, No. 1, 1934, p. 37.

[27] Elsie Parker, "Teaching Pupils the Conscious Use of a Technique of Thinking," *The Mathematics Teacher*, Vol. xvii, No. 4, 1924, pp. 191–201.

from one field of experience to another is possible, she set up a controlled experiment in an effort to answer the question, "Can pupils of geometry be taught to prove theorems more economically and effectively when trained to use consciously a technique of logical thinking; and furthermore, does such training, more than the usual method, increase the pupil's ability to analyze and see relationships in other non-geometrical situations?" In connection with this problem she says, "The traditional method of instruction has been to let the pupil discover for himself a method of reasoning which he thereafter uses without, in many cases, being aware of the fact that he is using that mode of procedure." The pupils in the experimental group studied the thought process used in proving geometric theorems and gained some understanding of the nature of logical thought, while in the control group the theorems themselves were recognized as of major importance and little attention was given to the thought process involved in proving them.

In order to measure the results of such training original geometric theorems were given to the classes before and after their work with Miss Parker, and after presenting the results of her study she says, "These data would seem to offer conclusive evidence, in so far as one experiment can be considered to do so, that when pupils are taught to use consciously a technique of logical thinking, they try more varied methods of attack, reject erroneous suggestions more readily, and without becoming discouraged maintain an attitude of suspended judgment until the method has been shown to be correct. The data on the reasoning tests would seem to indicate that such training in logical thinking with the materials of geometry tends to carry over these methods of attack and these attitudes to other problem situations not concerned with geometry."

A second experiment related to learning in geometry is reported by Winona Perry.[28] In this experiment Miss Perry had two control divisions and one experimental division. In each control division the instruction was definitely guided by a textbook. In one of these divisions the book propositions were emphasized as of primary importance, while in the other the emphasis was placed on the proving of original exercises. No attention was given in either group to any particular method of thinking, and in each case the class was conducted by the question and answer method. In the experimental division the development of a technique in reasoning about the exercises of geometry was of major importance. This technique in reasoning emphasized the "if-then" type of thinking as well as the analytic method and even though non-mathematical subject matter was not included in the course Miss Perry found, among other things, that in the experimental group "the ability to solve problems non-mathematical in character was markedly improved, following the period of training in the solution of exercises in geometry.

---

[28] Winona Perry, *A Study in the Psychology of Learning in Geometry.* Bureau of Publications, Teachers College, Columbia University, 1925.

This increased ability was most noticeable as resulting from those tests more nearly similar to the type of reasoning emphasized in demonstrative geometry in form and in content."

In dealing with the necessary training of teachers of geometry Dr. H. C. Christofferson[29] states, "There remains then the problem of securing still further professionalization of subject-matter with more emphasis on the fundamental pattern of teaching geometry as well as on the foundations of geometry, more actual contact with high-school geometry, and more attention to the system of formulated reasoning and its application to non-geometric as well as geometric situations." In order to suggest a solution for this problem he analyzes all the theorems and constructions in the Report of the National Committee on the Reorganization of Mathematics and on the basis of this analysis presents a list of "The Essential Constructions and Theorems of Geometry," "essential" being defined as "necessary for the proof of other propositions." Using these "essential" theorems as illustrative material, he then presents principles and methods of presentation which emphasize "the fundamental pattern of teaching geometry" and which, in his opinion, should result in the maximum of transfer to "non-geometric as well as geometric situations."

Now a teacher of geometry whose major interest is "to teach the nature of deductive proof and to furnish pupils with a model for all their life thinking" is not primarily interested in the factual aspects of the subject and has no fixed number of theorems which he feels must be covered. Furthermore, it is relatively unimportant for this purpose what theorems are covered since the deductive process by which they are established is illustrative of a method and the theorems are not important in themselves. However, in selecting these theorems from the large number available it would seem sensible to study those which contribute most effectively to an acquaintance with the important ideas of geometry and Dr. Christofferson's list of "Essential Theorems and Constructions" should prove particularly helpful for this purpose.

---

[29] H. C. Christofferson, *Geometry Professionalized for Teachers*, p. 2. George Banta Publishing Co., Menasha, Wis., 1933.

A Description of Procedures

Harold P. Fawcett

I T was the purpose of the writer to have an individual conference with each of the twenty-five pupils in the experimental group before the opening of school. It was not possible to arrange all these conferences and when the class met for its first session there were still eleven pupils with whom the teacher had not had an opportunity to confer. The conferences were arranged during the first week of school and there is little reason to believe that the few class sessions held in the meantime had any marked effect on the pupil's attitude. Through these conferences the teacher hoped to secure some understanding of each pupil's attitude toward mathematics in general and toward demonstrative geometry in particular.

The conference was most informal in character. No questionnaire was given to the pupil and no notes were taken during the interview. The teacher tried to make the pupil feel that he was definitely interested in helping him and that this help could be most effective only when the teacher understood the true nature of the pupil's attitude toward the work he was about to begin. The pupil was encouraged to talk freely and little direction was given to the conversation by the teacher until he felt that such an attempt to guide the discussion would not destroy the pupil's confidence in the situation. In the judgment of the teacher, there were only three instances where the pupil failed to talk honestly and frankly concerning himself and his attitude toward mathematics.

Immediately following the interview, the teacher made careful note of the pupil's comments which, in his opinion, gave indication of the pupil's attitude toward geometry. Although there was great variation in the form of these comments they fall into the general classification given in Table 5. This table also indicates the number of pupils in each classification. There is, of course, a considerable amount of overlapping.

Table 5

Comments Revealing Attitude of Pupils Toward the Study of Demonstrative Geometry

Kind of Comment Made by Pupils	Number Making Comment
If this work were not required I would not take it.	19
I need it to go to college.	18
I like the faculty. Since they require this course there must be some value in it.	12
I have heard that geometry is very difficult and I know I will not be able to do it.	10
Why should I know any more geometry? What I know already is of no use to me.	9
I have had enough geometry. I don't need any more.	8
I don't see any value in memorizing a lot of things which I will never use.	8
I have to take something and this will probably be as good as anything else.	3
I need geometry for later work which I want to take.	3
What does "demonstrative" mean?	2
I know I shall like geometry for I have always liked mathematics.	2

While one should be cautious in drawing any generalizations from these data, it is clear that at least nineteen of the twenty-five pupils were taking the work only because it was required. This is doubtless true also of the three who thought it would "probably be as good as anything else." Most of the pupils who had had any work in informal geometry considered it as a continuation of that sort of experience, and only two of them raised any questions as to the significance of the word "demonstrative." In general, it is probably safe to say that twenty-two of the twenty-five pupils had a negative attitude toward studying any more mathematics, although twelve of the twenty-two thought "there must be some value in it," otherwise the faculty would not require it. They "liked" the faculty and because of this apparently had some faith in their judgment.

Recognition of Need for Definition

As the first step in leading the pupils to understand something of "the nature of proof," it seemed important that they should recognize the necessity for clarity of definition in all matters where precise thinking is essential. To understand how the vagueness and ambiguity of ordinary words lead to serious errors in reflective thought is to appreciate the importance of clearly defined concepts in any technical vocabulary. It seemed advisable, then, to begin this work with a consideration of the importance of definition in matters which claimed the interest of the pupils, and the teacher gave careful thought as to what illustrations would be most helpful for this purpose. He recalled that during the preceding school year there had been a good deal of animated discussion concerning awards and that the question of whether or not awards were to be made for "outstanding achievements" in the school was still unsettled. Here, then, was a problem of real significance to the pupils about which there had been some controversy and in the consideration of which definition was likely to be an important issue. The teacher decided to give them an opportunity to discuss this problem, anticipating that in the argument which was almost certain to develop the necessity for clarity of definition would be recognized.

When the class met for the first time most of the pupils had their notebooks with them, some of the more thoughtful had brought their compasses and straightedges, and all of them expected to be given a text from which the work of the year would be taken.

There was thus considerable surprise when, after the usual routine of the opening day had been completed, the teacher said, "There is no great hurry about beginning our regular work in geometry and since the problem of awards is one which is soon to be considered by the entire school body I suggest that we give some preliminary consideration to the proposition that 'awards should be granted for outstanding achievement in the school.'" While it developed that within this particular group there were not many pupils who openly opposed the granting of awards, the opposition that was offered was thoughtful and intelligent.

Within a very few moments after the discussion started the question was raised as to whether a teacher's salary was an award. One pupil argued that everybody worked for an award of some kind. Another asked whether playing on the football team constituted an "outstanding achievement" while another believed that grades on academic achievement were awards. Vigorous disagreement developed, even among those who were supporting the general proposition that awards should be granted, and considerable time was spent in what might appear to have been useless discussion. However, when the teacher summarized the discussion, pointing out the differences that had developed and clarifying the real issues which had arisen one pupil said, "Most of this trouble is caused by the fact that we don't know what we mean by 'awards' or by 'outstanding achievement'"; and the evident agreement with this statement by other members of the class indicated that although the word "definition" was not used many of the pupils recognized in this situation the need for clearly defining these two ideas. The school body later decided to grant awards under certain conditions and one of the conditions was that the pupil receiving the award "must be a good citizen." In discussing this, one of the pupils promptly pointed out that before this award plan could be effectively administered "someone will have to explain what is meant by 'good citizen'."

In general, however, this explicit recognition of the need for definition seemed foreign to the thinking of the pupils. They were unable to select with any degree of accuracy the key words which need to be clearly defined before the real meaning of the statement in which they occur is evident. All of them agreed with the truth of the statement that "Abraham Lincoln spent very little time in school" and no one raised the point that the truth of this statement depends on how "school" is defined. However, when each pupil was asked to write his definition of "school" the results indicated that:

12 students considered "school" as a "building" set aside for certain purposes.
10 students considered "school" as a "place for learning things."
3 pupils considered "school" as "any experience from which one learns."

Before there was any discussion of the significance of these definitions the students were given the following short exercise:

Accepting the definition of "school" as "Any experience from which one learns" indicate your agreement or disagreement with the proposition:

"Abraham Lincoln spent very little time in school."

One pupil was absent when this exercise was considered, but of the twenty-four present all of them now disagreed with this proposition. A comparison of these results with those previously obtained from the same pupils served to emphasize the importance of definition and illustrated how a changed definition does affect conclusions. The teacher also used this opportunity to point out that when two people are discussing "school," and one of them defines it as a "building" while the other defines it as "any experience from which one learns," there is almost certain to be disagreement since the concept of "school" does not mean the same thing to each person.

Through such considerations the pupils began to recognize the need for clarity of definition, and as this recognition developed they began to suggest illustrations which were interesting to them. Among these were such points as the following:

1. Is the librarian a teacher?
2. What is an aristocrat?
3. What is one hundred percent Americanism?
4. How do I know when I am tardy?
5. What is a "safety" in football?
6. What is a "foul ball" in baseball?
7. What is the labor class?
8. What is an obscene book?

These suggestions, together with the many others that were made, not only reflect the interests of the pupils but also suggest how concepts and ideas can be developed "in connection with the field to which transfer is desired." While it was neither possible nor advisable to give detailed consideration to all of these points, numbers 4, 5 and 6 seemed to the teacher to have possibilities which should not be overlooked. Each pupil was familiar with the fact that when tardy for any class he could not be admitted without an "admit slip" which he secured at the office. One pupil made the point that he was considered tardy for one class and not for another, although the circumstances were the same. The teacher pointed out that the respective teachers involved had different definitions of "tardy." He then suggested that the pupils define "tardy" for that particular class and after some discussion, to which the teacher contributed as well as the pupils, they defined a pupil as tardy "when he reached the classroom after the door was closed." The teacher then pointed out that each time a pupil met the requirements of this definition, the rule concerning admittance to the classroom implied certain conclusions as to what that pupil should do, and this served to illustrate something of the nature of implication.

A consideration of "safety" and "foul ball" brought an eager response from the pupils. They quickly pointed out that in baseball a "foul ball" had been defined, and it was the responsibility of the umpire to determine whether or not a ball which had been struck satisfied this definition. Similarly a pitched ball is a strike only when in the judgment of the umpire it meets the requirements of the definition of a strike. This whole question of games proved to be a most fruitful field, for in addition to the excellent illustrations of the importance of definition the rules served to illustrate the importance of agreements among people and how these agreements determined conclusions. No difficulty was met in leading the pupils to recognize that these rules were nothing more than agreements which a group of interested people had made and that they implied certain conclusions relative to the activities of the players.

When the question was raised as to what was the effect of changing one of these agreements, numerous pupils were ready to point out that the activities of the players were changed. Many illustrations were offered. One pupil explained how a new rule concerning the forward pass in football had changed the game in certain respects, while another discussed how the game of handball had been improved through a change in one of the rules on which the game was based. The teacher used such illustrations as these to introduce the pupil to the idea that conclusions usually depend on a set of rules to which people have agreed, or which they accept. These agreements may well be called assumptions and through numerous illustrations the pupil learns that even the most elementary activities of life depend on certain assumptions. The teacher of algebra assumes that his students understand arithmetic, the man who deposits his money in a bank assumes that the bank will not fail, the patient assumes that his doctor can cure him.

It is important to recognize that in this introduction the thinking of the pupils was concerned with situations which were interesting and familiar to them. Most of the illustrations used were suggested by the pupils and reflected their interests. However, it was the responsibility of the teacher to guide the discussion in such a way that the attention of the pupils was focused on the important principles, common to all illustrations, and not on the illustrations themselves. He suggested that these principles be made explicit and after considerable discussion the following summary was made:

1. Definition is helpful in all cases where precise thinking is to be done.

2. Conclusions seem to depend on assumptions but often the assumptions are not recognized.

3. It is difficult to agree on definitions and assumptions in situations which cause one to become excited.

It should be pointed out that this summary is not the work of the teacher alone. It represents the joint thinking of the teacher and pupils. The statements when originally made were not in this form. They were awkwardly expressed and the ideas were none too clear. Some pupils objected to these original statements and suggested improvements. The teacher felt free to contribute in the same manner as other members of the group, and it was through such discussion and joint thinking that the original statements were refined and ultimately accepted, as expressed above, by all pupils.

Introduction to Space Concepts, Undefined Terms, Definitions and Assumptions

Following this introduction, which with this particular group required approximately four weeks, the thinking of the pupils was definitely guided to a consideration of space where the "ideas studied are devoid of strong emotional content" and "the pupils' native ability to think is not stifled by prejudice or bias." The pupils had already realized that "it is difficult to agree on definitions and assumptions in situations which cause one to become excited" and to illustrate this it was only necessary to remind them of the vigorous controversy which developed in our attempt to define "aristocrat" and "labor class." There was general agreement that it would be interesting to make definitions and assumptions about concepts which did not stir the emotions and to proceed to investigate their implications. Many suggestions were made as to the content with which these definitions and assumptions would deal. Some pupils suggested "government," some suggested "religion," while others suggested "war." All of these suggestions were rejected by the group since they did not satisfy the criterion that the concepts involved should be "devoid of strong emotional content." The fact is that no suggestions were offered which did satisfy this criterion until the teacher raised the question whether or not one was likely to "become excited" in thinking about space. This question apparently reminded the pupils that they had originally expected to study geometry, and the discussion which followed resulted in an agreement to "build a theory about the space in which we live."

There was no question about the interest of the pupils. All of them had participated in the discussion to some degree and all were anxious to begin to "build" this theory. However, there was some difficulty in knowing just how to begin. Their background had led them to believe that they should begin by agreeing on certain definitions, but there was no agreement as to what should be defined. One pupil said that space involved "great distances," and that perhaps "We should begin by defining distance." Another said that he could "fill up the space in the room with cubes," and he thought it would be well to begin by defining "cube." This suggestion received feeble support. Another suggested, however, that he could "fill up the space in the room by piling a lot of flat surfaces one on top of the other and so we might just as well begin by defining 'surface' as 'cube.'" This suggestion also received feeble support. Another said he would like to begin by defining "triangle" for he knew that "triangles had something to do with geometry." Someone else suggested that "squares also had something to do with geometry" and that "we might begin by defining them." A large majority of the pupils made no suggestions whatever, although some of them supported either the "cube" or the "surface."

The teacher wrote on the board all the suggestions that had been made and asked the pupils to select one of these concepts and define it. No one attempted to define "cube" or "surface" although these were the only suggestions that received any support when made. Two pupils attempted to define "distance," while the remainder selected either "triangle" or "square." From the discussion which followed it was evident that degree of familiarity with the concept was the basis of selection rather than anything else. The teacher examined these definitions carefully and selected for discussion at the next meeting of the class those which he considered typical. These were written on the board. They varied widely in meaning and precision of statement, and the pupils were asked to suggest any improvements they thought desirable. Comparatively few responded and the suggestions that were offered had no particular significance. The general inability of these pupils to recognize looseness of expression was particularly noticeable. After some discussion which resulted in minor improvements, the pupils were willing to accept each of the three definitions which follow:

> A *triangle* is a figure with three lines as sides.
> A *square* is a figure having four sides of the same length and four right angles.
> The *distance* between two points is the length of the line joining them.

The teacher pointed out that acceptance of these definitions really implied that the words used in them had the same meaning for each of the pupils. Attention was thus directed to a consideration of these words and the question was raised as to what is meant by "figure," "line," "side," "angle," "point" and "length," these being the only words questioned. In the course of the discussion which followed one pupil said, "Everyone knows what a line is," and when another pupil asked whether he meant "straight line" or "curved line" he replied that "It made no difference for everyone knows what a straight line is and everyone knows what a curved line is." The teacher considered this to be a most important statement and wanted the class to appreciate its significance. He asked whether they agreed with this pupil. Is the concept of "straight line" the same in the thinking of all people? Does everyone know "what a straight line is"? Do people in Moscow, Rome, Paris and Berlin have the same concept of "straight line" as the people in Columbus? All the pupils agreed that "straight line" did mean the same thing to all people and that no useful purpose would be served in attempting to define it.

Here then was a concept concerning the meaning of which there was apparently no vagueness or ambiguity *in the minds of the pupils*. It was accepted as meaning the same thing to all of them, and the question was raised as to whether the other significant words in the three accepted definitions were of the same nature. Each of them was carefully considered and after some discussion the pupils agreed to accept "point" and "angle" as concepts which were unambiguous and without vagueness. There was disagreement concerning "figure," "side" and "length," which raised some doubt whether the original definitions should now be accepted. However, no attempt was made to revise them at that time for it seemed evident to the teacher that the thinking of the pupils had turned in a direction which offered more promising results. Some of them recognized at once that instead of beginning the study of space by defining certain concepts it was much better and, in fact, necessary to begin with the selection of a few concepts about the meaning of which there was no disagreement and for which explicit definition was unnecessary. While the real significance of this was not recognized at that time by all pupils, the importance of and the necessity for these "primitive concepts" became increasingly clear to them as the work developed.

Up to this time there had been no general consideration of a textbook. While some pupils had wanted to know "What text is to be used," it was now generally recognized by all of them that to use any text would spoil to a large extent the opportunity for the kind of thinking they had been doing. Each pupil was thus encouraged to develop his own text and was given freedom to develop it in his own way. This procedure is consistent with the assumptions in the preceding chapter, for while group discussions usually resulted in common agreement concerning the undefined terms, the definitions and assumptions, each pupil was given opportunity to express his own individuality in organization, in arrangement, in clarity of presentation and in the kind and number of implications established. Most of the pupils called this text "A Theory of Space" and they now recognized that to build this theory it was essential to select a few concepts which were without ambiguity and which meant the same to all of them. They decided to call these "The Undefined Terms" and the first section of the text was to be used for the purpose of listing these terms as they were agreed to in the development of the work. This list began with "point," "straight line" and "angle," which had already been accepted. The teacher, however, reminded the pupils that they had agreed to build a theory of space and the question was raised as to whether or not these three primitive concepts belonged in the construction of such a theory. Had "point, " "straight line" and "angle" any relation to space? In answer to this one pupil said, "I think space is full of points," and through questions and suggestions from both teacher and pupils the following agreements were reached:

There are points in space.
A line can be drawn through any two points.

These were recognized as assumptions and it is important to observe that they were made *by the pupils* and considered *by them* to have originated in their own thinking.

To illustrate the second of these assumptions the teacher placed two points on the board and asked any pupils who felt they could draw a line through these two points to volunteer for this purpose. One was selected from the large number responding and he drew a straight line through the two points. The teacher then asked another pupil to draw the line he had in mind and he said it would be "the same line." Other pupils who had volunteered were invited to "draw a line through these two points," but the general reply was that "the line had already been drawn." Now there is a big difference between "a line" and "the line." This difference was emphasized by the teacher, who then drew a curved line through these same two points. Immediately the pupils said that they meant "a straight line" and that there was "only one such line" through these two points. The suggestion was made that "line" should always mean "straight line" and the assumption was then revised by the pupils to read "One and only one line can be drawn through any two points."

The teacher asked the pupils how long this line was. Some of them replied that its length was determined by "the distance between the two points," while others said that it was just "as long as you want to make it." This led to considerable discussion out of which grew the idea that "a line could be extended in either direction just as far as you want to extend it," and this was finally accepted by all pupils as a third assumption. No pupil recognized that certain properties of space were implicit in this assumption, and when the teacher raised this question one pupil did point out that "we are assuming that space has no end." With one or two exceptions, however, this point had at that time no significance to the pupils. The assumption seemed reasonable to them and was validated by their own experience.

The discussion which led to the preceding assumption also focused the attention of the pupils on the portion of the line between the two points. They referred to this as "a piece of the line," and pointed out that the length of this "piece" was the distance between the two points. The concept of "line segment" was thus introduced, and the pupils made the following definitions:

A *line segment* is a piece of a line.
The *length of a line segment* is the distance between its two end points.

Considerable emphasis was given to the fact that these two concepts, "line segment" and "length of a line segment," were defined in terms of other concepts which were undefined. When asked what these undefined concepts were the pupils replied that they were "point" and "line." No question whatever was raised concerning "distance," although this was one of the concepts for which definition was earlier attempted. It is significant that later on when the thinking of the pupils was more critical both "distance" and "between" were selected as terms which needed defining. However, after an effort was made to define them they were placed among the undefined terms.

As an exercise which offered promising possibilities the following suggestions were given the students:

On a piece of paper locate any two points, *A* and *B,* and draw the line *AB.* Then locate two different points, *X* and *Y.* What are all possible relations which line *AB* can have to line *XY?* Considering *only* lines *AB* and *XY,* write down all the properties of the resulting figure which you either know or are willing to accept.

Different pupils expressed "all possible relations" of these two lines in very different language. The relations which they tried to express are given in Table 6, which also indicates the number of pupils recognizing the possibility of each kind of relation.

Table 6
Possible Relations between Any Two Lines Recognized by the Pupils

Relations Recognized by the Pupils	Number Recognizing Relation
Intersecting lines	25
Parallel lines	11
Perpendicular lines	9
Skew lines	3

The properties of the figure either "known" or "accepted" were also expressed in a variety of ways. However, on analyzing the statements it was clear that the following points had been made:

Lines *AB* and *CD* can be extended indefinitely in either direction.
When two lines intersect four angles are formed.
Vertical angles are equal.
The sum of the angles about a point is 360°.
If the lines are perpendicular each of the four angles is 90°.
A straight angle contains 180°.
Parallel lines will never meet.
Skew lines are not in the same plane.
Two lines intersect in a point.

These results reflect the retention of ideas to which some of the pupils had been introduced in a study of informal geometry. However, although they showed some familiarity with the facts, there was evident need for a clarification of the concepts. The statements of the pupils were classified and samples were selected for criticism by the entire class. These were written on the board and as a result of the general discussion which followed "plane" and "equal" were added to the list of undefined terms, definitions were written for the following concepts:

vertical angles
right angle
straight angle

intersecting lines
perpendicular lines
parallel lines
skew lines

while there was general agreement that the following statements should be accepted as assumptions:

Vertical angles are equal.
Two lines can intersect at one and only one point.
One and only one plane can be passed through two intersecting lines.
One and only one plane can be passed through two parallel lines.
One and only one plane can be passed through a line and a point not on the line.
One and only one plane can be passed through three points which are not in the same line.

When the pupils later became familiar with the idea that certain factors "determine" certain other factors and when they understood the full significance of this concept, they revised the preceding five assumptions to read as follows:

Two intersecting lines determine one and only one point.
Two intersecting lines determine one and only one plane.
Two parallel lines determine one and only one plane.
A line and a point not on the line determine one and only one plane.
Three non-collinear points determine one and only one plane.

The word "collinear" had been accepted to replace "in the same line," while "non-collinear" had been accepted to mean "not in the same line." Similarly, the word "coplanar" was accepted to mean "in the same plane."

It is well, at this point, to make explicit certain procedures which are illustrated in the preceding discussion and which have a strong influence on the content.

1. *With respect to the undefined terms.*

   a. The terms that were to remain undefined were selected and accepted by the pupils as clear and unambiguous.

   b. No attempt was made to reduce the number of undefined terms to a minimum.

2. *With respect to the definitions.*

   a. The need for each definition was recognized by the pupils through discussion. Definitions were an outgrowth of the work rather than the basis for it.

   b. Definitions were made by the pupils. Loose and ambiguous statements were refined and improved by criticisms and suggestions until they were tentatively accepted by all pupils.

3. *With respect to the assumptions.*

   a. Propositions which seemed obvious to the pupils were accepted as assumptions when needed.

77

b. These assumptions were made explicit by the pupils and were considered by them as the product of their own thinking.

c. No attempt was made to reduce the number of assumptions to a minimum.

d. The detection of implicit or tacit assumptions was encouraged and recognized as important.

e. The pupils recognized that, at best, the formal list of assumptions is incomplete.

In addition to the section for "the undefined terms," each pupil reserved in his text a section for "definitions" and recorded in this section all definitions made in the development of the theory. Similarly, another section was reserved for listing the "assumptions" and new assumptions were added to the list as the need for them was recognized.

While "angle" was accepted as one of the undefined terms, the teacher suggested that the pupils think of an angle as being generated by the rotation of a line about a fixed point in a fixed line. This was illustrated by using the hands of a clock and many additional illustrations were given by the pupils, such as a pair of compasses, a pair of scissors and the like. The concept of angle was thus enlarged to include "amount of rotation," and in developing this idea it was pointed out that the rotation might be in either a clockwise or a counter-clockwise direction from the given fixed line. While the pupils felt that they should agree as to the direction of rotation, they also felt that this was a point of minor importance. One pupil pointed out that this situation resembled the number scale to some extent, since from any given point on the number scale one could go "either in the positive or negative direction." This was an important contribution to the group thinking, and by emphasizing the resemblance suggested it soon became apparent to the pupils that if the rotation in one direction is considered as generating a positive angle the rotation in the other direction may be considered as generating a negative angle. The pupils were unanimous in believing that the angle should be positive when the rotation was clockwise and negative when it was counter-clockwise. They were surprised to learn that mathematicians had agreed to consider counter-clockwise as generating a positive angle, but accepted this convention "since it is not important anyway."

As an outgrowth of the preceding discussion "fixed" and "rotation" were added to the list of undefined terms, while the following concepts were defined:

vertex	obtuse angle
initial side	reflex angle
terminal side	adjacent angles
acute angle	

Perhaps it should again be emphasized that definitions were made by the pupils *after* a recognition of the characteristics of the concept to be defined. For example, the definition of adjacent angles was an outgrowth of the discussion which resulted from a consideration of the following points:

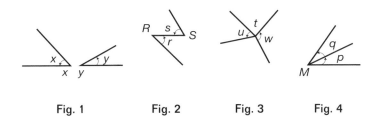

| Fig. 1 | Fig. 2 | Fig. 3 | Fig. 4 |

In Figure 1, what is the vertex of angle *x*? What is the vertex of angle *y*? What is the initial side of angle *x*? What is the terminal side? What is the initial side of angle *y* and what is the terminal side? Do these angles have any elements in common? If so, what are they? The angles in each of the other diagrams were examined in a similar manner and such an examination served to emphasize the characteristics of angles *p* and *q* which make the relation between these angles uniquely different from the relation between the two angles in each of the other diagrams. When the contributions of the pupils were of such a character as to indicate that they recognized these characteristics, the teacher told them that the name given to angles thus related was "adjacent angles." He then asked for a definition of adjacent angles and there was unanimous agreement that:

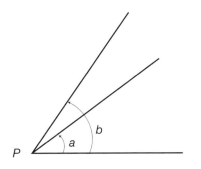

> *Adjacent angles* are angles that have a common vertex and a common side.

The teacher then drew angles *a* and *b* as in the accompanying figure and asked if they did not meet the requirements of this definition. He then asked if the relation between *a* and *b* was the same as the relation between angles *p* and *q* in Figure 4. There was an immediate response to this and all the pupils were ready to accept the corrected definition given by one of them:

Adjacent angles are angles that have a common vertex and a common side between them.

It is interesting to note that the pupils originally agreed that "An acute angle is one which is less than 90°." Later, however, it was pointed out by one of the pupils that "According to that definition −30° is an acute angle." No one disagreed with that statement, but when another pupil asked if −100° was an acute angle it became clear that the original definition included angles which were not meant to be included. After some discussion it was changed to read, "An acute angle is an angle which is greater than 0° and less than 90°."

The teacher frequently reminded the pupils that they were building "A Theory of Space" and that wherever possible their thinking should extend beyond the plane into three dimensions. Reminders of this nature soon became unnecessary, and many concepts of space were developed. The question was raised, for example, as to whether it was possible to rotate one plane about another plane just as a line could be rotated about another line. Through consideration of this problem, for which numerous illustrations were given, the concept of dihedral angles was developed, while the following assumptions were suggested and accepted by the pupils:

> If two planes intersect, they can intersect in one and only one line.
> Vertical dihedral angles are equal.

The teacher discouraged any attempt by the pupils to memorize the definitions and assumptions accepted. On the other hand, each pupil was encouraged to use his text freely and to refer to whatever definitions and assumptions he needed in the development of his work. This served to emphasize the importance of his text and was a strong factor in encouraging him to keep it neat, well organized and always up to date. As new definitions and assumptions were made they were written in the text with numerous illustrations and supplementary comments, depending on the interests and abilities of the individual pupils.

Misuse and loose interpretations of these basic concepts were unusually rare, largely, in the opinion of the writer, because of the fact that they were an outgrowth of the thinking of the pupils. When a term was used incorrectly, in either oral or written discussion, the pupil was referred to his text to check on his statement for the definition of that term. It was by such means and not by memorization that the pupils became familiar with the technical language they were helping to create.

Definition in Non-Mathematical Situations

The interest of the pupils was at all times particularly noticeable. They participated actively in discussion and there was increasing evidence that they were learning to think together. The emotional controversy which had marked earlier efforts to define "aristocrat" and "labor class" was noticeably lacking in their efforts to define mathematical concepts, and the teacher pointed out the desirability of thinking about the pressing problems of democracy in this same objective manner. To provide opportunity for this kind of activity exercises similar to the following were prepared and in many instances the content of such exercises was suggested by the pupils:

> In the administration of certain Ohio Laws it became necessary to know just what a restaurant is, and in connection with this problem the following editorial appeared in one of the Columbus papers:
>
> "What Is a Restaurant?"
>
> "The art of precise definition is not so easy as it may seem. To produce an acceptable definition of a dog calls for ability in the art of lucid and exact statement. . . .

* * * * * * *

80

"What is a restaurant? Well, a harder order to fill would be: What is a drug store? To say it is a place where commerce in drugs is carried on is as far afield as to say restaurants are places where people can rest their aunts.

"Well, the Ohio State Restaurant Association has tackled the job, has drafted a definition and will ask the Ohio Legislature to give them the sanction of their approval when they say that a restaurant is a place of business where 50 per cent or more of the gross sales accrue from the sale of food-stuffs consumed on the premises."

Now, in view of the way in which "restaurant" has been defined by the Ohio State Restaurant Association, let us consider the following questions:

1. The major part of the gross sales of the Atlantic and Pacific Tea Company is of food-stuffs. Can these stores rightly be called restaurants? Why?

2. The gross sales of one of the White Castles in Columbus is approximately $12,000 a year, all of this coming from the sale of food. Some of this food is eaten where purchased while the remainder is eaten elsewhere. If the amount eaten elsewhere is $6,185.40, is the White Castle rightly called a restaurant?

3. In 1934 the gross sales of a business amounted to $18,439.27. These sales were distributed as follows: drugs, $9,203.12; ice cream, $3,297.65; candy, $1,069.60 and lunches, $4,868.90. In the light of these data discuss the problem as to whether this place of business is a restaurant.

4. How would you decide whether or not a combined ice cream parlor and soda fountain which also serves light lunches is a restaurant?

5. Discuss the problem as to whether or not a place which sells beer and liquor and also serves sandwiches as well as other types of food is a restaurant."

This proved to be a most profitable exercise. It stimulated thoughtful discussion, and the point was very definitely made that before this suggested definition could be effective in distinguishing restaurants from other places of business in Ohio, "foodstuffs," "consumed" and "on the premises" must also be carefully defined.

An amendment to the constitution of Ohio which removed the sales tax on "food" also presented an interesting illustration of the importance of definition and this situation is used in the following exercise.

On November 3 the voters of Ohio approved the constitutional amendment which provides that "on and after Nov. 11, 1936, no excise tax shall be collected upon the sale or purchase of food for human consumption off the premises where sold." What words in this amendment must be clearly defined in order to make effective the administration of the law?

A number of questions similar to those which are sure to arise any time after this law is put into effect are suggested below. Consider each of these questions and in view of the above amendment answer them in accordance with your best judgment. In each case point out the factors on which your answer depends:

1. Mr. Carter went into an A and P store and purchased three pounds of beef. He happened to say he was buying this meat for his dog and the clerk insisted that a tax be charged. Was the charge legal?

2. Mrs. Page purchased a bottle of cod liver oil at the drugstore. She expected to pay a tax on this purchase but the clerk told her that was unnecessary. Was the clerk correct in requiring no tax?

3. Fred wanted to buy a box of chocolates that was priced at $1.50. He had only $1.50 in his pocket and he hesitated to ask for the chocolates because he feared he would have to pay a tax on the purchase. Were his fears justified?

4. A twelve-year-old girl went to the store to buy some baking chocolate for her mother. The clerk charged her a tax on the purchase. When the mother learned of this she said the clerk had made an error. She telephoned him and asked that the tax be refunded. Do you think she should get this refund?

5. Mr. Smith asked his friend to dine with him at the Statler Hotel in Cleveland. He had this dinner served in his room on the tenth floor. Was the waiter justified in charging Mr. Smith a tax on the amount of his bill?

6. Joe lived on the tenth floor of an apartment house and bought his groceries from a store located on the first floor of the same apartment house. He argues that his groceries were consumed "off the premises" and that therefore they were not taxable. Do you agree with him?

7. Tom bought fifty cents worth of apples at a fruit store and paid no tax on them. However, when the fruit merchant saw Tom eating these apples while seated in his car which was parked directly in front of the fruit store he asked him for the tax. Should Tom pay it?

8. Discuss in general the difficulties which are likely to occur in administering this law. How can they be avoided?

For the purpose of further emphasizing the necessity of clearly defining "significant words and phrases" before the statements which contain them can have any real meaning, the educational planks in the state platforms of the Republican and Democratic parties of Ohio in the 1934 election were selected and the exercise which follows below was prepared.

This kind of exercise had a most helpful effect on the thinking of the pupils. To compare the way in which the governor might handle the educational problems of the state with the educational plank of the platform on which he was elected, deeply impressed the pupils with the real need for clear definitions of such terms as "adequate," "fair," "sufficient" and "proper" in all such documents.

Here are two statements relating to the education of the boys and girls of Ohio. Statement One represents A's viewpoint on this problem while Statement Two represents B's viewpoint. These statements have little meaning until certain significant words or phrases are defined. In your opinion, what are these words or phrases? List them on the lines below the statements:

A's Viewpoint	B's Viewpoint
1. The education of childhood and youth is a fundamental obligation of state government. Therefore, we pledge ourselves to provide, without further delay and with safeguards which will preserve local control, financial support of our public schools that will establish and maintain a reasonable minimum standard of education throughout the state, to the end that every boy and girl in Ohio may secure an elementary and high school education.	2. We pledge an adequate program of state revenues, to be distributed by the state to the elementary and high schools of Ohio, in sufficient amounts to make up the deficiency in local revenues as determined by a fair foundation program and based upon proper economy of operation, fair and adequate salaries, and the maximum amount of local self help. The schools of Ohio must be kept open. The constitution guarantees education to the children of this state. We cannot afford nor can we tolerate any backward step in the education of the citizens of the future.

Let us now consider the following assumptions:

     1. That A and B are candidates for governor of Ohio.

     2. That education is the only issue between them.

     3. That you favor a strong and effective program of public education.

Under these circumstances for which of the two candidates would you vote?

Another exercise which led to much thoughtful discussion and which again emphasized the far-reaching importance of definition follows:

Not many years ago the Supreme Court of the United States rendered a decision concerning the relationship of the state to the schools of the state. In a unanimous decision the court ruled that the state had power over all schools in respect to the following matters and that it was the responsibility of the state:

     1. To require "that all children of proper age attend some school."

     2. To require "that teachers shall be of good moral character and patriotic disposition."

     3. To require "that certain studies plainly essential to good citizenship must be taught."

     4. To require "that nothing be taught which is manifestly inimical to public welfare."

Now before the state of Ohio or any other state can meet its responsibility as outlined in this decision it is essential to know just exactly what the decision means. Its meaning depends on certain significant words or phrases which it contains and the extent to which any state government will control the schools of the state depends, among other things, on how it defines these words or phrases.

Read this decision carefully and thoughtfully. What, in your opinion, are the words and phrases which need careful definition in order to make the decision clear? List them in the space provided below:

_____

_____

_____

This Supreme Court decision applies to all of the forty-eight states and presents the powers and responsibilities which each state has concerning the schools of the state. Now the educational opportunities of some states are definitely superior to those of certain other states and if we assume that each state government is equally effective in meeting its responsibilities for educating the young people of the state how may these wide differences in educational opportunity be explained?

The interpretation of papers of state is also an excellent illustration of the fundamental importance of definition. A part of the New Deal program, for example, was held to be unconstitutional because of the definition given to "Interstate Commerce." An illustration of this, which was very meaningful to pupils, occurred in the administration of the State government of Ohio and the following exercise was built around this situation:

The constitution of Ohio imposes certain obligations on the governor as to just how he is to handle a bill presented to him for his signature after it has been passed by both houses of the general assembly. One section of the constitution, for example, reads as follows:

"If he does not approve this bill, he shall return it with his objections in writing to the house in which it originated, which shall enter the objections at large upon its journal, and may then reconsider the vote on its passage. If three-fifths of the members elected to that house vote to repass the bill, it shall be sent, with the objections of the governor, to the other house, which may also reconsider the vote on its passage.

"If a bill shall not be returned by the governor within ten days, Sundays excepted, after being presented to him, it shall become a law in like manner as if he had signed it, unless the general assembly by adjournment prevents its return; in which case, it shall become a law unless, within ten days after such adjournment, it shall be filed by him, with his objections in writing, in the office of the secretary of state."

The second general appropriation bill, commonly known as "the revised budget bill," originated in the senate. It was approved by both houses of the general assembly and the governor received it for his signature on January 28. While awaiting the action of the governor the legislators took a recess until February 25 with a definite obligation to reassemble on that date. The governor did not wish to sign the bill as it was presented to him and vetoed certain items to the extent of $3,002,734. Then on February 7, in view of the fact that the legislature was not in session, he filed this bill together with his objections in the office of the secretary of state. Because of this action the legislators, when they reassembled on February 25, were unable to act on the governor's vetoes.

Some people believe that the governor did not comply with the obligations imposed upon him in the section of the constitution quoted above and therefore say that his vetoes are illegal. It is possible that this issue may be referred to the supreme court of Ohio for decision where it will be studied "without prejudice and in the clear light of logic." Considering the facts presented here:

> "Do you believe that the governor handled this bill in accordance with the provisions of the constitution? What are the significant words in the constitution which must be defined before an opinion can be reached? How would you define these words?

> "Present your analysis of this situation in logical form and state your decision as to the constitutionality of the way in which the governor handled this bill."

In their analysis of this situation most of the pupils were led to two different conclusions, depending on the way "recess" and "adjournment" were defined, and the exercise served to emphasize the great significance of definition in matters of far-reaching concern.

Exercises of this nature were not all given at one particular period. They were interspersed throughout the work of the year and were supplemented by thoughtful consideration of the points raised for discussion. The transfer value of this kind of exercise was almost immediately noticeable, for in other classes these pupils were asking that vagueness in the use of words be replaced by well defined terms, and they voluntarily contributed many illustrations from other school experiences and from out-of-school situations wherein their own thinking had been clarified because of their improved ability to select "significant words and phrases" which needed definition before their meaning was clear.

Implications of Definitions and Assumptions

At the beginning of one class session the teacher had drawn two intersecting lines on the board, as h and h' in the accompanying figure. Concerning this diagram he had written:

Assume that h is fixed in position and that h' revolves about O in a counterclockwise direction. State all the properties of the resulting figure that you are willing to accept.

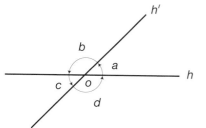

While these "properties" were stated in a variety of ways the ideas were clear and, in general, well expressed. Two pupils were absent. The responses of the remaining twenty-three are summarized in Table 7.

Table 7

Relations Recognized by Pupils When One Line Rotates about a Fixed Point in Another Line

Response of Pupils	Number Making Response
h and h' can be extended indefinitely.	23
Vertical angles are always equal.	21
h and h' determine a plane.	23
Angles a and c get larger while angles b and d get smaller.	18
There is a time when the four angles formed are equal.	17
There is a time when h is perpendicular to h'.	15
When the four angles are equal h will be perpendicular to h'.	5
Angle a is never smaller than 0°.	3
Angle a is never larger than 360°.	3

These results were presented to the entire class. They were discussed, critically examined and clarified. One pupil pointed out that the last two were based on the assumption that h' started its rotation from the position of h and that its rotation was limited to one complete revolution. This discussion also led to the agreement that:

> If two lines intersect in such a manner that the adjacent angles thus formed are equal then the two lines are perpendicular.

Originally perpendicular lines had been defined as "two lines which make angles of 90° at their point of intersection," and the limitations of this definition were now apparent.

The properties of the accompanying diagram were similarly explored. In this case, the pupils were given the following suggestions:

> *Assume* that w and h' are two lines intersecting at O' and that w and h' are fixed in position. Let O be any point on w other than O' and let line h rotate about point O.
>
> What properties of the resulting figure are you willing to accept?

While many of the properties listed resembled those in the preceding table, there were in addition other properties which related not only to the position of h with respect to w but also to the position of h with respect to h'.

Attention was focused on the three that follow:

> There is a time when h is parallel to h'.
> There is a time when angle a equals angle a'.
> When angle a equals angle a' then h is parallel to h'.

While these properties were not listed by all the pupils, the last being given by two pupils only, they were accepted without objection and some pupils expressed annoyance that they had overlooked them. The teacher now referred to the original definition of parallel lines, which was "Parallel lines are lines that will never meet even though extended indefinitely in either direction," and pointed out that it would be difficult to show that two lines met the requirements of this definition just as it was impossible to show that two lines made angles of 90° at their point of intersection, since measurement is not permitted in demonstrative geometry. He further suggested that since perpendicular lines had now been defined in terms of "equal adjacent angles" it might be possible to define parallel lines in terms of equal angles. The response was prompt and one pupil gave the following definition:

> Parallel lines are lines that have the same rotation from another line.

This definition was criticized and was soon corrected to read:

> Parallel lines are lines that have the same amount of rotation from a line which intersects them.

The pupils now seemed willing to accept this definition and the teacher apparently agreed also. He then proceeded to draw different diagrams, inserting numerical values for the sizes of the angles, and asked if the conditions were such as to make the lines parallel. Among these diagrams was one similar to the figure at the left and the usual question was asked, "Under the assumed conditions, is h parallel to h'?" This development served to make explicit that which the pupils had been assuming and illustrated most effectively the importance of "tacit assumptions." Many pupils then volunteered to correct further the definition of parallel lines and in its final form it was:

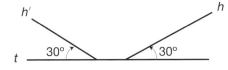

> Parallel lines are lines which have the same amount and direction of rotation from a line which intersects them.

Some question was raised as to the necessity for defining "direction," but there was general agreement that this concept belonged among the undefined terms.

Many of the implications of this definition were recognized at once. The teacher placed the accompanying diagram on the board and asked the pupils to assume that h and h' were parallel and that angle a equaled 80°.

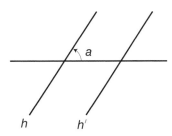

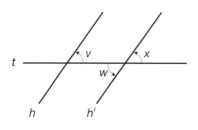

Without any apparent difficulty *all* pupils then gave correctly the size of each of the other angles in the figure. Later the problem was made more general and the pupils were asked to study the properties of the figure at left in which h and h' are assumed to be parallel. No one failed to recognize that under these conditions $x = v$ and $w = v$. However, more than half the pupils limited their analysis to these two statements because (it was later learned) it was assumed, since only these angles were marked, that no others were to be considered. The other pupils, however, inserted additional letters and gave most of the familiar relations between the angles formed when two parallel lines are cut by a transversal. These relations, however, were stated in terms of the particular angles in this diagram, and when the teacher suggested that they be generalized to cover all parallel lines the struggle to state these relations in general terms was found to be due to the difficulty of explaining just what angles were meant. For example, one pupil, in attempting to generalize the statement that $x = v$, expressed himself as follows:

> When two parallel lines are intersected by another line any outside angle is equal to the non-adjacent inside angle.

and when another pupil pointed out that "non-adjacent inside angle" might mean any one of the three "inside" angles that were non-adjacent to x, he revised his original statement as follows:

> When two parallel lines are intersected by another line any outside angle is equal to the non-adjacent inside angle on the same side of the line which intersects the two parallels.

The pupils recognized the awkwardness of such a statement and were ready for the suggestion of the teacher that a name be given to "the line which intersects the two parallels" and to any two angles so located with respect to this line as are angles x and v. As an outgrowth of this discussion the following definitions were accepted:

> A *transversal* is a line which intersects two or more other lines.
> When two lines are cut by a transversal any outside angle and the non-adjacent inside angle on the same side of the transversal are *corresponding angles*.

Before accepting this second definition, however, the question was raised as to how "outside" and "inside" were defined. After some discussion it was decided to place these among the undefined terms. The brevity and conciseness with which it was now possible to state the familiar theorem concerning "corresponding angles of parallel lines" impressed the pupils with the convenience of carefully defined terms. The idea of "converse statements" also developed from this discussion, and the point was made that when one accepts a definition he also accepts that definition when turned around. This was finally stated as "A definition when turned around is acceptable authority" and was regarded as an assumption.

88

Through a continuation of such processes the theorem concerning alternate interior angles of parallel lines was established and the class extended this to cover alternate exterior angles as well. Some of the pupils felt that these relations should be accepted as assumptions because "anyone can see those angles are equal," while others worked out deductive proofs for them because they "enjoyed it." Of the twenty-five pupils, however, no one suggested the relation between the two interior angles on the same side of the transversal, and the teacher directed their attention to this relation in the following manner:

Assume: h and h' to be two parallel lines cut by the transversal t.

Do these conditions imply any relation between angles a and c?

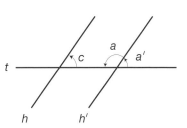

In their study of this situation some pupils gave a definite value to the size of angle a and then proceeded to show that for that particular value angles a and c were supplementary. They were encouraged to give other values to a and to see if the same relation held between a and c. Other pupils discovered this relation without using angles of definite size and presented a satisfactory proof for their conclusion. Five pupils made very little progress in the study of this figure until such questions were asked as "What is the sum of angles a and a'?" and "Do you know anything about angles a' and c?" Finally, both approaches to the proof of this theorem were discussed and examined. While the words "inductive" and "deductive" were not used, the essential difference between the two types of proof was emphasized and all pupils felt that the "deductive proof" was more convincing. It was pointed out, however, that this proof depended on an assumption which had not been stated explicitly and this led to an acceptance of the assumption that:

A quantity or magnitude may be substituted for its equal.

The first statement of this assumption was awkward, loose and inaccurate. This statement was criticized, improved and finally accepted as above expressed. However, in the course of this discussion numerous pupils asked how "quantity" and "magnitude" were defined and in the end these were placed among the undefined terms. The teacher then asked the pupils to generalize the conclusion, and an attempted generalization, followed by criticism, suggestion and refinement led to a statement of the *general* theorem which was accepted by all the pupils. There had already been agreement that "the converse of a definition is acceptable authority" and following these theorems on parallel lines the teacher raised the question whether the converse of a theorem was also acceptable authority. Many pupils believed that the converse of theorems should be accepted, but the converse of such propositions as:

If a man lives in Columbus then he lives in Ohio.
All right angles are equal angles.
If a man is rich then he can buy a car.

89

impressed all of them with the necessity of proving a converse before it can become an acceptable authority. The converse of each theorem already established was stated and most of them were proved.

These theorems were regarded by the pupils as *implications* of the definitions and assumptions they had made. Each pupil reserved a section of his text for "Implications of Definitions and Assumptions" and in this section placed all the implications which he established, giving the proof and generalized statement for each of these theorems. Wherever possible, corresponding three-dimensional concepts were considered and many properties of Euclidean space were thus accepted. The real purpose of this extension into space was to give the pupils an opportunity to visualize three-dimensional figures and, in general, no proofs were given. However, three of the pupils became greatly interested in this three-dimensional work and did prove a large number of theorems concerning the properties of space.

The teacher had a definite purpose in directing the thought of the pupils to the relation between the two angles on the same side of a transversal which intersected two parallel lines. With Figure 1 on the board—

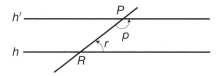

Fig. 1

he asked the pupils to assume that h and h' were parallel and to write down all the result-ing implications. Among these was the statement that angle p + angle r = 180°. He then asked them to assume that h is fixed in position but that h' rotates around the fixed point P. Much emphasis was placed on the fact that the slightest rotation of h' to either the right or the left brings into the finite plane and under control a point which, when the lines are parallel, is beyond control. It is the point at which h' will intersect h, and although it may be at a very great distance from R the assumption that non-parallel lines in the same plane intersect defines its existence. The pupils participated in this discussion and to a large extent guided this development by their questions and contributions. The teacher suggested that they study what happens when the rotating line h' takes a position such as that indicated by the dotted line h'' in Figure 2.

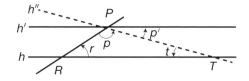

Fig. 2

90

The point T was thus defined and a pupil pointed out that PRT is a triangle. Another pupil said, "How do you know? Triangle has not yet been defined." The definition of triangle earlier accepted but never used was then recalled, and when a pupil pointed out that a diagram consisting of two parallel lines cut by a transversal might be considered "a figure with three lines as sides" the original definition was changed to read:

A triangle is a closed figure with three lines as sides.

The teacher then raised the question of what happened to angle p as the line h' rotated around P in the direction indicated. The pupils promptly agreed that the angle p was made smaller by the amount of rotation but that with the rotation there appeared in the figure a new angle t whose size was always equal to the amount by which angle P was decreased. Thus, if angle p is decreased by p' then $t = p'$ because h'' is a transversal cutting the parallel lines h and h' and the alternate interior angles of parallel lines are equal. Since angle r is constant and since the decrease of angle p is exactly balanced by the increase of angle t, there was unanimous approval of the statement that angle r + angle RPT + angle $t = 180°$.

After a further study of this figure, including the changes that occur as h' makes one complete revolution about P, the pupils were asked to draw any triangle and see what they could discover about the sum of the angles of that particular triangle. All of them felt that the sum was $180°$ but no one knew, at first, just how to proceed to demonstrate this fact. However, after thoughtful study followed by a suggestion from the teacher that reference to the diagram of the preceding discussion might prove helpful, there was increasing evidence from all parts of the room that discoveries were being made, and before the class period was over seventeen of the twenty-four pupils present had worked out an "acceptable" proof for the theorem concerning the sum of the angles of a triangle.

The only real weakness in each of these seventeen proofs was found in the statements relating to the drawing of the auxiliary line through one of the vertices. All the pupils felt the need for such a line and all of them drew it. However, twelve of them said nothing whatever as to where this line came from or how it was drawn, although in the remainder of the proof it was assumed to be parallel to the side opposite the vertex through which it was drawn. Four of the pupils did say that the line was drawn parallel to whatever the name of the side happened to be, but did not say that this line was to be drawn through any particular vertex; nor did they feel the need for any authority to support this construction. In the proof of only one pupil was the statement "Draw a line through A in such a way that it is parallel to BC" and as an authority he gave "any helping line can be drawn."

These proofs are called "acceptable" because in each of them the basic mathematical ideas are acceptable. Their weakness lies in lack of precise and accurate statement. All of them, however, served as excellent illustrations of the way in which unrecognized assumptions can creep into one's thinking and definitely affect conclusions. The pupils were impressed and, in the judgment of the writer, greatly helped by the discussion on this point and wanted to make explicit those assumptions which were implicit in their proofs. It was thus that the famous parallel postulate of Euclid was approached. The various statements of this assumption that were made before it was expressed in a form acceptable to all are interesting. The original statement, which is the first of those given below, was criticized, improved and further improved as indicated by those statements which follow it:

A line can be drawn parallel to another line.

Through any point a line can be drawn parallel to another line.

Through any point not on a given line it is possible to draw a line parallel to the given line.

Through a given point not on a given line it is possible to draw one and only one line parallel to the given line.

Enough of the history of Euclid's assumption was discussed to stimulate and retain the interest of the pupils and also to give them some background for the later discussion of the assumptions which led to non-Euclidean geometries. As a generalization of this theorem one pupil volunteered the statement:

The angles of a triangle equal 180°.

and a number of the pupils agreed with this statement. Other pupils, however, suggested that it be revised to read:

The sum of the angles of any triangle is 180°,

and as an illustration of the critical manner in which the pupils were thinking still another suggested that "in order to be more accurate" this general statement should be:

The sum of the interior angles of any triangle is 180°.

To this there was unanimous agreement and the complete proof of the theorem, followed by the general statement, was placed in each textbook in whatever form seemed logical to the individual pupil concerned and best suited his abilities.

At this point, it will be helpful to summarize the general principles and methods by which "implications of definitions and assumptions" or "theorems" are discovered. If the pupil is to have the opportunity "to reason about the subject matter of geometry in his own way," no theorem should be stated in advance; for such a statement fixes, to some extent, the direction of his thought and deprives him of discovering for himself the mathematical relations which control a situation. While in some cases assistance is certain to be needed the teacher should consider himself nothing more than a guide who directs *toward* the discovery and develops within the pupil increasing power to discover for himself. The general principles and methods which were followed in developing this sense of discovery and which are inherent in all preceding illustrations are summarized below:

1. No formal text is used. Each pupil writes his own text as the work develops and is able to express his own individuality in organization, in arrangement, in clarity of presentation and in the kind and number of implications established.

2. The statement of what is to be proved is not given the pupil. Certain properties of a figure are assumed and the pupil is given an opportunity to discover the implications of these assumed properties.

3. No generalized statement is made before the pupil has had an opportunity to think about the particular properties assumed. This generalization is made by the pupil after he has discovered it.

4. Through the assumptions made the attention of all pupils is directed toward the discovery of a few theorems which seem important to the teacher.

5. Assumptions leading to theorems that are relatively unimportant are suggested in mimeographed material which is available to all pupils but not required of any.

6. The major emphasis is not on the statement proved, but rather on the *method of proof*.

7. The extent to which pupils profit from the guidance of the teacher varies with the pupil and the supervised study periods are particularly helpful in making it possible to care for these variations. In addition individual conferences are planned when advisable.

Teachers who are "concerned not merely by the objective goals reached by the pupils, but quite as truly with the actual searchings themselves," will recognize that when the suggestions in 2 and 3 are translated into actual classroom practice the opportunity for discovery, to which so many teachers of mathematics lend verbal allegiance, is preserved for the pupil. Also, the suggestions in 4 and 5 make it possible for each pupil to develop "whatever sequence will give him the greatest sense of accomplishment," and while, as indicated in 4, there is a small number of theorems which constitute a common background for all pupils and which serve as illustrations of what proof really means, provision is made for original work commensurate with the abilities and interests of each individual pupil. Frequently, the original work of one pupil is of such a character that it has very real value for all others, and in such cases it is presented by the pupil to the group for criticism and discussion.

In order that the principles discussed in the preceding paragraph may be still further clarified, an actual illustration may be helpful. However, to appreciate the significance of this illustration it is essential that the experiences of the class preceding this illustration be explained. Parallel lines had been defined as "lines having the same amount and direction of rotation from a line which intersects them," and the following theorems were common to the experience of all pupils:

> If two parallel lines are cut by a transversal, the alternate interior angles are equal.
> If two parallel lines are cut by a transversal, the two interior angles on the same side of the transversal are supplementary.
> If two lines in the same plane are cut by a transversal and if the alternate interior angles are equal, then the two lines are parallel.
> The sum of the interior angles of a triangle is 180°.

A number of pupils had established many other properties of parallels but only the four theorems above had been proved by all pupils. Congruence had been discussed and defined and a consideration of the conditions which made triangles congruent had led to the acceptance of the following assumptions:

> If two triangles have two angles and the included side of one equal respectively to two angles and the included side of the other, then the triangles are congruent.
> If two triangles have two sides and the included angle of one equal respectively to two sides and the included angle of the other, then the triangles are congruent.

With only this background in common the pupils were working during a period of supervised study on whatever seemed most important to them at that particular time. Some were bringing their textbooks up to date, some were analyzing certain types of non-mathematical material which claimed their interest, while some were investigating one set of assumptions and some another. Among the situations available was one which related to the accompanying figure. The suggestions given were as follows:

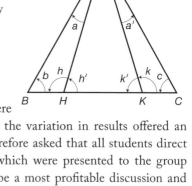

1. Assume that angle a = angle a'. What are the resulting implications?

2. Assume in addition that angle b = angle c, and study the figure for any added implications.

3. Now combine with the two preceding assumptions the additional assumption that $BH = KC$ and try to discover what additional properties are implied.

The teacher observed that a number of pupils were working on this exercise and he also observed that the variation in results offered an excellent opportunity for general discussion. He therefore asked that all students direct their attention to a consideration of these results, which were presented to the group by the individual pupils concerned. This proved to be a most profitable discussion and revealed in a rather impressive manner the cumulative effect of additional assumptions. Following the assumption in 2, some pupils reported that triangles ABH and AKC had the same shape, but no one made any attempt to show any line segments equal until the third assumption was available. One pupil, however, who had given no previous thought to this set of assumptions, raised the question whether it was necessary to prove any triangles congruent to show that $AH = AK$. He said he believed this followed at once from the assumption in 2 for that assumption did imply that angle h' = angle k' and "if those angles are equal, the sides opposite them would have to be equal." Other pupils supported this statement but all felt that it must be proved before it could be accepted. No one suggested that it be recognized as an assumption.

At the close of the period three pupils went to the teacher's office and worked independently in an effort to prove this conclusion. Two of them were successful, and the next session of the class opened with one of these pupils presenting his proof to the entire group. He met some resistance, particularly from those who themselves had been unable to discover a proof, and the nature of their criticism reveals the quality of their thought. When the bisector of angle *HAK* was drawn the question was raised whether an angle could be bisected and whether or not this bisector had to intersect *HK*. These criticisms resulted in healthy discussion, and as an outgrowth of this discussion the following assumptions, which were implicit in the proof but previously unrecognized, were accepted:

> It is possible to draw a line bisecting an angle.
> A line bisecting an angle of a triangle must intersect the opposite side.

Here, then, is an illustration of how a question raised by one pupil led to the proof of the general theorem that:

> If two angles of a triangle are equal, the sides opposite those angles are equal,

even before the "isosceles triangle" had been defined. The converse of this theorem was suggested almost immediately and was proved without serious difficulty by most members of the class.

Inductive Proof

The insertion of the word "interior" in the general statement of the theorem dealing with the sum of the angles of a triangle led many pupils to ask if a triangle had "any exterior angles." Illustrations of exterior angles were thereupon given to them and a definition of this concept was derived from a study of these illustrations. The question was immediately raised: "What is the sum of the exterior angles?" and this led to a consideration of what was meant by extending the sides "in succession." The pupils then discovered the sum of the exterior angles to be 360° and most of them made the following generalization:

> The sum of the exterior angles of a triangle is twice the sum of the interior angles.

The idea of studying the sum of the angles of a figure of more than three sides apparently occurred to none of the pupils, but since the teacher desired to use this opportunity to emphasize the nature of induction, he raised the question as to how the sum of the interior angles of a four-sided figure might be found. No further suggestion was needed. The pupils recognized the possibilities, and figures with various numbers of sides were drawn on the board. The teacher suggested that in studying the sum of the angles of these different figures it would be helpful to arrange the work in tabular form, and outlined a table for this purpose. While some pupils generalized after only two cases, there were some who failed to recognize the nature of the process and did not generalize at all. Individual conferences with these pupils seemed advisable for the purpose of clarifying this method of reaching generalizations.

As an outgrowth of this work "diagonal" and "polygon" were defined as well as some of the more commonly used special polygons, such as "quadrilateral, pentagon, hexagon" and the like. No pupil had apparently given any thought to the sum of the exterior angles of these polygons, and in reply to a question concerning the sum of the exterior angles of a quadrilateral practically all of the pupils indicated that this sum would be *more* than the sum of the exterior angles of a triangle.

Further questions revealed that this conclusion was related in their thinking to the idea that "the greater the number of sides, the greater the sum of the angles." As they gave more thought to this problem, however, some became more cautious while others became more certain. This latter group argued that:

> Since the sum of the exterior angles of a triangle is twice the sum of the interior angles, then the sum of the exterior angles of a quadrilateral must be 720°.

Discussion led to an explicit statement of the assumption on which this generalization was based, and the teacher used this as a further illustration of how hidden assumptions affect conclusions. Some of the more careful thinkers had by this time found the required sum and all were surprised that it was the same as that for the exterior angles of a triangle. Following a similar investigation by each pupil of the exterior angles of a pentagon, most of them were ready to accept the generalization:

> If the sides of any polygon are extended in succession, the sum of the exterior angles thus formed is 360°.

The pupils realized that here was a "new way of reaching generalizations" and the nature of this new process was examined in detail. Many illustrations of induction were given by different pupils, and the teacher added others which were drawn from various fields of thought but particularly from the sciences. Among these illustrations were some which showed the failure of induction, and the dangers and limitations of this new method of thought were recognized. As an outgrowth of this discussion the following general principles were accepted:

> *Induction* is the process of reaching a general conclusion through the study of particular cases. No one can be certain that conclusions established by induction are true. They are only probably true and the probability increases with an increase in the number of cases for which the conclusion is shown to be true.
> The validity of the generalization is destroyed if only one instance can be shown where it does not hold.
> A conclusion established by induction depends on the assumption that all cases which have not been studied are just like those which have been studied.

While the pupils were fairly cautious in accepting the absolute certainty of conclusions established by deduction, all of them felt that they were "much more certain" than those established by induction. This is, it seems to the writer, a most significant point, for in the course of the same discussion in which the above principles were abstracted from the general process there was agreement with the pupil who said, "Our assumptions are established by induction," which means that these assumptions have only probability in their favor. It seems, then, that at this time the pupils failed to realize the logical relation between conclusions reached deductively and the assumptions which imply these conclusions, for how can conclusions reached deductively be any "more certain" than the assumptions on which they depend?

Argument by induction is not limited to the field of mathematics, and if this method of thought is to be learned "in connection with the field to which transfer is desired," the pupil should be provided with an opportunity to analyze such arguments in non-mathematical situations. It often occurs that an editorial writer uses induction in an effort to establish his major thesis. The following exercise illustrates one way in which editorials using this "method of proof" can be used to advantage:

Following the world war the government of the United States gave to every eligible soldier who applied for it an adjusted compensation certificate which is commonly known as "the bonus." The certificate provided that this bonus would be paid in 1945, but ever since it was made available to the veterans the Congress has been subjected to almost constant pressure that this bonus be paid without further delay. On numerous occasions bills have been passed by both houses of Congress, conforming to the veterans' request for immediate cash payment, only to be vetoed by the President. However, both the House of Representatives and the Senate recently passed such a bill over the President's veto and on January 27, 1936, immediate cash payment of the bonus became the law of the land. It is believed by many people that a large number of representatives and senators voted to pay this bonus even though such a vote was against their own personal belief as to what was best for the country as a whole. They felt that their own political future depended on retaining the good will of the soldiers and they did not wish to antagonize this highly organized group. In a recent editorial, however, a distinguished writer states the following proposition:

"Senators who voted against immediate payment of the bonus are not usually defeated when running for reelection."

and in an effort to establish the truth of this proposition in the minds of the readers says:

"1. The interesting experience of two senators who have twice voted against the bonus, once with an election right ahead of them, seems to prove the proposition and also that members in Congress with nerve enough to stand up do not really suffer at the polls.

2. One of these senators is a Democrat, Edward Raymond Burke of Nebraska. The facts about Senator Burke are these: In 1934 he was a member of the House and the only man in the Nebraska delegation to vote against the bonus.

3. It was his first term and he was warned by his colleagues what would happen to him. After the session was over Mr. Burke went back to Omaha and one day met a professional leader of the veterans on the street.

4. 'All right, Burke,' he was told, 'you went back on us, didn't you? Well, you're through. You can't go back to the House.'

5. 'Isn't that interesting?' said Senator Burke. 'In that case I shall run for the Senate,' which he did, was elected, voted against the bonus again two weeks ago, and also voted to sustain the veto of the President. The most interesting part of this incident, however, is that not one of Senator Burke's four Nebraska pro-bonus colleagues of 1934 is now in public life. One was beaten in the primaries, one in the general election, one retired and the other made an unsuccessful fight for governor.

6. The story of Republican Senator Arthur H. Vandenberg of Michigan is equally interesting. In 1934, just ahead of his campaign for reelection, Senator Vandenberg voted against the bonus and also to sustain the 1934 veto. He was never asked a question until his final meeting, held in Detroit.

97

7. Then a veteran arose and asked him why. Senator Vandenberg gave his reasons and the veteran replied, 'Well, Senator, I don't agree with you but I must say you gave me an honest answer.' Senator Vandenberg thinks he gained votes rather than lost them on this issue.

8. At any rate, he was reelected and is now recognized as a Republican aspirant for the presidency. In spite of this he recently voted against the bonus and also upheld the presidential veto.

9. The 'scare cat' senators who vote entirely through fear and against their own convictions might well ponder on the political experiences of Senators Burke and Vandenberg in voting against rather than with the organized minorities."

The writer of this editorial apparently believes that he has established the truth of his proposition. Do you agree with him? Assuming that the facts he has presented are reliable, is his conclusion justified? By what process of reasoning did he arrive at his conclusion? If you do not feel that his argument is convincing, can you point out the weakness in it? Do you think this argument would lead a senator to vote according to his convictions in a similar situation? Discuss the argument and in your discussion consider these points which have been raised. The paragraphs are numbered for your convenience if you wish to refer to them.

The pupils were quick to point out the limitations of the argument used in this editorial, although they felt that it might be quite convincing to any reader who did not understand the nature of an inductive proof. One thing which emphasized the weakness of this argument more than anything else was the suggestion of one pupil that if "Senator Burke" and "Senator Vandenberg" were replaced by the names of two senators who had voted against immediate payment of the bonus and who had been defeated when running for re-election, the argument would be just as potent in proving the opposite proposition.

A number of editorials concerning the controversy over the suggested enlargement of the Supreme Court of the United States are excellent illustrations of an attempt to prove a broad generalization by induction. One writer, for example, states as his main proposition that:

Men do not deteriorate at the age of 70.

and evidence which he presents in support of this proposition consists of the following statements:

Senator Norris is a power in the senate at the age of 76.
Senator Glass has great national influence at the age of 75.
The Vice President of the United States is 75.
Elihu Root became a world figure after he was 70.
Stanley Baldwin, that steady rock of England during these perilous days when clear vision is demanded, will be 70 in August.
Pope Pius, whose view of facts certainly is not blurred, carries on successfully at the age of 80.
Benjamin Franklin's greatest work was done after he was 70.

Gladstone was called to the prime ministership of England three times after he was 70.
Von Hindenburg was drafted to save Germany in her darkest days when he was elected president after he had passed 80.

An argument of this type offers many possibilities to a teacher who is interested in leading his pupils to understand the nature of proof, and can be used to particular advantage with relation to induction. Does this writer really prove the proposition he apparently wants to establish? Are the statements in his argument likely to be accepted as facts by all people? Is there any weakness in the argument? If so, what is it? A consideration of questions of this sort will emphasize important points in connection with any argument, the pupils will be interested and their "reflective thinking" improved.

Another writer of some note presented an argument in support of the proposition that:

Once an amendment to the constitution has been submitted to the people, the time required for ratification is slightly over a year.

When stripped of verbiage, much of which was irrelevant to the proposition, his argument consisted of the following facts:

The eighteenth amendment was ratified in 13 months.
The nineteenth amendment was ratified in 15 months.
The twentieth amendment was ratified in 11 months.
The twenty-first amendment was ratified in 9 1/2 months.

After generalizing from these four cases the writer mentions the child labor amendment, which up to the present time has failed of ratification, and refers to it as "the exception which proves the rule" whereas actually it is the exception which destroys the generalization. Arguments of this sort, which can be found on almost any editorial page, offer an excellent opportunity to show young people how the kind of thinking which is applied to idealized concepts in mathematics can become distorted when applied to concepts which tend to stir one's prejudices.

Detecting the Factors Which Determine Conclusions

While the ability to gather pertinent evidence in support of a proposition, and to present it clearly, logically and effectively is one mark of an educated person, it is equally important to be able to analyze evidence presented by others in support of conclusions one is pressed to accept. To understand the nature of proof as defined earlier in this chapter is to know that these conclusions are "true" only within the limits of the assumptions on which they depend and to be able to detect these assumptions is an important attribute of "reflective thinking." At the beginning of each college year, for example, Professor Harold Hotelling[1] of Columbia University

---

[1] Harold Hotelling, "Some Little Known Applications of Mathematics," *The Mathematics Teacher*, Vol. xxix, No. 4, 1936, pp. 157–169.

presents to his class in mathematical economics a mathematical demonstration of the proposition that "if everyone is left to himself and will just pursue vigorously his own maximum profit, then everybody will be as well off as possible." His purpose in doing that, he says, "is not to make people believe in the proposition but to show what definitions and what assumptions have to be made in order to make a mathematical proof possible. By the time a person has understood the definitions and assumptions involved in these proofs, he is quite willing to reject the result."

As an illustration of the way in which the analysis of evidence serves to make explicit the basic factors on which a conclusion depends, the teacher guided the pupils in an examination of a proof for the theorem that "if two parallel lines are cut by a transversal, the two interior angles on the same side of the transversal are supplementary." The assumptions and definitions which determine this conclusion were explicitly stated and the undefined terms involved were recognized as essential to the proof. With this illustration as a guide the pupils were asked to analyze the evidence supporting the theorem that "The sum of the interior angles of a triangle is 180°," which the teacher selected for a definite purpose. While the factors which determine what the sum of these angles will be were presented in various ways, the following arrangement suggested by a pupil perhaps indicates better than any other the nature of the relation between the conclusion and the factors which imply it:

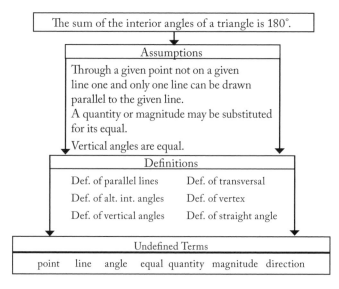

This analysis clearly revealed that one proof of this theorem, so important in the field of mathematics, actually depends on three assumptions, six definitions and seven undefined terms. The question was then raised as to who made these assumptions, who made the definitions and who selected the undefined terms? From these considerations the pupils realized that the sum of the interior angles of a triangle was not fixed by some divine power, for it was they who made these definitions and assumptions and it was they who selected the undefined terms. They disagreed with the statement of Edward Everett who wrote in 1870 that "In the pure mathematics we contemplate absolute truths, which existed in the divine mind before the morning stars sang together, and which will continue to exist there, when the last of their radiant host shall have fallen from heaven."[2] They saw that it was their own minds and not "the divine mind" that had manufactured this so-called "truth" about the sum of the angles of a triangle, and they recognized that this "truth" was relative to the factors which imply it.

The way in which the pupils had been introduced to the nature of assumptions was helpful in leading them to see that a change in any one of the three assumptions on which this conclusion depends would be likely to change the conclusion. In order to relate this to the space in which we live the teacher raised such questions as the following for the pupils' consideration:

"Is each of these assumptions validated by your own experience of actual space?"
"Which of these assumptions do you consider to be the most complex?"
"Do any of them have definite implications concerning the nature of space?"

There was no question in the mind of any pupil concerning the validity of these assumptions. The pupils felt that they squared with their own experience and "they must be valid because they work." There was general agreement that the first was the most complex "because it contains the greatest number of ideas," and there was uncertainty and doubt as to the implications of any of these assumptions "concerning the nature of space." The teacher then recalled the assumption that "A line could be extended in either direction just as far as you want to extend it," and raised the question whether this assumption would be valid in a finite space. Furthermore, if space is finite and parallel lines are lines in the same plane that do not meet in the space in which they are drawn, might it not be possible to draw through a point more than one line parallel to a given line?

The history of the parallel postulate was now considered in greater detail and the teacher acquainted the pupils with the nature of the work done by Saccheri, Lobatchewsky, Bolyai and Riemann.[3] They learned what it meant to "challenge an assumption," and other important results which had been derived from questioning what had seemed "obvious" were discussed, such as the far-reaching consequences of Einstein's challenge of the axiom of the simultaneity of events.

Through such considerations, it became increasingly apparent to the pupils that the assumptions which they had selected and which had seemed "obvious to them" were not inherent in the nature of space. They realized that in their choice of assumptions they were

---

[2] Eric T. Bell, *The Queen of the Sciences*, p. 20. The Williams and Wilkins Co., Baltimore, 1931.

[3] References helpful to the pupils are:
Vera Sanford, *A Short History of Mathematics*. Houghton Mifflin Co., 1930.

David Eugene Smith, *History of Mathematics*, Vol. 2. Ginn and Co., 1925.

Lillian Lieber, *Three Moons in Mathesis*. Brooklyn, 258 Clinton Ave.

Edwin E. Slosson, *Easy Lessons in Einstein*. Harcourt, Brace and Co., 1920.

Eric T. Bell, *Men of Mathematics*. Simon and Schuster, 1937.

really defining space, and this was a most surprising idea to them. They became conscious of the important fact that there was no way of telling whether their world actually corresponds to the assumptions they had selected or to those of Lobatchewsky or Riemann. To know which set of assumptions is "true" is relatively unimportant, even if that were possible, but it is very important for the pupil to recognize that these assumptions, whether they be those of Euclid, Lobatchewsky or Riemann, are in fact nothing more than agreements about an abstract space and that it is not possible to establish by logical proof any properties of that space which are not contained in the assumptions.

In the judgment of the writer, this analysis of the proof for the theorem concerning the sum of the interior angles of a triangle, followed by a consideration of the important developments connected with the parallel postulate, was most helpful in broadening the understanding of each pupil as to the real nature of deductive proof. It revealed in an impressive manner how a change in only one assumption can change a conclusion, and the "truth" of a conclusion was now a question of *consistency* rather than one of absolute verity. It extended the thinking of the pupils beyond the limits of the Euclidean world and introduced them to the non-Euclidean worlds of Lobatchewsky, Riemann and Einstein. It stimulated their imagination and liberated their thinking just as the work of Lobatchewsky revolutionized the thinking of the early nineteenth century.

The Recognition of Assumptions in Non-Mathematical Arguments

While a very large majority of the people in the United States will never be faced with the necessity of analyzing the evidence presented in support of a mathematical theorem, no thoughtful citizen of a democracy can avoid the necessity of examining the evidence in support of the great variety of conclusions he is pressed to accept. There has probably never been a time in the long history of human thought when the ability to detect the hidden assumptions in an argument and to recognize the "weasel words and phrases" was more to be desired than at present. No evidence is available to show that this generalized ability will be developed through the study of mathematics alone. It must be developed "in connection with the fields to which transfer is desired." To provide for this need many exercises similar to those which follow were given the pupils:

> The fact that Thomas Jefferson was one of America's great mathematicians is known to comparatively few people although one might recognize his mathematical talent from a study of the Declaration of Independence of which he is the author. While this document, as we know it, was accepted and approved by fifty-six men, they did not change the general form which closely resembles that of a mathematical treatise. Will you analyze this document, using the following questions as guides in your consideration of it:
>
> 1. What are the assumptions to which the fifty-six men who signed this document agreed?
> 2. What generalization is made about George III, who at that time was the British king, and what sort of argument is used in an attempt to "prove" this generalization?

3. What conclusions are reached? Are these conclusions consistent with the assumptions?

4. Are these conclusions reached by induction or by deduction?

The kind of thinking which the pupils had been doing from the very beginning of this work was definitely reflected in these analyses and it was still more apparent in the discussion which followed. Such points as the following were raised by the pupils:

"Would the history of the United States have been changed if the assumption had been that 'All white men are created equal'?"

"How is 'equal' defined?"

"I do not accept the assumption that 'All men are created equal'."

"Was 'men' defined so as to include the Negro?"

"In the phrase 'certain inalienable rights' the word 'certain' should be defined to explain what these rights are."

"To assume that governments 'derive their just powers from the consent of the governed' implies democracy."

"Who is going to define what 'just powers' are?"

These illustrations, some of them taken from the papers of pupils and some from oral discussion in class, do not exhaust all the points made but they do offer some indication of the extent to which the thinking of the pupils had been affected by their growing understanding of the nature of proof.

The kind of advertising to which students are daily exposed is another fertile field in which to find situations which involve the idea of proof. Behind every advertisement is a multitude of assumptions. Once these assumptions are stated explicitly the advertisement loses a large part of its appeal. An illustration of this type follows:

While reading a magazine Helen's attention was drawn to the picture of a beautiful girl with an attractive smile who was represented as saying:

"I'd wished a thousand times for a brighter smile. One tube of Colgate's gave it to me. It was so annoying to see other girls with lovely smiles get all the dates. Then I tried Colgate's. Now my smiles are bright too."

This is really an argument for using Colgate's Dental Cream. The argument is based on certain assumptions. What are these assumptions? List them in the space below:

(Space was left here for the statement of the assumptions.)

Rate this argument as excellent, good, fair, poor, very poor, using whichever word in your opinion best describes it:

_____ Rating

Numerous exercises of this kind, varied somewhat to suit the type of advertisement considered, have impressed the pupil with the need for evidence "in support of any conclusion

103

he is pressed to accept"; have helped him to "analyze that evidence," "to recognize both the stated and the unstated assumptions" essential to the conclusions, and "to evaluate the argument." There follows another illustration of the type which involves the added feature of "evaluating the assumptions, accepting some and rejecting others":

> While reading a magazine one evening, John's attention was drawn to the picture of a fountain pen. Underneath the picture he read these words:
>
> > "A brain harassed by a pen that runs dry loses its brilliance, power and expression. Hence we have created a pen with 102 per cent more ink capacity and visible ink supply."
>
> These statements seemed to impress him. He had always been irritated when his pen went dry but he had never before realized the extent to which this affected "the brilliance, power and expression" of his brain. He determined he would own this new pen "with 102 per cent more ink capacity" and on the next day he purchased it. What assumptions are involved in the argument which led him to this conclusion? Write them below, placing a plus sign before those which you accept and a minus sign before those which you reject:
>
> (Space was left here for the statement of the assumptions.)
>
> Using these assumptions as authorities, prepare in logical form the argument suggested in the above advertisement for the purchase of this pen:
>
> (Space was left here for the presentation of the argument.)
>
> Would this argument alone influence you to buy the pen?
>
> Answer: _____

Evidence that this kind of exercise had a helpful influence on the ability of the pupils to detect assumptions and to make them more resistant to the power of advertising and other types of propaganda is given in Chapter V of the Thirteenth Yearbook.

It very often occurs that different people who use the same laws of thought and who reason clearly and logically about a given problem reach different conclusions. The point is raised in the following illustration which in other respects is similar to those that precede it:

> By a decision of 6 to 3 the Supreme Court of the United States recently declared the Agricultural Adjustment Act of the New Deal to be unconstitutional. This created wide discussion among people in various walks of life. The viewpoints of a distinguished group of gentlemen who discussed this decision have been summarized by a local paper in the following proposition:
>
> > "It is regrettable that the United States Constitution in its present form is inadequate to meet the needs of present day agriculture and government procedure."
>
> Let us consider that this summary is dependable and that it expresses their real belief. They have reached this belief by making a number of assumptions. What are some of them? Give as many as you can in the space below:

(Space was left here for the list of assumptions.)

It is possible that if these gentlemen could examine the assumptions on which their belief depends they might reject some of them and thus change their conclusions. Do you accept all the assumptions which you have listed? If not, which do you reject? Indicate those you reject by placing a minus sign before them.

It is worth while to note that three of the justices considered the A.A.A. constitutional while six of them did not. As one writer says, they reached their decision "without emotion, without prejudice and through the clear light of logic." One might expect "the clear light of logic" to yield a decision which would be unanimous. How do you account for the fact that three of these distinguished gentlemen found the act constitutional while six of them ruled otherwise? Discuss this on the other side of the paper.

In addition to the other values already considered, this type of exercise serves to emphasize the point that different assumptions are likely to lead to different conclusions when these conclusions are reached "without emotion, without prejudice and through the clear light of logic."

Many people who exhibit great mental power when dealing with mathematical concepts fail to think clearly when controversial issues are under consideration. A local problem involving such an issue and one which claimed the eager interest of all pupils is summarized in the following exercise:

A group of citizens in a certain Columbus precinct met all the legal requirements to have a local option vote on Tuesday, November 5, 1935, the day of the general State election. The purpose of this vote was to find out whether or not the majority of the citizens in that precinct favored the continued sale of intoxicating liquor in the district. The "wets" charged that "the election should be prevented" and appealed to the County Judge for an injunction which would stop it. The argument they presented ran somewhat as follows:

Argument	Statement of Fact	Assumption
It was necessary to file a petition to hold this election. The last day this petition could be legally filed was October 4.		
A citizen must register if he is to qualify as a voter.		
The "drys" did not file their petition until October 4.		
By October 4 all the "drys" had registered.		
After filing of the petition only a day and a half remained in which the "wets" could register. Because of this lack of time all the "wets" were unable to register.		
Nothing should interfere with a citizen's right to register.		
Therefore the injunction should be granted and the election stopped.		

It is probable that no other issue is more conducive to prejudice than the liquor issue and when prejudice replaces reason, conclusions are not likely to be reliable. Lay aside any prejudice you may have on this particular issue and consider the argument by the "wets" only on its merits. By checking in the proper column indicate which of the statements in the argument you believe to be facts and those which you believe to be assumptions. Do you find that the argument is sound? In your opinion are the assumptions justified? It is not necessary to be a county judge to determine the validity of this argument. Suppose you were in the position of the judge. Considering the argument alone, what would be your answer to this request for an injunction to stop the election? In the space below present your decision as the conclusion of a logically developed argument, the conclusion being either "Therefore the injunction is granted" or "Therefore the injunction is not granted."

In evaluating the argument presented by the "wets" the pupil not only must distinguish "facts" from "assumptions" but he must be able to detect the important assumptions which are essential to an honest conclusion and which are not explicitly stated in the argument. He then places himself in the position of the county judge, evaluates these assumptions, either denies or grants the injunction and presents a "logically developed argument" supporting his position. This proved to be one of the most profitable exercises considered.

The purpose of an editorial writer, in general, is to convert the reader to a definite point of view. To do this he usually states a major proposition and then presents evidence which, in his opinion, should lead the reader to accept the proposition. Editorials of this kind are excellent material for the consideration of any class interested in the study of "proof." An exercise built around such an editorial follows:

During the latter part of October, 1935, the corn-hog farmers of the United States were asked for an expression of opinion as to whether or not they desired the continuation of the Agricultural Adjustment Act. The result of this referendum seemed to indicate that they did favor this program by a majority of about 8 to 1. In considering this, however, a distinguished writer recently stated the following proposition:

"It is surprising that any votes were cast against the continuation of the policies of the A.A.A."

and in attempting to establish this proposition he presents the following discussion:

"1. Today there is country-wide wonder that any votes were cast in opposition to the referendum which brought an 8 to 1 victory to the A.A.A.

2. For the Roosevelt administration did not present both sides of the argument to the corn-hog farmers, but merely asked whether they would continue to favor crop control and a subsidy.

3. The consumers who have to pay, in the cost of living, higher prices for foodstuffs, were not asked to vote.

4. None of the arguments that might be made on the faulty economics of the A.A.A.'s policies was presented to the corn-hog farmer. It was a one-sided election among a group who were really being asked whether they would like to have a better price for their product by limiting the output of their farms and making their products relatively scarce.

5. When steel manufacturers used to get together and try to agree among themselves on limitation of production and prices, the government used to call it a trust and apply the anti-trust laws. But farm organizations of all kinds are exempt from the operation of anti-trust laws and the A.A.A. is really a substitute for the old-fashioned trust or monopoly, but with governmental control.

6. The spectacle of a government-managed election among the members of a minority group, whose decision now is to effect the prices paid by the majority, is still too novel for widespread appreciation of its implications. Minority by minority, the New Deal offers money benefits in the form of processing taxes or subsidies and the result is to build up a cumulative weapon of blocs for the presidential elections. Minorities swing national elections because they move from one party to the other, while the straight ticket voters remain indifferent.

7. If each powerful minority group is to be appealed to on grounds of direct benefit to it with funds either taken out of general taxation or by levying assessments on the cost of living, the chaotic consequences will hardly be called 'planned economy.' It will be difficult for any opposition political party to win in the near future on the simple truths that century-old experience has proved. It only means that New Deal economy may have to run its full course, bringing in its wake the friction that has always arisen between classes and the concussion that has always come from government control of production and price-fixing, no matter where it has been tried in human history."

All of this discussion is not pertinent to the argument, for in it there is a good deal of the writer's own philosophy concerning government policies. What are the principal statements which in your opinion he uses to "prove" his proposition? Arrange these statements in what you believe to be a logical order and reconstruct his argument in the space below. Do not attempt to support these statements by any authorities but by placing a check mark in the proper column indicate which of them you believe to be statements of fact and those which you believe to be mere assumptions.

Argument	Statement of Fact	Assumption

(Sufficient space was left here for the presentation of the argument.)

In view of the argument which you have presented do you consider the conclusion justified? If you do not consider the conclusion justified by the argument, discuss your reason for this in the space that follows:

(Sufficient space was left here for the presentation of reasons.)

There are paragraphs in this discussion which appear to have no direct relation to the proposition which the writer really wanted to establish. Some of them are irrelevant to the argument, some are generalizations about the result of the referendum and some are inserted for other purposes. In the space below will you discuss briefly the nature of each of these paragraphs? They are numbered for convenient reference. . .

(Sufficient space was left here for the discussion.)

An exercise of this sort, while combining most of the features included in the preceding illustrations, has additional values. To analyze any discussion for the purpose of determining which statements support the major proposition and which are irrelevant is critical thinking of the most helpful sort. The reconstruction of the writer's argument, omitting all irrelevant statements, also proved to be a most revealing and profitable activity.

The exercise which follows is built around an event which had been discussed in the local papers and which had been given wide publicity:

Mrs. Lewis Seymour was recently struck by an automobile and instantly killed. The driver of the car did not stop and while a man saw the accident he failed to see the number of the license plates on the car. However, he did notice that the right headlight was broken and that a tire blew out at the time of the accident. He reported these facts to the police and twelve hours later they found a car with a flat tire and with the right headlight broken. This car was parked behind the house of Hezekiah Berry and belonged to him. Numerous conclusions considered by the police are stated below. Place a plus sign in column one opposite each conclusion which you will accept from a consideration of only the facts given above:

	1	2	3	4
a. It is certain that the car which struck Mrs. Seymour belonged to Hezekiah Berry . . . . *a.*				
b. The given facts are irrelevant to the problem of discovering who owned the car that struck Mrs. Seymour . . . . *b.*				
c. It is certain that Hezekiah Berry was not driving the car that struck Mrs. Seymour . . . . *c.*				
d. Other facts are needed before it can be definitely proved that Hezekiah Berry was the driver of the car that struck Mrs. Seymour . . . *d.*				
e. It is probable that the car that struck Mrs. Seymour belonged to Hezekiah Berry . . . . *e.*				
f. It is certain that the car which struck Mrs. Seymour did not belong to Hezekiah Berry . . . . *f.*				
g. It is certain that Hezekiah Berry was the driver of the car that struck Mrs. Seymour . . . . *g.*				

Through further study of the problem the police found that the glass at the scene of the accident was of the same pattern as that in the broken headlight on Mr. Berry's car. Using this added fact in connection with those already given indicate in column two which of the preceding conclusions you would accept.

It was also established that when Mrs. Seymour was struck she was carrying a quart of potato soup. Some of this soup was found at the scene of the accident while traces of the same kind of soup were found on Mr. Berry's car. Does this added fact change your judgment as to which of the preceding conclusions you would accept? Indicate in column three those conclusions which you believe are now definitely established by the known facts.

The police also found strands of hair on the broken headlight of Mr. Berry's automobile. Now considering all of these facts will you indicate in column four which of the conclusions you believe to be justified.

Exercises of this type reveal the effect of cumulative evidence on the thinking of the pupil and indicate just how much and what sort of evidence is needed to change his acceptance of a conclusion from probability to certainty. Such work, when followed by thoughtful discussion, has a definite tendency to increase the caution of the pupils in reaching conclusions.

As an outgrowth of this kind of thinking some of the pupils became conscious of the fact that underlying their beliefs were many assumptions, and out of this came a request that the examination of these assumptions become a matter for group consideration. Many topics for such study were listed and from these an exercise similar to the following was prepared:

Underlying the beliefs of an individual are numerous assumptions, and anyone who accepts a conclusion regarding any issue at the same time accepts the assumptions on which that conclusion depends, even though he may not know what some of them really are. For this reason it is well that we often examine the assumptions behind our beliefs so that we may be fully conscious of just what our position on any important issue involves. Will you then state your present position on the following topics and in the space provided for analysis, list the statements on which in your opinion this position depends? By checking in the proper column indicate which of these statements you consider to be facts and those which you regard as assumptions. It will also be well for you to define any words or phrases which will help to clarify your position:

1. Racial Superiority.
Many people believe that the white races are superior to the colored races.
My present belief concerning this is:

Analysis of my belief	Statement of Fact	Assumption

109

2. Compulsory Education.

We have in this country certain laws which compel all young people up to a certain age to attend an organized school. Some people believe that such laws are most desirable, while others feel that their operation accomplishes little if anything in improving the quality of our citizenship.

My present belief concerning this is:

Analysis of my belief	Statement of Fact	Assumption

3. A Citizen's Obligation to His Government in Time of War.

Many people in the United States have recently stated that they would never bear arms in any war. Others feel that it is their duty to support their government at all times whether in peace or in war.

My position on this important issue is:

Analysis of my belief	Statement of Fact	Assumption

4. Awards.

Some pupils in the school feel that there should be a set of fixed criteria for honors and awards, and once a student has satisfied these criteria, he should be granted an award. Others feel that the problem of awards is more complex, and that any set of fixed criteria cannot possibly take into account all factors in a given situation.

My present position on this problem is:

Analysis of my belief	Statement of Fact	Assumption

Discussion of such questions as these was noticeably unprejudiced. All of the results had been mimeographed and each pupil had before him the work of the others. Contrasting positions were considered and the assumptions underlying each of these positions were examined. The objective way in which the pupils went about this was particularly impressive. Many of them stated that such discussions were especially helpful in clarifying their own thinking and it is believed that all of them learned many valuable lessons of tolerance in their thoughtful consideration of these complex problems.

4

From *The Learning of Mathematics: Its Theory and Practice,* Twenty-first Yearbook (1953)

NCTM's Twenty-first Yearbook was the first entry in the series since the very first one in 1926 that was not edited by W. D. Reeve, the editor of nineteen consecutive yearbooks. Howard F. Fehr, editor of the 1953 volume, also wrote its initial chapter, which NCTM past president Henry Kepner (2008 to 2010) recommended we include here.

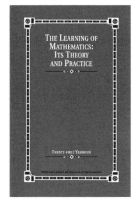

"**Theories of Learning Related to the Field of Mathematics**" presents a thorough treatment of learning as considered for mathematics. In this chapter, Fehr examines changes in behavior and intelligence as well as learning situations and the products of learning. Connectionism and field theories are discussed, and in all cases the discourse connects directly to mathematics examples that reflect elementary- and secondary-level mathematics. Fehr's discussion of "areas of agreement" provides a possible foundation for a most interesting theory of learning. Readers interested in learning and particularly in historical perspectives related to learning will, like Hank Kepner, find this chapter to be a valuable resource.

Theories of Learning Related to the Field of Mathematics

Howard F. Fehr

Ways of Studying Learning

THERE are a number of ways to study the learning process of human organisms. One is purely physiological, that is to study learning as physical reactions of the brain, the nervous system, the glands, and the muscles, as they are acted upon by physical stimuli. Another method, partly physiological and partly observational, is to study the way the organism reacts in various situations so as to abstract common elements called laws of learning. It is recognized that physical changes are taking place in the organism, and that some of these physical changes can be ascribed to certain actions and reactions of the organism in particular situations. But the general explanation of the reaction of the organism is given in terms of the situation, and not in terms of physical changes within the organism. This procedure is followed by psychologists. A third method is to ignore all internal physical changes and to describe learning purely in terms of introspection and logical considerations. All three methods have provided and are providing new insight into human behavior, but recently the psychological investigations have given the most promise of help to the teacher.

What do we know about the physical behavior of the brain? In the first place, it is composed of more than 10 billion nerves which are connected by an exceedingly complex network. These neurons consist of a center or cell body, from which run fibers of two types, axons which are single strands of various lengths, and dendrites, which are ramified short fibers. Impulses travel along these nerve fibers at rates which have been measured to vary from three feet to 300 feet per second. The impulse is relayed from one neuron to another by a synapse, and the flow of the impulse is in one direction only. The response of each fiber is an "all or none," that is, if it is not sufficiently agitated there is no response, but at a certain degree of stimulation the whole response goes forward. The amount of stimulation necessary for the "all or none" response varies also from neuron to neuron. Hence any overt action of an individual, and all actions, involves many, many nerve fibers, and is dependent upon the number of stimulated neurons.

We also know that certain areas of the brain are related to certain functions such as sight, information storage, control of the sympathetic nervous system, and emotional behavior, and that damage to these parts of the brain interferes with the corresponding functions. There is also evidence that in time, certain parts of the brain can take over the functions of other damaged parts. But how the physical behavior of this vast network of nerves in the brain and nervous system produces the response $a^2 - b^2$ is $(a - b)(a + b)$ is totally unknown. How the cells get their information, how they transfer it from the sign of $a^2 - b^2$ to one area of the brain, to another area, to an ultimate response from the organism of $(a - b)(a + b)$ is a deep, dark secret. Further, any attempt to study the physical behavior meets with many obstacles. To open the brain to observation is usually accomplished by destroying the very nerves we would study. Further, the nerve cells, axons, and dendrites are exceedingly minute objects, and to see a synapse at the end of a nerve fiber is exceedingly difficult. At present, explanation of human actions in terms of physical phenomena within the brain seems very, very remote (23).[1]

Hence psychologists have resorted to experimental and observational procedures to explain what the human brain is, and what it does. They create certain situations and observe under as controlled conditions as possible the behavior of the organism, and describe the operation of human learning by the various behaviors that take place. Thus human learning is defined as a change in behavior acquired through an experience. The learning is usually directed toward specific goals through organized patterns of experience. In order to clarify our concept of change in behavior, we give several examples from the mathematical field.

Examples of Change in Behavior

When a student enters a beginning algebra class and is asked, "What are the two numbers of which the sum is 6, and the difference is 1?" he *behaves* as follows: Try 1 and 5; the difference is 4, no. Try 2 and 4; the difference is 2, no. Try 3 and 3, no. Maybe there is no answer; try again; 2 and 4; 3 and 3. Oh! Maybe I can use fractions; $1^1/_2$ and $4^1/_2$, no; $2^1/_2$ and $3^1/_2$; there it is. He has solved the problem, he has reasoned, and he has exhibited a type of behavior in a given situation, but it is not the goal behavior you will ultimately expect from his instruction in algebra. Now let us repeat the same problem three months later. If the student reacts in the same manner as above, he has not learned anything new in this situation. If, however, he behaves as follows: Two numbers x and y; sum, $x + y = 6$; difference $x - y = 1$; add $2x = 7$, $x = 3^1/_2$ and $y = 2^1/_2$, then his behavior has decidedly changed; he has learned a new mode of action. His mind proceeds in a manner entirely different from before. We should set up our goals of learning in the mathematical field, in terms of all desired changes in behavior with reference to numerical, spatial, quantitative, and logical situations.

Another example. At the start of the year in plane geometry, you give the following hypothesis: A triangle has sides 2 in., 3 in., and 4 in. The middle points of sides 2 in. and 3 in. are joined by a straight line segment. Then you ask, "How long is this segment?" The

---

[1] In this chapter the symbol $(x{:}y)$ will be used to refer to page y of reference x in the numbered list at the end of the chapter.

student responds by using his ruler, a pair of compasses, and paper, actually constructing the triangle and segment and measuring the latter. Assuming careful work, the student responds, "I measure it to be 2 inches." His behavior in this case is a result of his past experience. Three months later you confront the student with the same problem. If he has learned, his response now is solely the result of an inner brain reaction. He says, "It is 2 inches, since it must be one-half the length of the third side." Thus he has had a complete change in behavior from one involving perceptual-motor skills to one involving purely concept-relationship.

One task in education is to create such experiences and situations that will enable a student to reconstruct his behavior towards goals desired by both himself and the teacher. When we have accomplished this, we shall have improved our instruction.

Learning thus becomes a developmental process. It is change in behavior brought about through brain action or thinking. It comes about through facing situations that call for making discoveries, abstractions, generalizations, and organizations in mathematics. It is problem-solving, for without a problem felt by the organism, and motivation toward the solution of the problem, there will be little learning of mathematics. On this most psychologies of learning agree. The disagreement arises in the theoretical explanation of how the solution comes about.

The Nature of Intelligence

To what extent is it possible for human beings to change their behavior? It is quite common in academic circles to hear such expressions as, "He does not possess enough intelligence to learn mathematics," or "He is a highly intelligent individual." Intelligence as used in these expressions is that quality which permits an individual to adapt himself successfully to a given situation. This was one of the earliest definitions of intelligence. However, if a dog adapts himself to a household in a manner to get good care, we do not say the dog is intelligent (in the sense we apply the word to human behavior). There is more to intelligent action than mere adaptation.

Binet in his early work on testing used the ability to make judgments as the best description of intelligence. To this end he constructed many tests devised to measure the ability to make judgments or choices. This has culminated in the construction of many types of mental tests, and we could describe intelligence as that faculty, or quality, or characteristic which is measured by the intelligence tests. Intelligence would then be the ability to perform mental tasks, to remember, to make generalizations, to form relationships between concepts, and to deal with abstract ideas. The amount of intelligence would be measured by the degree of difficulty of tasks completed, of their complexity, of their abstractness, and of the speed and lack of interference with which the tasks are completed.

Dewey in all of his writings has concerned himself with the nature of human intelligence. A brief summary of his concept would be: Intelligence is acting with an aim; it is purposeful activity. The activity must at all times be controlled by a perception of all the facts in a given situation and their relationship to each other. Even more important in intelligence is the capacity to refer present conditions to future desired goals and conversely to refer the goal to the present conditions. Thus to refer counting to acquiring of addition facts (as an elimination of a needless time-consuming process) and to refer the facts back to counting is intelligence. If we act without knowing the consequences of our acts or even considering them, we are unintelligent. A shot in the dark is not intelligent action. (This is not to be confused with acting on a hunch; a hunch is usually related to the goal.) If we make a guess at an answer as a loose stab, and not as a related action, we may be exhibiting some intelligence (goal-directed action) but it is very imperfect. If, however, we act with an aim toward changing our behavior to a new desired pattern which is perceived as desirable, we are making intelligent action. "Intelligence is the power to understand things in terms of the use made of them" (3).

It is in this sense that intelligence is the ability to solve problems—to think—to learn. And this is more than merely an ability to think in terms of abstractions which is one kind of intelligence. It is also the ability to grasp relations in physical or concrete setups (situations) and to see how to readapt these for more useful purposes. This has been referred to as a practical or mechanical aspect of intelligence. A technologist has a different type of intelligence than a theoretical scientist. He foresees future conditions in terms of concrete situations rather than abstract relationships. His type of intelligence is very important in modern society and should be developed. It may be characterized in one way by a space-perception activity as contrasted with a deductive propositional activity. Another type of intelligence recognized by Thorndike (16) is social intelligence. This is the power to understand people, to get along with them and to lead them. It involves personality traits and actions between humans which relates present conditions to future desired states of happy, cooperative living, and vice versa.

In the learning of mathematics, the power with which an individual can make generalizations, abstractions, logical organizations, and relate these to purposeful action, determines his ability to progress. As teachers of mathematics, we are interested in this phase of intelligence. However, as teachers of children, of young men and women, we are certainly interested in the mechanical and social type of intelligence, and hence must consider all these types in our study of learning.

A Learning Situation

To study how we learn, consider your solution to the following problem: A man in a department store noticing the escalator in motion, raises the question, "How many steps are there in the escalator between the floors?" He walked down the escalator as it was in motion, timing the distance between floors. When he reached the lower floor having walked down 26 steps, it took him 30 seconds; similarly, when he walked down 34 steps, it took him 18 seconds. What is the answer to his question?

What answer did the man find? How did he find it? If you, the reader, are interested in these questions, if you really want to find the answers, you are in a learning situation. All your past experience in mathematics has created a mental set and the type of problem gives sufficient motivation to send you into action toward the solution. You now use your previous learning to find a solution.—(Before reading further, stop and seek your solution, keeping a diary of every move you make. Then you can study your method of obtaining a solution or how you learned in this situation.)

You may have gone directly to the solution of the problem on your first trial by applying a technique previously learned. In this case you did not learn, you did not need to learn, you merely recalled a previous learning. But you may have proceeded in one of the following manners: The difference between the numbers of steps walked during each of the two trials was 8; the difference in times was 12 seconds; but what relation has this to the problem? Here many a student would cease learning because he would have reached a block in his reasoning without sufficient drive to go on. However, another student would say, "Is there any relation between the difference in times and difference in steps traveled? Oh yes, there is a relation to the motion of the escalator and the steps move at a rate 8 steps per 12 seconds or $2/3$ steps per second. Now does this help me?" Again the student may be blocked or he may return to the problem and think, "In 30 seconds then, the escalator will move $30 \times 2/3$ or 20 steps and the man move 26 steps. Aha! There are 46 steps in the escalator between floors." Now many a student would stop having secured satisfaction with the answer. But, if he is to have better learning he will check and analyze his thinking as follows: "Let me see if this is so. If there are 46 steps and I go down 34 of them, the escalator must move 12 steps. At a rate of $2/3$ steps per second, it takes $12 \div 2/3$ or 18 seconds and that's right." At this point, having checked a hypothesis, the student may again cease his learning. He has all that he desires. But a still better learner will say: "Now let me see how I solved the problem. First I found the rate at which the escalator moved, then I found the number of steps the escalator moved in 30 seconds, and I added to this the number of steps walked. This is the number of steps in the escalator. In situations where I know two distances, and two corresponding times, I had better try first to determine a rate."

This whole learning leads to a change in behavior. When confronted with a similar situation the student will now act differently from what he did in solving this problem. Of course, there are several ways of solving the problem besides this method of arithmetic. This method was shown to illustrate how we learn.

The accompanying diagram (shown on the following page) similar to that given by Dashiell (2) can be taken as the starting point in the study of how we learn.

At the start of learning or readjustment of behavior, there must be a situation in which the student feels a *need*. A need is a feeling of the organism for something which is absent, the attainment of which will tend to give satisfaction. The situation is such that the student is motivated to satisfy the need. This creates tensions and drive within the organism which impel it towards its goal. Thus the learner is spurred to physical and mental action, or making a response. The first response often does not lead to the goal; he runs against a barrier. If the motivation to learn is strong enough, the learner seeks another response or series of responses. One after another of these responses may fail to lead to a

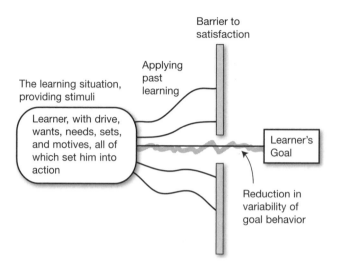

solution, but finally he selects a path of action that reaches the goal. He has solved the problem; he is ready to readjust his total behavior in this situation. He may go over the solution, to make the meaning and structure more precise, and his formulation more articulate; to make the whole situation more highly differentiated from previous learning, and more generalized, until he has developed a new pattern of behavior that will function in new problems containing the same or similar situations. He has learned.

Each of the several psychologies of learning has its own explanation of the way the learning goes from need to goal. That these theories do conflict at a number of points should not concern us too much. If the application of one of the conflicting theories proves more useful for our purpose in a given situation, and application of another theory in another, we shall use each as it fits the occasion. Until psychology develops into a more significantly unified, scientific theory we must do this. Physicists do this in the study of light where both the wave theory and corpuscular theory are applied, giving consistent results in some instances and contradictory results in others. Finally the areas in which all of the theories are in agreement will be especially important for our study of learning. Here we shall briefly examine what three theories, association, conditioning, and field psychologies, say about learning.

End Products of Learning

In the learning of mathematics, a student is expected to do everything from handling concrete objects in counting to making abstract logical deductions with the use of symbols. There exists a sort of hierarchy of end products of learning in which we strive for the highest level. One of the simplest types of human learning is a *sensory-motor skill*. The response is practically automatic once it is learned. This is illustrated in teaching a child how to use a pair of compasses to draw a circle.

On a slightly higher level we have *perceptual-motor skill* learning. This can be characterized as learning which is applied immediately to a perceptual pattern. It is illustrated in the learning to use a protractor to measure an angle and a ruler to measure a line segment, or in drawing a geometric figure.

The next type of learning which occupies a large part of the school activities is *mental association*. This is the type of learning which gives the child his store of number facts, names of algebraic terms such as exponent, coefficient, binomial, the names of geometric figures. It includes vocabulary learning.

While a student may learn to recognize an exponent, coefficient, or median of a triangle, he may not fully comprehend these objects of thought. For this purpose he must *learn concepts.* How concepts are learned in mathematics is the topic of an entire later chapter. When a child has a mental image of a thing and can relate it to other things through definitions, laws of operation, application, or generalizations, he does a great deal more than mere identification through association.

A final end product of learning, of concern to mathematics teachers, is *problem-solving* as illustrated in the example in a previous section. Here all of the other end products are brought to bear in making hypotheses, judgments, organizing evidence to give solutions, and forming structures of knowledge such as pure mathematics. The chapter in this Twenty-first Yearbook on problem-solving is particularly concerned with aiding the mathematics classroom teacher to develop this type of learning on the part of his students.

It should be noted that while the end products appear quite distinct in form, yet their learning has in common the elements in the learning diagram. In learning to use a pair of compasses for example, there must be motivation, there will be motor movements which will not give the desired circle, then a correct use of the hand and fingers comes forth, finally the learner will try this successful technique until the variability has been reduced to a desired level of manipulative skill.

Animal Learning

Conditioning as a theory of learning grew out of laboratory studies of animal behavior. In most of these experiments, the animals were restricted or constrained so as to be unable to avoid stimuli. The reward or punishment was in most cases the same, an electric shock or food, respectively. Since for most animals food is a strong incentive, it was easy to condition the animals and to have them react in a given manner to a given sign as a stimulus. The animal thereafter did what he had been stimulated to do and it was said that the animal had learned.

Cole and Bruce (1) characterized two levels of freedom in animal learning, (a) when the animal is almost totally restrained in a harness and free to move only one or more legs, and (b) when the animal is confined in a cage or maze but free to move about within it. In the first case the animal learns by responding to a stimulus; in the second by selecting from random activity those responses that lead to satisfaction. In neither case can the animal *explain* why he behaves the way he does.

119

On a higher level of animal learning, Köhler (9) described how, confined in a cage, an ape could piece together two sticks inside his cage, and reach outside the cage to scrape food to within reach of his arm. The ape had previous experience with using a stick as a scraper, but not with putting two sticks together. The ape learned the latter by accident or by random error, but having learned it, he had a flash insight as to its use in getting food.

Animal trainers use the method of conditioning in training their subjects, using a lash and food for punishment and reward. Even fleas and worms can be shocked into behaving as we would have them behave. The questions for the teacher are: Shall we use the techniques of the animal trainer in our classes, imposing the necessary restraints, with accompanying punishment and reward for failing or successful responses? Or shall we permit freedom of learning experience? Or are both techniques of value depending on the time and the nature of the learning.[2]

In the following description of theories of learning, it will be well to recall at all times the limitation of applying animal learning theory to the learning of mathematics. Thorndike, and many others since, in their experiments on animals, conclude that they do not learn by reasoning or by social imitation. They learn *only* by physical doing. They do not develop a culture, or transmit their culture from generation to generation. They do not speak, they do not use symbols, and they communicate only by signs, where by sign is meant the stimulus for a fixed response. In contrast, mathematics is learned by reasoning, by the use of symbols, and by the transmission of cultural patterns. We recognize that the learning of infants by reproof and reward is the same as the conditioning of animals. We also recognize that much that we learn in the early stages of mathematics is learned by doing, and to this end we should examine what conditioning theory has to offer.

Conditioning

To enter into a detailed discussion of the theory of the various schools of psychology is beyond the purpose of this Twenty-first Yearbook. The interested teacher can obtain this by studying the literature in the bibliography appended to this and successive chapters. We shall state the main principles and characteristics of each of the psychologies and illustrate these with applications to mathematics learning.

Conditioning, with its emphasis on stimulus and response, was one of the first psychological theories carried over to human learning and still either consciously or unconsciously guides the teaching patterns in many of our classes. Since we cannot tell the difference, by examining the brain and nervous system, between a boy who gives a correct response to a problem and one who does not, we resort to predicting from observing what each boy does (his responses) and the situation that brings about his responses (his stimuli). Evidently the boy with an incorrect response is in a situation with unfavorable stimuli, which is different from the situation or stimuli acting on the boy who gives the correct

---

[2] See film "Willy and the Mouse," 16 mm., 11 min., black and white. Teaching Films Custodians, 25 West 43 Street, New York City.

response. If the outside situations are the same, then we can predict the inner (inside the organism) stimuli are different.

The fundamental principle of conditioning as given by Guthrie (13:23) is: *a stimulus pattern that is acting at the time of a response, will, if it recurs, tend to produce the same response.* According to this theory we learn only what we do in a given situation. We learn only our reactions, our responses. To bring about learning we, as teachers, must induce our students to follow certain mathematical patterns or behaviors, at the same time that they are confronted with stimuli. These stimuli become the signals for the mathematical behavior, and when this has been done, the signal replaces the inducement, that is, the response tends to come forth with the signal thereafter and learning has taken place.

Thus the learning of the addition facts, under this theory, can be brought about by having children combine groups, for example placing 3 chairs and 4 chairs in a row. At the same time the children are confronted with the stimulus 3 + 4 and the response 7 is given. Thereafter any stimulus pattern similar to 3 + 4 will tend to evoke the same response 7. Thus the child is learning what he is doing at any given time.

We can learn incorrect responses as well as correct ones, and in this case it is necessary to break down the incorrect response. If an algebra student says $(a + b)^2 = (a^2 + b^2)$, we must remedy the situation. To do this, conditionists use *associative inhibition* in the following manner: present the signal $(a + b)^2$ and along with it a stimulus for inhibiting the incorrect response (a teacher's disapproval—no, no; or the correct response, $a^2 + 2ab + b^2$ are all inhibitory stimuli). After sufficient repetition the incorrect response will be forgotten. In this case forgetting is failure to respond to a signal and it is due to new associations formed, that is, learning to do something else that is more desirable.

In general, conditioning has as its basis for learning:

1. The making and breaking of habits, the acquisition of skills.

2. The response to a pattern of stimuli is conditioned. We learn what we are doing. We learn incorrect responses (errors) as well as correct responses.

3. New responses result from conflict and inhibitory stimuli.

4. Learning occurs normally in one conditional response. The need for repetition in skill learning is that a skill is not simple, but it is a large collection of habits.

5. Learning best takes place when a desired response is associated with appropriate signs, gestures, mathematical symbols, and words that act as stimuli for the desired action.

6. Since we learn what we do, we must be free to act. Hence learning takes place best in a free situation, not in a forced and harnessed activity.

It is easy to see under these principles how rote learning and fact learning can be brought about. It is rather difficult to see how these principles aid in learning to solve an original problem in geometry or algebra. For this latter purpose the theory takes recourse to trial and error which is explained better under connectionism.

Connectionism

The fundamental characteristic of connectionism is the bond established between a situation and the response made by the organism. These bonds become unified and patterned through selection (trial and error) according to certain laws of effect, exercise, readiness, and analysis.[3] The degree of learning to which the organism can aspire is largely determined by its inherited qualities. As the organism matures, it develops connections (habits and skills) which must be practiced to achieve permanence. The more complex the acquired bonds can become, the greater is the capacity to learn mathematics.

The law of effect has particular interest for learning mathematics. It says: *A bond is strengthened or weakened according as satisfaction or annoyance attends its exercises, and reward upon success is the most potent factor for insuring learning.* If this is so, we learn, practice, and have an interest in those things which are pleasurable. Thus, the first experiences a pupil has in mathematics should be simple enough to insure successful results and should be accompanied by reward in the form of praise or encouragement. *Start right, and practice.* Under connectionism some adherents say it would be detrimental to learning to allow a student to flounder about or to make mistakes. Others, patterning their belief on animal experiments, say the initial experience should be entirely free to permit a choice of any trial, no matter how blind. Out of this experience should come direction of later efforts. When a correct bond has been established, as for example $a^n \cdot a^m = a^{n+m}$, immediate and frequent repetition will strengthen the bond and give greater probability of its functioning in a similar educational situation at a later time. So we should drill repeatedly on the thousands of facts throughout the sequential learning of mathematics.

The law of exercise says: *When a connection is made between a situation and a response, the strength of the bond is increased; when the connection is not made over a period of time the strength of the bond is decreased.* Thus in learning to solve a quadratic equation by the use of the formula, the oftener the equation is accompanied by the proper use of the formula, the stronger the bond, and in later appearances of quadratic equations, the formula is more apt to come as a response. Further, the sequence of operations in applying the formula—equating the function to zero, determining the coefficients, substituting, simplifying the result, checking—form a *belonging sequence,* the repetition of which accompanied by success or other reward promotes learning.

The law of readiness says: *When a bond is ready to act, to act gives satisfaction, not to act gives annoyance. When a bond is not ready to act, and is made to act, annoyance is caused.* Thus to attempt to make a child form addition facts or learn the multiplication table, when his organism is not ready to act, is to cause dislike, and to interfere with later learning of the arithmetic. If a child cannot substitute 10 pennies for a dime in a practical subtraction of 23 cents minus 15 cents, then he is not ready to do subtraction involving borrowing

---

[3] These laws are stated by Sandiford (13:111). While in the early formation of his theory of learning, Thorndike states these as specific laws. He and his followers later amended these statements to serve as descriptions or characteristics of learning rather than as laws. They should be thought of in this latter aspect in this chapter.

(changing a ten to ten ones) and to force him to do the abstraction would interfere rather than aid his later learning of the process.

The law of analysis (similarity and dissimilarity) says: *When a given response has been connected with many different situations which differ in all respects except one common element, the response becomes bound to that element.* In later situations totally different from the previous situations, the presence of this common element will tend to evoke the given response. This law is closely related to trial and error learning or problem-solving. A child is confronted with three blocks and one block put into a single group, and hears or gives the response *four.* Then two apples and two apples, with the same response; then various patterns of four objects of various kinds, all with the same responses. In all cases the situation is different except for the fourness. The law of analysis says that in any later complex situation the recurrence of fourness will evoke the response *four.* This is how a child learns *four.* Thus analysis, in the sense used here, fosters learning.

Under analysis, trial and error is not a befuddled, blind chance affair. On a given trial we do not get a desired response so we discard the mental path we used and select another path to our goal. We do not return to unsuccessful paths (errors). Thus trial and error is deliberate choice. Each succeeding trial takes less time until finally we solve the problem. A permanent bond is then formed between the stimulus and the goal and in later different situations in which the stimulus occurs (along with many other elements) this goal response will come forth. Thus the $S \rightarrow R$, $a^2 - b^2 \rightarrow (a - b)(a + b)$, as a desired learning should be taught in many different situations involving $a^2 - b^2$ as a common element but always with the same response $(a - b)(a + b)$. Then when the right triangle occurs with hypotenuse r and one side x, the situation $r^2 - x^2$ should evoke $\sqrt{(r - x)(r + x)}$ as the remaining side.

The law of analysis indicates that all complex learnings should be analyzed into simple elements, and then taught or presented in a sound, pedagogical order. If you wish to learn how to add a/b to c/d, analyze every step involved—the definition of a fraction, changing a fraction to an equivalent fraction, defining a common denominator, finding common denominators, changing fractions to common denominators, the rule of adding numerators—then drill on the process until it is mastered. This is the way much of our mathematics is to be learned.

The recent war and the present mobilization are focusing our attention on knowledge as a tool. The goal of learning is performance. To this end we have stressed the learning of facts and skills and connectionism has been the prevailing psychology. Under this theory our whole program in mathematics has been largely concerned in getting students *to do* their operations quickly and accurately whenever they occur. Problem-solving in mathematics is reduced to a method of steps to be followed in a belonging sequence, from step one (read the problem) to step n (check in the original problem) which, if mastered, will automatically lead to the solution of the problem. Drill is the keynote to achievement. If the multiplicity of facts and skills becomes so great that it finally overwhelms the learner, he has reached the limit of his inherited capacity. With sufficient capacity one can learn all the mathematics as a set of sequentially ordered and related facts.

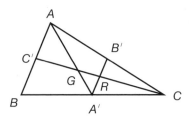

The following geometrical original may be used to illustrate how connectionism leads to a solution. In the triangle shown, AA' and CC' are medians. (These are stimuli which evoke the response A' and C' are mid-points of the sides.) AA' and CC' meet at G. (This evokes the response that AG is $2/3\,AA'$ or GA' is $1/3\,AA'$.) B' is the mid-point of AC. $A'B'$ meets CC' at R. What part of $\triangle ABC$ is the $\triangle A'GR$?

The student begins his goal-seeking by trial and error: $\triangle A'GR$ is a part of $\triangle GA'C$ which in turn is a part of $\triangle AA'C$. But $\triangle AA'C$ is $1/2$ of $\triangle ABC$. (This is a transfer of an identical element in many previous problems in geometry.) What relation has $\triangle GA'C$ to $\triangle AA'C$? Here the response $A'G$ is $1/3\,AA'$ may be forthcoming (and if it is not, the student may be given a cue that suggests this previous learning). Then $\triangle A'GC$ is $1/3$ of $\triangle AA'C$ and thus it is $1/6$ of $\triangle ABC$. (The student evokes the conditional response $1/2$ of $1/3$ is $1/6$.) Now how can I find what part $\triangle A'GR$ is of $\triangle GA'C$? Since vertex A' is common, he thinks of bases GR and GC. This evokes drawing an altitude from A', but this seems to complicate the figure and this trial is rejected. Finally the student says, "I can't find any relation." The relation of GR to GC and RC does not come forth because he has never had this response in his past learning. The goal is at his door but he fails to make the connection.

What usually happens is the abandonment of this procedure and a search for another way. He says, $\triangle AC'G$ is also $1/6\ \triangle ABC$. Are $\triangle A'GR$ and $\triangle AC'G$ related? $A'B'$ is parallel to AB. (The past learning of a line joining the mid-points of two sides of a triangle evokes this response.) Immediately the pattern of similar triangles $AC'G$ and GRA' is recalled since this is a very common learning. Similar triangles can bring forth many responses but the common element here is area and this recalls the fact that areas of similar triangles are in proportion to the square of their corresponding sides. Side GA corresponds to GA' and their ratio is 2 to 1. Squares bring forth 4 to 1. Then $A'GR$ is $1/4\ \triangle AC'G$ which is $1/6\triangle ABC$. The numbers $1/4$ of $1/6$ finally give $1/24$ as the solution. Under connectionism, the answer $1/24$ would be given at the beginning of the problem, and it is the path from the given to the conclusion that is sought, not the discovery of the relation $1/24$.

Of course, in the above, other unsuccessful trials may have been made and then discarded, until the path to the goal is made. The student then repeats his solution several times, each attempt taking less time until he has made the solution readily available for further use in future learning situations.

The reader, no doubt, can supply many similar examples from arithmetic, algebra, trigonometry, or mathematical analysis. The principal characteristics of this theory of learning are:

1.   Thinking back to similar situations to find a particular response that worked previously; the transfer of identical elements.

2.   Trial and error, discarding unsuccessful paths (responses); avoiding wrong responses.

3.   Each complex situation is to be broken up into a series of simple elements arranged in a sequential order. Each simple element is mastered separately. The seriated set of mastered elements make up the whole.

4.   After the whole solution is obtained, repeat and drill until the solution is sufficiently strengthened (conditioned) for later recall.

5.   Reward successful learning of desired goals.

It should not be assumed from the foregoing discussion that connectionism was not concerned with organized systems of related knowledge. Quite the contrary, Thorndike consistently insisted on organization and interrelatedness in learning. He said: "Arithmetic consists not of isolated, unrelated facts, but of parts of a total system, each part of which may help to knowledge of other parts, if it is learned properly. . . . *Time spent in understanding facts and thinking about them is almost saved doubly*" (17).

"Knowing should be not a multitude of isolated connections, but well ordered groups of connections, related to each other in useful ways . . . a well ordered system whose inner relationships correspond to those of the real world . . ." (18).

"Every bond formed should be formed with due consideration of every other bond that has been or will be formed; every ability should be practiced in the most effective possible relations with other abilities" (19).

If connectionism is held to be not adequate, it is not for its objectives, but in its means used to secure the objectives. Through its emphasis on the detailed analyses of every mathematical process into a large number of serially related bonds to be practiced, the ultimate outcome of fundamental concepts, generalizations, and organizations frequently failed to materialize. "The forest could not be seen because of the trees." Recently psychologists are developing a more adequate explanation of learning through which understanding and well-organized patterns of knowledge come to the fore. The results of their study are now considered.

Field Theories

A major desired outcome of school education is an ability to solve problems. We are determined that this ability will be permanent and grow stronger. While we learn many facts and skills, it is the developing of the process by which they were learned that is as important as the material learned, for it is this procedure that will enable us to "go learning," to solve new problems. The goal is thus *to learn how to learn*.

It is in this aspect that field theories differ most from other theories of learning. In conditioning, the ceiling of learning is dictated by the inherited capacities of the

125

organism. In gestalt theory, the inherited capacity is increased (modified) within limits by training. There is a body of mathematical knowledge that, regardless of the capacity of the learner, could never be acquired without appropriate prior physical symbolic and linguistic experience. When this experience is acquired the learning ability rises. Thus experiencing and vocabulary building can increase the inherited capacity of the organism to learn, and they become important elements of the learning process.

Experiencing a situation and finally understanding the situation calls for a study of the *whole* of it, rather than a detailed study of the individual elements of the situation. It is only as the relation of a part to the whole is sensed that a solution of a problem can emerge that will be permanent. This is one of the fundamental principles of field theory: *Always consider the whole situation in responding.* It is not how many facts you know about a situation (a geometric original or a problem in installment buying) but how much *relatedness* in all possible ways there is between the facts and the whole of a situation. For example, a median is not only a bisector of the opposite side of a triangle, but of every line segment in the triangle parallel to this side.

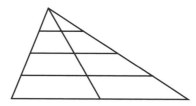

When the various elements of a situation are grasped in their relation to the whole situation, *insight* occurs. Insight is thus the final outcome (behavior) of a given situation. Before insight leading to the solution of a problem occurs, there may be trial and error, but it is not the type explained by connectionism which rejects false leads, but rather an *analysis* of the relationship of parts to the whole and the seeking of those relationships which give the complete understanding of the problem. According to connectionism the rejection of each false lead (error) brings one closer to the level of his goal; according to gestalt each analysis helps, but there is not necessarily a closer approximation after each analysis, only a larger number of relations. When all the relations discovered shape into an organized pattern, there is insight and the problem is solved. The following diagrams illustrate these two points of view:

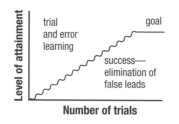

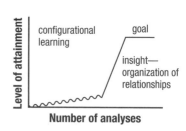

Examples

A study of two problems may aid in clarifying the concept of total configuration and insight. Consider the geometry original given on page 124. The problem is, "What relation has the area of $\triangle GA'R$ to the whole $\triangle ABC$?" A gestaltist would not give the answer $1/_{24}$, but expect the student to discover it. According to gestalt, the first emphasis is to be on the whole figure, and its related parts. We are to study the relationship of the whole

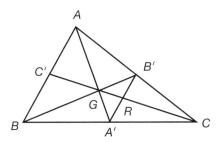

to its parts, and of the parts to the whole. Looking at the figure this way, we draw the remaining median BB' and note:

$$\triangle AGB = \triangle AGC = \triangle BGC = 1/_3\triangle ABC$$
$$\triangle B'GA' \sim \triangle AGB \text{ (They look similar, and } A'B' \parallel AB)$$
$$\triangle B'GA' = \sim 1/_4\triangle AGB = 1/_{12}\triangle ABC$$
$$\triangle GRA' = \triangle GRB' = 1/_2\triangle GA'B' = 1/_{24}\triangle ABC.$$

As soon as the pattern $\triangle GRA' = 1/_8\triangle AGB = 1/_{24}\triangle ABC$ emerges (or a similar pattern), insight has occurred.

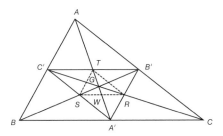

Another approach is to draw all the medians and then analyze the figure. In this case the parts are in closer relation (in a narrower field) than in the above figure, and almost by flash you see

$$\triangle A'B'C' = 1/_4 \triangle ABC; \triangle A'GR = 1/_6 \triangle A'B'C'.$$

It is also significant that the latter figure has within it elements of a new learning situation, which was not apparent in the other figures. Real creativeness lies not so much in the solution to a problem, as it does in the significant new problems that emerge out of a solution. This is in part the element of *generalization* which gestalt psychologists hold essential for permanency of learning and transference to new situations. In the figure, the relationships evolved suggest drawing SR, ST, TR, and continuing on in this manner. Then $\triangle GWR$ has the same relationship to $\triangle A'B'C'$ as $\triangle GRA'$ had to $\triangle ABC$. This immediately suggests two infinite series of areas, $1, \frac{1}{4}, \frac{1}{16}, \frac{1}{64}, \ldots$ and $\frac{1}{6}, \frac{1}{24}, \frac{1}{96}, \frac{1}{384}, \ldots$ This aspect of *discovery* and *extension by generalization* is a major aspect of configurational learning.

Finding the rate of interest charged on an installment loan illustrates several of the types of learning. Consider the problem: A television set can be bought for $300 cash, or for $60 down and 6 monthly payments of $45. What is the approximate rate of interest paid on the installment loan? At times a formula is given and the students substitute. They are conditioned to do this by the recall of the formula, say

$$= \frac{2fI}{P(n+1)}, \quad i = \frac{2(12)(30)}{240(6+1)} = 42.86\%,$$

and the knowledge of what each letter represents. They substitute numerical quantities, carry out the operations and give the result. Thus the desired end response of learning (a formula) is *given* with certain stimuli and the association is made. It is then practiced in sufficiently different situations to secure its desired strength.

Those who believe in conditioning would analyze the problem into its separate parts and arrange them in a belonging sequence, perhaps (a) finding the total installment price, (b) finding the cost of the installment plan, (c) finding the amount borrowed, (d) finding the size of each monthly principal payment, (e) finding the total time of the loan, (f) finding the interest rate. Each of these elements would be practiced separately until known, and then in the order given until known. A student who had finally learned would make the proper series of related stimulus-response actions and get the answer. His actions would probably be as follows: (a) $60 + 6 \times 45 = \$330$, the total cost; (b) $\$330 - \$300 = \$30$, the cost of the plan; (c) $\$300 - \$60 = \$240$, the principal borrowed; (d) $\frac{240}{6} = \$40$, each principal payment; (e) $\frac{1+2+3+4+5+6}{12} = 1\frac{3}{4}$ years, the time of the loan; (f) $30 = 40 \times i \times \frac{7}{4}$ or $i = 42.86\%$, the interest rate.

In field theory, the words configuration or gestalt (form) are constantly used. For this reason, the psychology using field theory is referred to as *gestalt*. A configuration is a pattern of all the elements entering into the perceptual field of the learner. If the elements of the field are reorganized, a new pattern or configuration is formed. While the elements may be abstract, it was frequently found helpful to represent the configuration by a geometric form, in which each of the geometric elements symbolizes an element of the field, and the positions of the geometric elements indicate the relationships of the field elements to each other and to the total situation. More generally, however, a gestalt is to be considered the total situation with which the learner is confronted.

The field theorist would try at the start to bring all the elements of the problem together. He would not hesitate to use a geometric drawing to aid the mental organization of the elements. The following figure (derived with the help of the students) is the initial step. (Several other configurations are possible.)

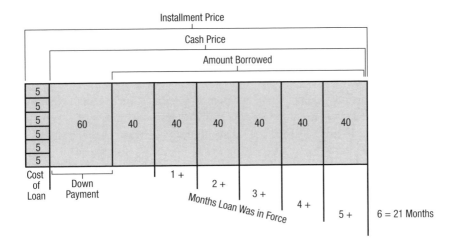

A study of the diagram allows a simpler rephrasing of the problem. If $30 is paid on a loan of $40 for 21 months, what is the interest rate? A student may now recall the simple interest formula and solve the problem. Having done this problem, and several like it, he would then generalize (pattern, structure, organize) his solution and obtain as the end result, the formula. The formula is the result of his learning, not the starting point.

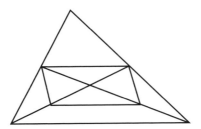

Further Aspects of Field Theory

When a field theorist says "The whole is greater than the sum of the parts," he is not referring to the physical characteristics of a situation, but the concepts and relationships involved. One can learn (a) a median is . . . (b) a line joining the mid-point of two sides of a triangle is . . . (c) a parallelogram is . . . (d) diagonals of a parallelogram . . . as four relationships. Now put them together in a whole and we also have (e) medians meet

129

two-thirds. . . . Thus any whole learning has within it concepts and potentialities for further learning that are greater than the sum of each of the elements that make up the whole. Further, it is easy to recognize that if learning is to be extended, each whole is only a part of a greater whole. By total configuration, then, is meant all the elements that come within the perception of a given situation. Learning is the integration and reorganization of the elements of a given situation into a mental pattern. When the pattern is finally organized, it is done swiftly, in a flash. The rest is a matter of progressive clarification or smoothing the performance, which is the *function of drill or practice.*

In summary, the various studies in field psychology explain learning by the following characteristics:

1. Initial learnings come from experience (physical and mental experiments), constructive methods, not from definitions. It is the dynamic aspects of events that aid learning. Whatever is to be learned must have its roots in some challenging, problem-presenting situation.

2. All parts related to the learning situation must be brought into focus to see the problem as a whole. Scattered elements or isolated details prevent insight.

3. The analysis and obtaining of relations of parts to whole and whole to parts, the recalling of past patterns of learning, and blending of the given elements permit the restructuring of the elements into a new pattern. When this occurs the student has *insight.* It is here that abstractions and generalizations come to the fore. It is the analysis and insight that give meaning to arithmetic, algebra and geometry. Lack of variety of previous experience, and over-preoccupation with fixation of specific habits, operate to prevent insight.

4. After insight, the student practices the solution to smooth and clarify the new learning (structure). The more sharpened and systematized the knowledge is, the less chance is there for forgetting.

5. A whole (configuration) is always a part of a greater whole. The relationships in one configuration (e.g. congruent triangles) appear and are generalized in later configurations (e.g. similar triangles). The relationship of relations is organized into a structure of knowledge through analysis, synthesis, and deductive logic. A system of knowledge must be built. We draw from the system (and not from a multiplicity of isolated facts) for further learning. Thus project learning and systematic courses are not contradictory but of a different level of maturity in the learning process.

Some Examples of Mathematics Learning

There are several methods of teaching the multiplication facts. One common procedure is to present the facts, simpler facts first, and drill on responses until they are automatic and correct. Thus the learner is conditioned, or makes the connection-stimulus 6×9—response 54. Then the pupil is taught how to apply this to solving problems. He goes from given specific facts to experience. Another procedure less common is to have pupils build their own facts out of experience, and then organize them into related tables. Which of these is the best learning procedure?

Multiplication

	0	1	2	3	4
1	0	1	2	3	4
2	0	2	4	1	3
3	0	3	1	4	2
4	0	4	3	2	1

Consider the table shown of 5 numbers only, and call it a multiplication table. Make flash cards to show the items to be mastered: e.g., 2 × 3 = 1; 3 × 2 = 1; 4 × 3 = 2; etc. Now drill. The reader can assure himself that this table of multiplication facts can be learned in five minutes. With appropriate drill of five minutes each day for a week it can be made fairly permanent. With continued drill at spaced intervals over a year it can become as permanent as "hickory, dickory, dock." But what does it mean? To most elementary-school teachers, and to many readers, the answer is, "Nothing." It is merely an association of meaningless responses to given stimuli, which can be learned.

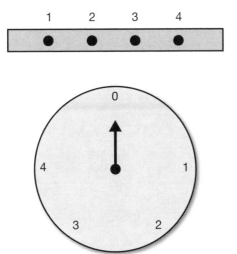

Now consider the dial and its control, as shown in the figure. Each time the button 1 is pressed the dial moves one space, clockwise. It is easy to see that 2 times 1 is 2, 3 times 1 is 3, and 4 times 1 is 4. If the button 2 is pressed, the dial moves through 2 spaces, clockwise. Thus 2 × 1 = 2; two moves of 2 spaces is 4; and three moves of 2 spaces places the dial at 1 and hence we say 3 × 2 = 1. Similarly, 4 × 2 = 3. If the button 3 is pressed 3 times, the dial makes 3 moves, each through 3 spaces and ends at the mark 4, hence 3 × 3 is 4.

From this experience it is easy to build up the table and establish meaning for it. If we are to use the table of dial multiplication efficiently, we had better practice the table to make the results automatic. But we can interpret any result in the table readily. The use of the dial to establish the table and then apply it is the less common way of teaching the usual multiplication facts in school, but it would appear to be a more satisfactory, need-fulfilling method than specific conditioning.

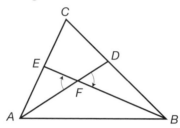

Luchins performed an experiment in solving originals in plane geometry. His subjects were high-school, tenth-grade students. The students were drilled on proving lines and angles equal by using corresponding parts of congruent triangles. They developed a set or were conditioned so that equal angles evoked the response parts of congruent triangles. This response acted as a stimulus to get sufficient parts of the triangle equal, to establish congruence—then the final response, corresponding parts of congruent triangles are equal. A series of originals on proving angles equal was given to the class after the above learning. In the set, the following original occurred: In isosceles triangle *ABC*, the bisectors of the base angles, *AD* and *BE*, meet at *F*. Prove ∠*AFE* = ∠*DFB*. So strong was the conditioning that all but one of the students proved the triangles *AFE* and *BFD* were congruent in order to arrive at the equality of the angles, whereas the solution should have been apparent at once by the use of vertical angles. A mind-set procedure (conditioning, transfer of identical elements) was a hindrance to learning.

The following sequence of problems was presented to a large group of teachers of mathematics, all with the same result.

Area = 60

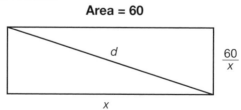

1. Represent the altitude of a rectangle of area 60 as a function of the base. The problem was a simple recall of an area pattern and the answer given was $h = \dfrac{60}{x}$.

2. In the same rectangle, represent the diagonal as a function of the base. Continuing from problem 1, most students responded by

$$d = \sqrt{x^2 + \frac{3600}{x^2}} \quad \text{or} \quad d = \frac{1}{x}\sqrt{x^4 + 3600}$$

3. In the same rectangle, represent the diagonal as a function of the perimeter. Continuing from problem 2, all students responded by placing $p = 2x + 2\dfrac{(60)}{x}$ or $x^2 - \dfrac{px}{2} + 60 = 0$. They obtained $x = p \pm \dfrac{\sqrt{p^2 - 960}}{4}$ and attempted to substitute this in the expression for d in problem 2 with various results from the involved algebraic manipulation.

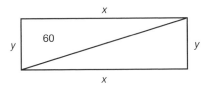

The students were then confronted with the rectangle, each part labeled with a unique symbol, and asked to study the relationship of the parts to the whole. They wrote $x^2 + y^2 = d^2$, $xy = 60$, $(x + y) = \dfrac{p}{2}$, and it was only a very short period before the "ahs" came forth as insight gave the pattern $(x + y)^2 = x^2 + 2xy + y^2$ or $\dfrac{p^2}{4} = d^2 + 120$. A set belonging-sequence can help recall old patterns but it can interfere with the discovery of new patterns.

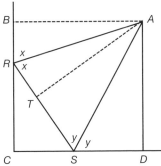

A geometric original that has caused much trouble for high-school teachers as well as their students is the following one.[4] In the figure shown we start with right triangle RSC and bisect the exterior angles at R and S. These bisectors meet at A. From A we draw perpendiculars to the sides RC and AD forming rectangle $ABCD$. Is it a square?

---

[4] I am indebted to Miss Barbara Betts of Boston for this example.

To prove the figure a square we must prove two adjacent sides equal. At once congruent right triangles is the response. An attempt is made to get angles equal through the use of complementary angles and angle sums. This fails. Even the relationship that x is not necessarily equal to y in this general configuration fails to bring out the response, "try something else." Despite the failure, the students continue to return to the unsuccessful path of obtaining congruent triangles. They fail to use the relationship of the given parts to the whole configuration. An approach that studies the whole configuration says "What is the pattern of angle bisectors?" This would suggest perpendiculars AB and AT for bisector AR, and perpendiculars AT and AD for bisector AS, and immediately insight is obtained, $AB = AD$. It is only in rejecting unsuccessful paths that trial and error can lead to a solution. It is in the relation of the whole configuration to its parts that an insightful solution emerges.

When we see an immediate relation between a given condition and a desired goal, that is, when the recall is forthcoming by simple association because of the simplicity of the problem, conditioning seems to be a good explanation for learning. When, however, the situation is complex, and we cannot see the path to the solution, the use of seriated bonds, the study of chain responses through a study of the parts of the situation seem to have much less value than a field approach of studying the whole problem, and making an analysis of the relationship of the parts to the whole, parts to parts, and whole to parts.

It is in the study of examples of learning and the various theories about learning that a teacher can gain a philosophy regarding a psychology that will work for him in his classroom activity. In the rest of this chapter we shall try to seek some common principles of learning that are concrete and acceptable for classroom management.

Areas of Agreement

Most of us, with a little introspection (and perhaps some rationalization) can recall experiencing most of the elements in all these psychologies of learning and the examples cited. In solving a geometric original, how frequently we stabbed at one route, then another, meeting block after block, coming back to the facts we started with, stopping and resting, and then going on until suddenly the solution appeared. Was it trial and error, or progressive clarification? Did we take separate steps in a sequential order or did we analyze the relationships of the parts of the figure to the whole configuration? Was there "insight" as the configuration became clear? Did each step we took put us in a frame of reference that made us take the next step? Perhaps all of these to some extent.

If you offer a mechanical puzzle to a class of students and they are not too homogeneous, you can observe a hierarchy of learning situations. Some students will merely shake, pull, and shove the puzzle by almost blind trial and error (no thinking) and by sheer accident they may solve the puzzle. Asked to try again they do the same stunt for hours and do not solve it. Others will attempt a deliberated trial and error procedure, remembering false leads and avoiding them in subsequent trials. Eventually a series of selected trials (belonging sequence) gives the solution, and then repetition of the successful sequence insures permanency of the solution. Still others will study the whole mechanism and the relationship of the parts to each other and the whole puzzle before any attempt at solution. They will try a certain movement, not necessarily to reject this movement if it does not succeed in solving the puzzle, but to see how it is related to other parts. After study, experience and trial with errors, the solution is obtained as an organized pattern. There is little need to practice each operation, but there is need to have the whole puzzle pattern clarified and verified.

The classroom teacher is desirous of having at his command a fundamental set of acceptable principles of learning upon which to organize his teaching so that the best possible learning takes place. At first glance the several interpretations of learning seem so different and in some respects contradictory, that confusion seems the outcome. But on further consideration, the teacher will find certain elements in each of the theories that appear to be important to him, and many other elements common to all the theories. We can derive from all the psychologies a theory of learning that is effective for ourselves and modify it as we ourselves learn more. The following elements may serve as a foundation of an effective theory of learning.

1. There must be a *goal* on the part of the student to learn. The learner must be aware of this goal. Thus a teacher must not only know *why* a student should learn to solve a quadratic equation, but he must know how to transform this *why* into a recognized goal on the student's part. Motivation conditions the quality of the learning. A pupil will stop counting and learn addition facts when counting becomes inadequate for him and he desires a more efficient method.

2. All cognitive learning involves association. The situation response may be simple or complex, it may be patterned, but it is an important aspect of learning. When we see a^3 we expect the response $a \cdot a \cdot a$. Even a relationship of one element to another is a form of association. The situation, similar triangles, is expected to bring forth the response—proportional sides and equal angles.

3. We recognize trial and error or analysis in most learning. If it is blind groping, then the learning situation is bad and the learner is very immature. If the trial and error is deliberate, then it can be better called approximation and correction, or analysis of relations, which continues until insight occurs. The learner should not be allowed to flounder. He should be guided in his experimentation and activity toward the final goal.

135

4. Learning is complete to the extent to which the relationships and their implications have been understood. These relationships sometimes are learned on an initial trial, especially if attended by an emotional response. More usually, practice is needed, that is, more study of the situation. This may be accompanied in some situations by studying the related set of responses, and in others by analyzing the whole situation. The more simple the pattern to be evolved, the less analysis is needed.

5. The learner must be in action, mentally and/or physically. In conditioning, the learner learns what he is doing. In connectionism, the learner must react correctly to a mathematical stimulus. In field psychology the learner experiments and organizes a pattern. It may or may not be a correct pattern of knowledge. Unless the learner is active mentally and physically, and his actions lead to success, he is not learning.

6. Intrinsic reward of success and awareness of progress toward a goal strengthens the learning and the motivation for further learning. Punishment is a deterrent rather than an aid to learning. Praise a successful response. Encourage students to make a new and different response when their first response is incorrect. With success, students raise their level of aspiration as well as their ability to solve new situations.

7. Discrimination of attributes (abstraction) and generalization are essential to effective learning. Thus all learning situations should be of the type where a relationship can be abstracted and a process can be generalized. This is only possible if the situation is meaningful.

8. New learning is in part a matter of transference of past learning. The degree to which this takes place depends on the degree of similarity of the new situation to the original learning situation, the learner's ability to analyze relationships, and the amount of varied experience in previous learning.

9. We learn facts and skills and we also learn how to learn. Our learning situations might well be changed from "topics" such as factoring, parallel lines, law of sines, and similar things, to problem situations involving the material to be learned.

10. We also learn feelings (attitudes). From unsuccessful experiences we learn to dislike mathematics and to shun the subject. We also learn to dislike teachers of subjects in which we have unfortunate experiences. We also learn to like mathematics from happy experience with it. There are many concomitant learnings that accompany the mathematics lesson that are outside the actual subject matter.

Introspection as a Guide to Learning

The task of our secondary schools is to establish within the minds of our students the fundamental bases for productive thinking. Unless we have left our students in a position where they can solve new problems—where they can go on with their learning independent of the teacher—we have accomplished but little in our mathematics instruction. The work of the teacher is to develop learning ability.

How we learn has been of interest to others besides psychologists. From the time of Plato, philosophers and educators have attempted to explain how we think. They have attempted this explanation through an introspection of how they themselves and how others have come to know whatever they do know, and to act however they behave. While this may appear to be an unscientific approach, yet the results of the thinking of these philosophers have had wide influence in establishing learning procedure.

Most famous of modern interpretations is John Dewey's *How We Think* (3) written in 1910. Dewey's interpretation of a complete act of thought (the solution of a problem) consists of five major phases: *problem-presenting situations, analysis, hypothesis, deduction, verification*. Each of these areas can be related back to some one or several of the psychological aspects of learning, and to do this can aid us as teachers in establishing our credo of learning.

A problem-presenting situation, or dissatisfaction, occurs when an individual is in a situation in which he is confused, or in which his previous knowledge does not give him satisfaction; he is not adjusted. The individual's previous ways of acting in a situation are inadequate. A student in setting up a problem obtains a quadratic equation which he does not know how to solve since his previous experience has been only with linear equations. He does not know what to do. He may experiment or flounder about, but he is dissatisfied and unhappy. We recognize here the concept of felt need, and of goal-seeking behavior, since no past learning exists to give immediate satisfaction. We also recognize that learning begins in a concrete problematic situation in which the answer is desired by, but unknown to, the individual.

A dissatisfaction causes the student to make a diagnosis of the situation, but only if the motivation is strong enough. If by trial and error, a student can find an acceptable answer to his quadratic equation, the puzzlement may cease, but a further analysis of the quadratic situation will not go on. New learnings are not desired when old ones suffice. We must make the old way of acting so inadequate that the motivation for new knowledge becomes strong enough to send the student on. A quadratic equation with no rational roots may do this.

The analysis is an examination, within the mind of the student, of the situation in which there is dissatisfaction. He discovers why he is dissatisfied and clarifies the goal that would give satisfaction. He recognizes or states his problem. The girl who, upon solving the quadratic $y^2 - 3y + 4 = 0$, obtained the answer $1/2(3 \pm \sqrt{-7})$ was in a state of perplexity because of $\sqrt{-7}$. "It just couldn't be," so she said, "for the number under the radical sign

137

should be positive." But it wasn't positive, and it had to be accounted for. Investigating past experiences in mathematics of creating negative numbers and irrational numbers, which was an analysis or diagnosis of the situation, she finally stated the problem: "I shall have to find an interpretation for $\sqrt{-7}$ to make it meaningful as an answer."

In some theories of learning little is said about the awareness of the problem. Most theories assume the existence of a problem. Thus Dewey has shown that analysis is used not only in the solution of the problem, but much earlier in the study of the difficulty, in the clarification of the desired goal. Unless the learner can detect his goal as a verbalized or unverbalized expression, he will flounder in his learning of mathematics. Thus a student meeting the quadratic equation for the first time should analyze his problem as "I must learn how to find the root of the equation when an x^2 is in it."

The third element of thinking (learning) is a search for hunches, promising leads, tentative hypotheses. This is related directly to trial and error, making responses, or analyzing relations. Testing a hypothesis is making a trial. Recalling the solution of linear equations, a student may try putting x terms on one side of a quadratic equation, and the constant on the other. If the equation is $x^2 = 25$, the hypothesis may prove fruitful, but if the equation is $x^2 + x = 25$, the trial fails (the hypothesis is not sustained). Then a new response is made. If the learner is active and purposeful, he may go back and make a further diagnosis to differentiate between the two types of quadratic equations. In forming hypotheses, the learner may need the help of "cues" to recall those past patterns of learning that will help. It may be suggested that the equation can be written in the form $x^2 - 4x + 3 = 0$ and the student asked to focus attention, for the time being, on the left-hand member (this is not to ignore the right-hand member). In what situations has he seen such expressions before? This may recall the factoring of a trinomial into two linear factors. Here is a recall of identical elements (learned patterns) and perhaps with it the response $x^2 - 4x + 3 = (x - 3)(x - 1)$. This pattern suggests two linear equations. What is the relation in the whole form $(x - 3)(x - 1) = 0$ between the left-hand member and the right-hand member? Another hypothesis (or trial) is, "Let each factor be zero." This hypothesis gives $x = 3$ or $x = 1$, and either answer satisfies. The student has broken the block and reached his goal. Note the importance of the recall of similar patterns (association). In framing hypotheses, an important question is, "What have I learned before that can be of help in solving this new situation?" Students should be imbued with this question.

The whole process of framing and testing hypotheses until a satisfactory route to the goal has been reached is the heart of the learning process. It is the most difficult part, and if a student is not successful after a few trials, he may deem the problem too difficult to solve. Many of the readers can recall "giving up" in their own learning, and allowing the problem to rest until more elementary forms were mastered, or until a further diagnosis of the problem could be made. A situation can be too difficult to be mastered, or the student may be too immature (not ready) for the task. It is important to note that the student must not be told how to solve the equation. He must be guided to make his own analyses, his own hypotheses, his own trials, and arrive at his own final solution. The more he does this, the better learner he will become. We recognize in this much of the field psychology point of view on learning.

The fourth element in Dewey's analysis of learning is deduction. By this is meant the organization of the solution of the problem into a logical frame of reference. To Dewey, this was the most important phase of the learning process, and most of his book, *How We Think* (3), is concerned with logical organization. He says, "Information . . . is not merely amassed and then left in a heap; it is classified and sub-divided so as to be available as it is needed" (p. 41). And further, "Only deduction brings out and emphasizes consecutive relationships, and only when *relationships* are held in view does learning become more than a miscellaneous scrap-bag" (p. 97).

Once insight has occurred, and the goal has been reached, it is seldom the case that the whole pattern is so clear and distinct that it will be recalled when needed in later learning. The organization of the pattern into a logical construct, after insight has occurred, can come about in several ways, (a) by going over and then generalizing the particular solution, (b) by taking many similar examples and abstracting the common elements of solution, (c) by making a logical chain of known theorems to the new result, and (d) by a mixture of these methods. The essence is this—the learning of the situation is not completed with the initial obtaining of a solution. The pattern of the solution must be organized.

The organization, according to conditioning, is organized through drill of many like situations. Actions that lead to success and satisfaction tend to recur. The more they recur the clearer the whole action. As seen above, Dewey insists on a logical deduction. Field psychology says that it is a meaningful analysis of the relations that will give the desired organization. Most mathematics teachers insist that all three procedures are necessary.

The student who obtained his solution to the quadratic equation would probably first verify his answers, then he would proceed to generalize his solution, then verify his generalization. His first task is to make some organization such as: The quadratic equation is related to the quadratic trinomial, equating the equation to zero I factor the one member, the product of two factors is zero only if at least one factor is zero, equating each factor in turn to zero I have transformed the quadratic into two linear equations, I know how to solve these, the two answers satisfy, a quadratic equation has two roots. It is only through such thoughtful organization that he can avoid such errors as equating each factor of $(x - 2)(x - 3) = 6$ to the number 6. This concept is entirely outside of the logical organization obtained from the original solution.

An organization sometimes reveals that the solution to the problem was a fortunate accident, and not a general pattern for solution. Suppose a student is confronted with a number of quadratic equations of the form $x^2 - a^2 = 0$. He writes $x^2 = a^2$. He takes square roots of both sides being very careful to use both positive and negative roots. Then he generalizes: Keep x^2 on one side, get the other terms on the other side, take the square roots of the latter side. This works until he is confronted with $x^2 = 4x + 16$. Then he is lost. The generalization did not really solve his problem. This suggests that for the most part we set up our learning situations to avoid particular solutions right from the start. The problem should lead to a generalization that has the greatest number of possible applications if it is to be of value in new learnings.

The fifth step of reflective thinking is verification, precising, and observation. If the learning has been accomplished, it is ready to be used in new experiences. C. I. Lewis (10) expressed this nicely when he said, "Knowing begins and ends in experience, but it does not end in the experience in which it begins." It is this application of what we have learned to new experiences that leads to creative learning. This is an important aim of mathematics teaching, but an aspect of learning about which most psychologists have little to offer. But if we have learned how to learn through the steps one to four, it would appear that this fifth step is to a large extent the reapplication of these four steps in a new problem with special attention to the use of the material learned.

The person who has learned and organized the general solution of a quadratic equation can now extend it in the new situation $x + bx^2 + c = 0$, or more generally, $x^{2n} + bx^n + c = 0$. This is an entirely different type of experience from that in which the original learning took place. These five steps suggest a modification of the learning diagram of Dashiell as shown below. The steps are illustrated by Roman numerals. The diagram aids in seeing the relationship between the psychological and philosophical explanations of learning. The reader can refer this figure back to the preceding discussion.

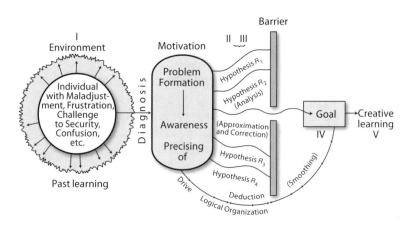

Creative Learning

It is essential to note that the logical organization or structuring of knowledge is the final step (never the first step) in developing a basis for future learning. It is also necessary to note that the steps in Dewey's complete act of thought do not necessarily occur in the order one, two, three and, four; nor are they necessarily complete learning. Dewey's explanation of thinking is a final structured explanation of what goes on in learning. A complete analysis must not be made before the learner is aware of the problem. Perhaps after he has an idea of his problem he finds further diagnosis helps further to clarify the problem, and when hypotheses fail, one after the other, he may return and reshape the problem. Further, the process of making a logical structure of the solution may result in the discovery of new and more fruitful hypotheses that give a better structure to the problem. There is a weaving back and forth across the whole learning situation with progressive clarification and elimination of unnecessary details until we finally arrive at a mature, sufficiently well ordered solution. Surely Dewey must have gone through some such procedure in arriving at his final structure of the learning process.

Regardless of how problems of world affairs appeared years ago, today to most people they seem more complex and more serious than ever. There are at least two widely separated theories on how we should behave. The one is to hold or to return to accepted and proved behaviors those that have stood the test of time. In this case our problem is not to unravel a new pattern of living, but to resolve new conditions to old patterns. The other theory is that a changing culture (new conditions) demands a new pattern of living. We are now in one of the ever emerging stages of maladjustment. We must therefore diagnose and clarify our problem, and make hypotheses. We must experiment, generalize, and deduce solutions; we must verify and "precise" our learning. We must be inventive and creative, whether it be in mathematics, social studies, government, physics, or art. We must seek solutions and take responsibility for whatever action our solutions initiate. This is the scientific attitude. This is the way to new knowledge. The key is learning how to learn.

Jacques Hadamard (5) in his *Psychology of Invention in the Mathematical Field* attempted to analyze creative thinking. As he saw it, it comprised four stages.

The first three steps of Dewey's analysis may be taken as the first or the *preparation* stage of creative thinking (step five). To create we must first solve many problems within the field. We must learn facts, skills, attitudes, habits, relationships, and be thoroughly versed in the mathematics. Then a new and unsolved problem may arise in the field. For example, a student may be asked to create a method of drawing a graph of a quadratic function for complex values of the argument which make the function real. It has never been solved so far as the learner is aware. He brings to bear all the past learning and procedures of learning, yet fruitful hypotheses are not forthcoming. To aid the student in this is the function of our teaching.

It is then suggested that for the time being the problem be allowed to rest from conscious attack. This can be looked upon as the latter part of Dewey's step three, and is called the *incubation* stage. Here the subconscious or unconscious, of which we know little, except that it plays an important part in our behavior and probably operates in the inner part of the brain, takes over. Our ideas (hunches), our past knowledge, our concept of the problem, our many unsuccessful hypotheses, and all our knowledge related to the problem, are in, or have had an effect on, the brain and the nervous system. What goes on we cannot explain except to say there is a reorganization, a restructuring of all these elements. The uncertainty and vagueness of action in the unconscious may be one reason why psychologists have, for the most part, omitted it in their theories on learning. Yet certainly something goes on in the unconscious. How do we know it?

141

By introspection, many, many creators in all fields of knowledge agree that there comes a certain period, frequently unannounced, maybe as we are relaxed before we go to sleep, or as we doze off in a chair, or on the bus or subway, when apparently without effort, insight or emergence of the solution suddenly occurs. This is called the *illumination* stage, when there rises from the subconscious to the conscious all the elements of the problem in an ordered total configuration. It is the beginning of Dewey's step four.

Having achieved insight, the discoverer immediately precises or sharpens his solution. He develops it into a neatly organized, logical, deduced pattern, placing it properly in the larger organization of mathematical knowledge. It is verified and tested as a new learning, as a new contribution to knowledge. This is called the *verification* stage. It is analogous to Dewey's steps four and five. Thus, *preparation, incubation, illumination,* and *verification* as stages in creative thinking have their counterpart in both the philosophic and psychological explanations of learning.

There is nothing new in this process of creative thinking that does not occur in any problematic approach to learning. All such learning is creative. There are successive and overlapping steps which occur in all fields of human endeavor that are making new contributions to knowledge. Insofar as learning how to learn, or problem-solving procedures are general, we can make a contribution in the mathematical field by stressing procedures of learning as much as we now stress the outcomes, without, however, detracting from the securing of necessary skills and facts. Perhaps our lack of creative learning may be traced to our lack of teaching so as to have students learn by problem-solving.

The Elements in Learning—Summary

How we learn is described in part by physical processes, by psychological aspects of behavior, and by philosophical considerations. In all descriptions there are present the elements of maladjustment, insecurity, dissatisfaction, motivation, drive, set, emotional disturbance, diagnosis, problem realization, preparation, recall, associations, trial and error (approximation and correction), analysis, hypothesis formation, incubation, solution, insight, goal attainment, illumination, structure formation, smoothing of goal route, precising, deduction, logical organization, and verification. This chapter shows in some manner how these terms are related and what they indicate for classroom practice.

The questions raised by this discussion are numerous. How is the organism motivated? How does it form its attitudes and habits? How are the senses related to learning? What is a concept and how are they formed? To what extent does language aid the learning of mathematics? Is learning achieved at one trial, or is practice a necessary requirement of learning? Does the procedure of learning transfer to later learning and learning in other fields? Just what is a problem, and how does the organism learn to solve problems? Do learners differ in ability to achieve, and if so, how do we provide for various rates of learning within a given class? Must learning be planned or is it a hit or miss affair? If answers to these questions are available, how can a teacher make sensible use of them in his daily instruction?

The rest of this Twenty-first Yearbook seeks some answers to these questions.

BIBLIOGRAPHY

1. Cole, Lawrence E., and Bruce, William F. *Educational Psychology*. Yonkers, N.Y.: World Book Co., 1950.

2. Dashiell, John. *Fundamentals of General Psychology*. Third edition. Boston: Houghton-Mifflin Co., 1949.

3. Dewey, John. *How We Think*. Boston: D. C. Heath and Co., 1910.

4. Gates, Arthur I., and others. *Educational Psychology*. Third edition. New York: Macmillan Co., 1949.

5. Hadamard, J. *Psychology of Invention in the Mathematical Field*. Princeton, N.J.: Princeton University Press, 1949.

6. Harlow, Harry F. "The Formation of Learning Sets." *Psychological Review* 51: 65–66; January 1949.

7. Hilgard, E. R. *Theories of Learning*. New York: Appleton-Century-Crofts, 1948.

8. Judd, Charles H. *The Psychology of High School Subjects*. Boston: Ginn and Co., 1915.

9. Köhler, Wolfgang. *The Mentality of Apes*. Harcourt Brace and Co., 1927.

10. Lewis, C. I. "Experience and Meanings." *The Philosophical Review* 43: 134; 1934.

11. McGeoch, John. *The Psychology of Human Learning*. New York: Longmans, Green and Co., 1942.

12. Murphy, Gardner. *Personality, A Biosocial Approach to Origins and Structure*. New York: Harper and Brothers, 1947.

13. National Society for the Study of Education. *The Psychology of Learning*. Forty-First Yearbook, Part II. Chicago: Distributed by the University of Chicago Press, 1942.

14. National Society for the Study of Education. *Learning and Instruction*. Forty-Ninth Yearbook, Part I. Chicago: Distributed by the University of Chicago Press, 1950.

15. Sherif, Muzafer. *An Outline of Social Psychology*. New York: Harper and Brothers, 1948.

16. Thorndike, E. L. *Human Nature and the Social Order*. New York: Macmillan Co., 1940.

17. Thorndike, E. L. *New Methods in Teaching Arithmetic*. New York: Rand-McNally, 1921.

18. Thorndike, E. L. *Principles of Teaching Based on Psychology*. Mason-Henry Press, 1906.

19. Thorndike, E. L. *Psychology of Arithmetic*. New York: Macmillan Co., 1922. p. 140.

20. University of Minnesota. *Learning Theory in School Situations*. Studies in Education No. 2. Minneapolis: University of Minnesota, 1949.

21. Wertheimer, Max. *Productive Thinking*. New York: Harper and Brothers, 1945.

22. Woodruff, A. D. *The Psychology of Teaching*. Third edition. New York: Oxford University Press, 1951.

23. Young, J. Z. *Doubt and Certainty in Science*. New York: Oxford University Press, 1951.

5

From *Historical Topics for the Mathematics Classroom,* NCTM's Thirty-first Yearbook (1969)

"The History of Mathematics as a Teaching Tool" by Phillip S. Jones, the first chapter of the Thirty-first Yearbook, was recommended by John Egsgard, NCTM president from 1976 to 1978 and the only Canadian ever elected to serve as Council president.

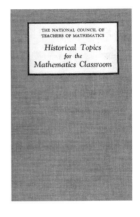

This chapter will be particularly valuable to readers who are interested not only in the history of mathematics but also in how this subject can be employed in the actual teaching of mathematics. The historical topics that Jones addresses include the development of numeration systems (with particular attention to base-and-place-value systems and the binary system), mathematics as an art, and non-Euclidean geometry. Jones also gives attention to both teacher and student use of the history of mathematics.

The History of Mathematics as a Teaching Tool

Phillip S. Jones

> *I have more than an impression—it amounts to a certainty—that algebra is made repellent by the unwillingness or inability of teachers to explain why. . . . There is no sense of history behind the teaching, so the feeling is given that the whole system dropped down ready-made from the skies, to be used only by born jugglers.*
>
> —Jacques Barzun, *Teacher in America*

EACHING so that students understand the "whys," teaching for meaning and understanding, teaching so that children see and appreciate the nature, role, and fascination of mathematics, teaching so that students know that men are still creating mathematics and that they too may have the thrill of discovery and invention—these are objectives eternally challenging, ever elusive.

The "sense of history" cited by Barzun will not, of itself, secure these objectives. However, properly used, a sense of the history of mathematics, coupled with an up-to-date knowledge of mathematics and its uses, is a significant tool in the hands of a teacher who teaches "why." There are three categories of whys in the teaching of mathematics. There are the *chronological*, the *logical*, and the *pedagogical*.

Teaching for Meaning

Chronological whys.—The first of these categories includes such items as why there are sixty minutes in a degree or an hour, and why sixty seconds in a minute; it looks for the sources of such words as "zero" and "sine" as well as "minute" and "second." These and many other historical facts, large and small, not only give the why of a particular question but also, in the hands of a skillful and informed teacher, trigger discussions about the necessity and arbitrariness of definitions and of undefined terms, about the psychological bases of mathematical systems, and about the development of extended definitions, such as those of "number" or "sine," which change with the development of new systems, being generalized and rephrased to include the new ideas as well as the old.

Logical whys.—The second category mentioned—the logical "whys"—may seem to be independent of the history of mathematics, but actually history may contribute greatly to developing the logical insights of students. Logical whys include an understanding of the nature of an axiomatic system as well as the logical reasoning and proofs that clothe the axiomatic skeleton with theorems. It is important that our students grow to understand this structure, but for many topics the direct and minimal statement of axioms and proofs is neither the way these ideas developed historically nor the way perceptions grow in the minds of many of our students. One of this century's leading mathematicians, Henri Poincaré, is quoted on this subject by Jacques Hadamard (1954, p. 104).

> To understand the demonstration of a theorem, is that to examine successively each syllogism composing it and ascertain its correctness, its conformity to the rules of the game? . . . For some, yes; when they have done this, they will say, I understand.
>
> For the majority, no. Almost all are much more exacting; they wish to know not merely whether all the syllogisms of a demonstration are correct, but why they link together in this order rather than another.

It is not true that a historical approach is the only way to communicate this sense of why theorems hang together as they do—this "mathematical motivation" for a proof. The historical approach is not even always the *best* way to communicate these insights. But it can often help tremendously! The authors of the article "On the Mathematics Curriculum of the High School" (see under title, p. 192) were on the right track, even though they may have slightly exaggerated the utility of what they called "the genetic method," when they made the following statement:

> "It is of great advantage to the student of any subject to read the original memoirs on that subject, for science is always most completely assimilated when it is in the nascent state," wrote James Clerk Maxwell. There were some inspired teachers, such as Ernst Mach, who in order to explain an idea referred to its genesis and retraced the historical formation of the idea. This may suggest a general principle: The best way to guide the mental development of the individual is to let him retrace the mental development of the race—retrace its great lines, of course, and not the thousand errors of detail.
>
> This genetic principle may safeguard us from a common confusion: If *A* is logically prior to *B* in a certain system, *B* may still justifiably precede *A* in teaching, especially if *B* has preceded *A* in history. On the whole, we may expect greater success by following suggestions from the genetic principle than from the purely formal approach to mathematics.

More recently John Kemeny (1963, p. 76) has pointed out that

> in the historical development of mathematics, it is usually, though by no means always, the case that a certain body of mathematical facts is first discovered, and then one or more people perform the very important task of systematizing this information by specifying a minimal number of axioms and deriving the other facts from these. It is, therefore, clear both that some acquaintance with axiomatic mathematical systems is an important part of mathematical education, and that mathematics is something over and above mere development of axioms.

148

A portion of the history of mathematics can often display Kemeny's "something over and above." Recapitulating all the errors of the past would be confusing and uneconomical. However, it is more than merely consoling to find that Descartes called negative numbers "false" and avoided their use; that Gauss had a "horror of the infinite"; that Newton wrote four approaches to his fluxions, probably because none of them quite satisfied him conceptually; that Hamilton really thought he was doing something quite different logically when he invented the abstract, modern, "ordered-pair" approach to complex numbers. Not only do these stories offer the consolation that great men once had difficulties with what today are fairly-well-clarified concepts. They also show how mathematics grows and develops through generalizations and abstractions; and they indicate that the mathematics of a few years from now will no doubt be different from ours while still including today's ideas, perhaps in altered form.

Pedagogical whys.—Pedagogical "whys"—our third category—are the processes and devices that are not dictated by well-established arbitrary definitions and do not have a logical uniqueness. They include such devices as "working from the inside out" in removing nested parentheses. This is not logically necessary, but experience and a little thought suggest that it is less likely to lead to errors. Similarly, there may be many logically sound algorithms for a particular process, say, finding rational approximations to $\sqrt{26}$. Pedagogically, I would choose the "divide, average, and repeat" process, sometimes called merely the "division" process, for the first one to teach. There are several reasons for this. It is closer to the basic definition of square root (in fact, thinking the process through calls for continually reviewing this definition); it is easier to understand; and it is easier to recall. Later the time-honored square-root algorithm may be taught in several contexts— for its connection with the square of a binomial or a multinomial, or for its connection with the expansion of a binomial with a fractional exponent into an infinite series, or for its connection with iterative processes.

Historical ideas may help in both the selection and the explication of such pedagogically motivated processes. For example, there is good evidence that the division process for finding square roots was used by the Babylonians. It is clear that the approximation to $\sqrt{a^2+b}$ as $a + (b/2a)$ was known to the early Greeks and can be motivated by reference to the square diagram associated with $(a + b)^2$ in Euclid's *Elements*, Book II, 4. The repeated use of this to get successively better rational approximations is to be found in the work of Heron (or Hero) of Alexandria. Heron, unlike earlier Greek philosopher-mathematicians such as Pythagoras, Euclid, and Apollonius, was also interested in mechanisms (he invented the first jet engine), surveying, and the related calculations. (The formula for the area of a triangle which bears his name, "Heron's formula," involves a square root and was no doubt used by him, though it is probably actually attributable to Archimedes.) The actual algorithm as we teach it involves an approximation and an iteration of that approximation. It also involves a process (marking off periods of two, etc.) that depends on the use of a place-value numeration system which, though partially achieved by the

Babylonians, did not develop to the extent of being ready for this algorithm until the rise of Hindu mathematics. Here is a prime example of a situation where a genetic approach does more than help the teacher to select a good pedagogical sequence and good teaching devices. Exploitation of this story, with its diagrams and processes, may really contribute to an understanding of such important concepts as irrational numbers, rational approximations to them, iterative processes, and base-ten numeration.

Recommendations for Use of the History of Mathematics

It is easy to collect testimonials about the importance and value, both for teachers and for students, of some study of the history of mathematics.[1] Recommendations for the inclusion of some study of history in teacher-training programs are to be found in many studies and committee reports in many countries.[2] A survey by John A. Schumaker (1961, pp. 416, 418–19) of trends in the education of secondary school mathematics teachers shows that the percent of teacher-education institutions offering such a course increased from 44 to 52 over the period from 1920/21 to 1957/58. In the latter year such a course was required of prospective teachers majoring in mathematics in 12 percent of the institutions. K. A. Rybnikov, chairman of the Department of the History of Mathematics and Mechanics at the University of Moscow, states that such a course is required of all mathematics majors there.

In spite of all these claims and recommendations, however, the history of mathematics will not function as a teaching tool unless the users (1) see significant purposes to be achieved by its introduction and (2) plan thoughtfully for its use to achieve these purposes. The remainder of this chapter will enumerate such purposes and illustrate methods of achieving them. We shall use historical anecdotes and vignettes or sketchily outlined sequences to do this. The aim is not to present the history of mathematics itself but to suggest in a concrete way how and for what purpose historical materials may be used. We believe that this concreteness is desirable even though some clarity may be lost because of the fact that the sketches generally illustrate more than one important objective.

The Development of Numeration Systems

The story of the development of different systems of numeration and of the binary system in particular is an excellent one to illustrate the reasons why persons do mathematics, as

---

[1] Several of these are to be found in an excellent article, "The History of Mathematics and Its Bearing on Teaching," which is the ninth chapter of *Teaching Mathematics in Secondary Schools* (see under title). This article includes, also, suggestions for using the history of mathematics as a teaching tool, as do articles by Jones (1957) and the Freitags (1957).

[2] See particularly AAAS (1960, p. 1028); Bunt (1963, p. 664); Dodd (1963, p. 313); Kemeny (1963, p. 76); MAA (1935, p. 276); and MAA (1960, pp. 635, 637, 643).

well as to illustrate the connections between mathematics and the physical world. The basic elements of a modern positional or place-value numeration system are these: an arbitrary integer greater than 1 selected as a base, b; a set of b distinct digits including 0; and multiplicative and additive concepts. (The multiplicative concept is that each digit is to be multiplied by the power of the base corresponding to the position in which the digit is written, and the additive concept is that the number represented by these symbols will be the sum of these products.) A final necessity is a decimal point or other scheme for marking the position of the "units digit" or the position associated with b^0.

Base-and-place-value systems.—The various elements composing the complete concept of a positional numeration system first appeared in history at different times and in different countries. Egyptians, Greeks, and Romans grouped their symbols by tens and in a sense were using the idea of a base. The ancient Babylonians made the first use of a positional system; but their system, using base sixty, was ambiguous because they used repetition in forming symbols and they lacked a zero and a "sexagesimal point." The symbol ▼▼ could represent 2, 2/60, $1 \cdot 60 + 1$, or even $1 \cdot 60^2 + 0 \cdot 60 + 1$. The Hindu-Arabic system's separate symbol for each of the whole numbers less than ten, the base, eliminated the first two sources of ambiguity. The Babylonians, however, anticipated by over three thousand years the development of place-value notation for numbers less than 1. It was not until the time of Simon Stevin of Bruges (1585) that the Hindu-Arabic numeration system was extended to decimal fractions with a scheme for marking the point separating representations of integers from representations of fractions. Such a historical analysis of the components of our numeration system, along with an investigation of its advantages and a realization of the dependence of our computational algorithms upon the system, can add substantial understanding as well as interest to the teaching of arithmetic.

The binary system.—Moreover, several very important concepts and insights can be simply demonstrated by detailing the development of the binary system. An anthropological study of the western tribes of the Torres Straits, published in 1889, tells of one tribe which had only two number words, *urapun* and *okosa*, for 1 and 2 respectively. Out of these two words, by grouping, they manufactured names for 3, 4, 5, and 6: *okosa urapun, okosa okosa, okosa okosa urapun,* and *okosa okosa okosa.* This was the totality of their number names, with the exception of *ras,* which was used for any number greater than 6. Although not so ancient in years, this primitive society resembles prehistoric ones; and its story suggests two things. It suggests that number names and systems of numeration undoubtedly existed in prehistoric times (prehistoric notched bones and bronze counters confirm this) and that no doubt this numeration developed out of the needs of daily life, commerce, taxation, and trade.[3]

---

[3] For details of the Torres Straits numeration system, see Conant (1931). For an account of genuine prehistoric mathematics, see Vogel (1958, I, pp. 7–20). For a theory of a ritual (religious) origin for both counting and geometry, see Seidenberg (1961 and 1961/62).

In spite of the antiquity of some of the components of a base-and-place-value numeration system, a general and abstract perception of its structures is relatively recent. The Torres Straits number words, grouping by twos, may suggest a primitive binary system to us, looking at it from the vantage point of today, but the first published description of a binary system appeared in 1703. This was written by Gottfried Wilhelm von Leibniz, the famous German philosopher and developer of the calculus, who was partially motivated by an attempt a to explain semimystical symbols found in an ancient Chinese work. Leibniz, however, was so intrigued by this notion that he wrote a generalized and extended discussion of the manner in which one could use any arbitrarily chosen number as the base for the development of a numeration system. His interest in religion and philosophy led him to feel a particular interest in the binary system, where any integers as large as you wish may be written using only two symbols, 0 and 1. Leibniz saw a parallel to the story of the creation of the universe in Genesis, in which we learn that God, whom he associated with 1, created the universe out of nothing, or a void, which he associated with 0. It is even reported that Leibniz caused a medal to be struck commemorating this idea.

Concepts illustrated by introducing the binary system.—At any rate, for our students these stories further illustrate several points that are now commonly accepted in mathematics, viz., that (1) connections are frequent between mathematics and philosophy and even between mathematics and religion; that (2) practical, social, economic, and physical needs often serve as stimuli to the development of mathematical ideas; but that (3) intellectual curiosity—the curiosity of the man who wonders, "What would happen if . . . ?"—leads to the generalization and extension and even abstraction of mathematical ideas; and that (4) perception of the development of general, abstract ideas, along with perception of the patterns or structure in mathematics, may be one of the most practical goals of mathematical instruction.

This fourth point is very well illustrated by the fact that, although Leibniz proposed no use whatsoever for the binary system, it has in recent times played a critical role in the development of high-speed electronic digital computers and in the development of information theory and data processing of all sorts. In this case the insights and perceptions associated with mathematical theory as expounded by Leibniz, and not the narrow, immediate application to counting as employed by the natives of the Torres Straits, led to recognition of the utility of the binary system in a context that had never been anticipated by its inventor.[4]

The concept of "model" in mathematics.—Another point illustrated by this historical sequence is basic in modern mathematics, namely, that (5) the concept of a "model" is important in both pure and applied mathematics. The model in this case is very simple

---

[4] For further details and references to the binary system, see Jones (1953).

both to describe and to understand. Electrical circuits are capable of two states: they are either open or closed. Electron tubes or electric lights are either illuminated or not illuminated: they are on or off. The binary numeration system uses only two symbols. In every position there is either a 1 or a 0. The parallel, when one thinks of it from the vantage point of history and of the perception of a general structure, is clear and obvious. If each position in the numeration system were made to correspond to an electrical circuit, and if the digits 0 and 1 were thought of as corresponding to an open or a closed circuit or switch, one would have a correspondence between a set of physical elements and a set of mathematical concepts and symbols. Each can be thought of as a model for the other. If the models are well chosen, any operation or state in the one system has a corresponding operation or state in the other. Hence one may draw conclusions by operating in one system and interpreting the conclusions in the other.

An understanding of the role of matched "models" (mathematical and physical or social or economic) in the applications of mathematics not only supports the importance of pure mathematics and the role of intellectual curiosity; it also helps students to develop a better understanding of the nature of mathematics itself. Mathematicians may be stimulated by physical problems and assisted by geometric diagrams, but the mathematics they create is an abstraction. This "model" concept itself is quite modern, but it has an extensive historical background, a study of which contributes to its understanding. For example, mathematicians for years thought of Euclidean geometry not as merely one possible model for physical space around us but as the true and ideal system abstracted from a system of physical points and lines. Its axioms were viewed as necessary and self-evident, probably on the basis of observations of the physical world and geometric diagrams. The source and role of axioms are viewed differently today, however.

Mathematics as an Art

Today mathematicians frequently liken mathematics and its creation to music and art rather than to science. J. W. N. Sullivan (1925, p. 7) wrote:

> It is convenient to keep the old classification of mathematics as one of the sciences, but it is more just to call it an art or a game. . . . Unlike the sciences, but like the art of music or a game of chess, mathematics is a free creation of the mind, unconditioned by anything but the nature of the mind itself. . . . There is nothing necessary in any of the fundamental postulates or definitions of any branch of mathematics. They are arbitrary: which does not mean that their choice was psychologically arbitrary, but that it was logically arbitrary.[5]

This view of mathematics certainly was not held by the ancient Egyptians or Babylonians, nor even by the Pythagoreans, who thought the basis of music, of philosophy, and indeed of the universe, lay in number. Not even Euclid, who organized the first successful

---

[5] See also Whitehead (1925). The second chapter begins: "The science of Pure Mathematics, in its modern developments, may claim to be the most original creation of the human spirit. Another claimant for this position is music." This interesting chapter is reprinted in Newman (1956, I, pp. 402–16).

axiomatic structure and did such a fine job that it persisted with little change for over two thousand years, had this view of his accomplishment.

If mathematics is an art, some appreciation of this fact, and of the relation of mathematics to the world of physical reality, can be as much a part of the liberal education of a doctor, lawyer, or average intelligent citizen as is some appreciation of the humanities.

Non-Euclidean Geometry

For persons whose exploration of modern advanced mathematics must be limited, a survey of the history of non-Euclidean geometry or of the relations between mathematics and man's view of his universe can be a vehicle for communicating a valuable, though partial, insight.

Non-Euclidean geometry was born of intellectual curiosity, without concern for practical applications; these appeared many years after its birth. In this respect it resembles an art. Its story, like others we are reciting, illustrates more than one of the major concepts and insights that discussions of historical background serve to convey. Dissatisfaction with Euclid's parallel postulate existed almost as soon as it was enunciated. Attempts to find better or simpler statements equivalent to it (as a postulate) accompanied attempts to prove it as a theorem dependent on other postulates. Although there was some earlier work by Arabic geometers, the first person to develop an extensive list of theorems in non-Euclidean geometry was Girolamo Saccheri, who never knew what he had actually done. He thought that his book *Euclides ab omni naevo vindicatus* ("Euclid Freed of All Flaws"), published in 1733, had proved the parallel postulate by assuming a different postulate and showing that it would lead to a contradiction. The "contradiction" he finally found was not a real contradiction but, rather, a theorem of non-Euclidean geometry which was different from the analogous theorem of Euclidean geometry. Without the generalized perception of the importance and arbitrary nature of postulates and the possibility of the existence of different geometries based on different sets of consistent, but arbitrary, postulates, Saccheri could not have had any real understanding of what he had accomplished.

Nikolai Ivanovich Lobachevsky and Janos Bolyai, who in the nineteenth century were the real inventors of non-Euclidean geometry, did not themselves comprehend the revolution in mathematics that was to grow out of their invention of another geometry. Out of their work developed (1) an understanding of the nature and importance of axioms, (2) the concept of the possibility of many different mathematical systems, and (3) attempts to axiomatize all branches of mathematics—attempts resulting in different algebras, different geometries, the need for proofs of consistency and independence, and finally in recent years the whole body of philosophical, logical studies comprehended under the new name "metamathematics."

Not only did the creators not comprehend these connections with philosophy and logic and the foundations of mathematics, but they certainly never anticipated the fact that one hundred years later the physicists, when formulating their theory of relativity, would find a non-Euclidean geometry just the tool they needed for rephrasing and simplifying the original statements of Einstein.

This fascinating story, with its range of connections all the way back to ancient Greek philosophy—notably, to Plato's understanding of the nature and importance of definitions and to Aristotle's work in logic—typifies again the connections of mathematics with other aspects of our culture, especially philosophy and the philosophical science of logic. It also illustrates two facts: (1) that often the applications of a mathematical concept or system are unforeseen by the inventors and may follow years later in unpredicted ways and (2) that over the years mathematicians' perceptions of their own subject change and develop.

The relations between mathematics and the physical world are an interesting and sometimes complicated two-way street. Physical needs and intuitions based upon ideas of such physical things as points and lines may serve well to stimulate the thinking and intuition of a mathematician; on the other hand, mathematics developed as pure theory or extended and generalized into abstract theory may at the hands of other persons serve as a vehicle for the discovery and extension of unforeseen physical or even social and economic theories.

Successive Extensions and Generalizations

The process of making successive extensions and generalizations has gone on over the centuries, as we have seen. Let us now look at this process as a pedagogical necessity and a pedagogical problem in the design and teaching of any modern mathematics curriculum. Although people are today urging that we begin at a much more advanced level than we have in the past, no one has suggested that we should begin in the kindergarten with discussions of the entire real and complex number systems. If we assume that it will be pedagogically desirable to study and teach positive integers before dealing with negatives, fractions, irrationals, etc., we automatically set up a situation in which we are going to be responsible for helping children to understand how number systems were extended repeatedly in order to overcome difficulties such as the impossibility of subtracting a larger number from a smaller one, of dividing 3 by 2 (in the domain of integers), of finding the square root of −4, etc. The pedagogical difficulty is that what has previously been deemed impossible suddenly is made possible and the same word that we have used in the past, "number," now has a new and extended meaning as well as new properties and uses.

A helpful example of this process of extension and generalization can be traced in the development of modern trigonometric functions. These grew out of Greek astronomy, where a conception of planets moving in circular orbits developed a need for determining the lengths of chords of circles if one knew the length of the corresponding arcs. Significant segments of early Greek geometry, especially those dealing with the construction of regular polygons, were useful (at least in part) in the computation of tables of chords by Ptolemy, who in this computation was extending work begun by Hipparchus. Hindu mathematicians of about A.D. 500 saw the greater utility of tables of half-chords and not only computed them but gave us, indirectly, the word "sine." This has come to us as the English version of the Latin word used to translate an Arabic word which, through an error, was used for the Hindu *jya,* "half-chord." In later stages the sine has been variously defined as "one side of a right triangle with a given acute angle and a hypotenuse of one," "the ratio of the side opposite an acute angle to the hypotenuse of a right triangle," "the ratio of the ordinate to the radius vector determined by an angle at the center of a circle," "the ordinate of a point on a unit circle determined by a line segment 'wrapped' around the circle," "the limit of an infinite series corresponding to a given real number substituted in the series," "the limit of an infinite series corresponding to a complex number substituted in the series," and "the inverse function of a relation determined by a definite integral." All the tables computed under these varying definitions of the sine are in some way equivalent to or derivable from one another. All these definitions and all these tables could be used to solve problems involving chords of circles and sides of triangles both right and obtuse. But the more abstract and general definitions, which do not depend on the notions of angles, triangles, or even necessarily of circles or arcs, are the definitions that particularly stress the special properties of these trigonometric functions. Amongst these special properties the most important is, perhaps, their periodicity.

The study of periodic phenomena such as vibrating strings, electrical impulses, radio waves, sound waves, and the development of physical theories such as the wave theory of light have made it important to have periodic mathematical models, functions, and relations. As the trigonometric functions were progressively through history separated from their dependence on circles and triangles and ratios, and as they were generalized and abstracted to be sets of ordered pairs of numbers matched with one another by such devices as summing infinite series, they became more useful. Their use was no longer restricted to situations with angles and triangles, but they became functions of real (or complex) numbers, which, in turn, could represent time intervals or distances or whatever was useful. These abstract functions were then studied for their intrinsic properties, such as their periodicity, rates of change, maxima and minima, to such an extent that they were well understood as separate entities and were ready for use in constructing a mathematical model of periodic phenomena. Alfred North Whitehead, in commenting on this story, wrote, "Thus trigonometry became completely abstract; and in thus becoming abstract, it became useful" (1925, p. 48).

It may be possible to exaggerate the importance of this idea that the generalizations and abstractions of mathematics are really the useful parts because it is they that make the mathematics applicable to physical situations as yet unknown and perhaps even unforeseen. However, this concept of mathematics is far too little appreciated, and the history of mathematics serves a valuable purpose if it makes this aspect of the subject understood by future citizens. It should be clear that an understanding of this role of abstraction and generalization is a goal to be sought, not a guide for initial pedagogical procedures. It is not our contention that the introduction to new mathematical topics should be general and abstract, especially at the more elementary levels; it is our contention, however, that as the subject grows and develops we should lead students not only to understand generalizations and abstractions but also to appreciate their importance and their role. A sufficiently concrete and detailed tracing of the history of the development of a generalized idea is one of the best ways to teach an appreciation of the nature and role of generalization and abstraction.

Intuition, Induction, and Analogy

However, balance demands that we also stress the importance of intuition, induction, and analogy in the conception and perception of mathematical ideas. The fascinating story of Srinivasa Ramanujan may help to illustrate this idea (as well as the internationalism of mathematics). Ramanujan was a poor, largely self-taught Hindu mathematical genius who lived from 1887 to 1920. His story, along with stories of Gauss, Galois, Euler, Archimedes, and many others, can further serve to stimulate interest when used as a subject for reading and class or mathematics-club reports. One of the remarkable things about Ramanujan is the fact that in almost complete isolation from other mathematicians he developed for himself substantial portions of mathematics. Some of his developments were awkwardly and incompletely done. Some of them were even incorrect. But, in spite of this, an almost unparalleled imagination and intuition led him to many remarkable discoveries. When he was being brought to England, his sponsor, the famous mathematician G. H. Hardy, wrote of a question that was bothering him:

> What was to be done in the way of teaching him modern mathematics? The limitations of his knowledge were as startling as its profundity. . . . All his results, new or old, right or wrong, had been arrived at by a process of mingled argument, intuition, and induction, of which he was entirely unable to give any coherent account.

> It was impossible to ask such a man to submit to systematic instruction, to try to learn mathematics from the beginning once more. I was afraid too that, if I had insisted unduly on matters which Ramanujan found irksome, I might destroy his confidence or break the spell of his inspiration.[6]

The caution here expressed illustrates the fact that all mathematicians recognize the role of imagination and intuition in leading to the conjectures they must later test for truth, then prove if found to be probably true. A story told by Hardy of a conversation with Ramanujan when visiting him in a hospital also illustrates the mathematician's natural inclination to ask, "What would happen if . . . ?", to generalize, and after each new idea to ask himself, "What more can I make of that? To what new idea does it lead me?" The story is that as Hardy was talking to Ramanujan, he remarked

---

[6] For this anecdote and more about the life of Ramanujan, see Newman (1956, I, pp. 366–76).

that he had ridden to the hospital in taxicab number 1729 and that this number seemed a rather dull one. "No," Ramanujan replied, "it is a very interesting number; it is the smallest number expressible as a sum of two cubes in two different ways." Hardy goes on to say, "I asked him, naturally, whether he knew the answer to the corresponding problem for fourth powers."

Interdependence in Mathematics

Another theme in the history of mathematics is illustrated by our brief reference to the development, simultaneously, of non-Euclidean geometry by a Hungarian and a Russian who had no connections with or knowledge of each other. The frequent occurrence of simultaneous discoveries in mathematics illustrates the growing and maturing nature of mathematical knowledge and the fact that frequently new discoveries are generated by earlier ones. Not only are new discoveries generated by earlier ones, but often earlier ones are so necessary as a preparation for the next stages that when the preparatory stages have been completed a number of persons will see the next step. The discovery of logarithms by the Scotsman John Napier and the Swiss Jobst Bürgi, as well as the development of a geometric representation of complex numbers by the Norwegian Caspar Wessel, the Swiss Jean Robert Argand, and the German Carl Friedrich Gauss, all at much the same time, furnishes additional illustrations of simultaneity and internationalism.

Mathematics advances by the interplay of many devices and approaches. In spite of what has been quoted previously about the bareness of mere statements of axioms and proof, the modern axiomatic method can be a tool and a motive for extending the bounds of mathematics itself. For example, Whitehead and Russell (1925–27, p. v), in the Introduction to their famous *Principia mathematica*, say:

> From the combination of these two studies (the work of mathematicians and of symbolic logicians) two results emerge, namely (1) that what were formerly taken, tacitly or explicitly, as axioms, are either unnecessary or demonstrable; (2) that the same methods by which supposed axioms are demonstrated will give valuable results in regions, such as infinite number, which had formerly been regarded as inaccessible to human knowledge. Hence the scope of mathematics is enlarged both by the addition of new subjects and by a backward extension into provinces hitherto abandoned to philosophy.

The variety of ways in which today's students can participate, both vicariously and directly, in the formulation of the axioms needed to advance a proof and in the fumbling for conjectures that follow from a set of assumptions, has been discussed in many articles; they do not necessarily have a strong historical connection. However, the examination of Euclid's axioms and their probable sources, and of some of his theorems and the gaps in their proofs occurring from the use of diagrams, is an instructive way to point out the nature and need of an axiomatic system.

Teacher and Student Use of the History of Mathematics

This then leads to the question, How do you use the history of mathematics to accomplish all the fine things suggested above—to increase the understanding of mathematics itself; of its relation to the physical world; of how and why it is created, grows, develops, changes, and is generalized; of the internationalism of its appeal; of the practicality of generalizations, extensions, and abstractions; and of the nature of structure, axiomatization, and proof?

There is no neat, simple, general answer to this. The age and background of the student and the ingenuity of the teacher join in determining the approach that may be used. It may be that a brief historical talk or anecdote is indicated. Or the class may be given—with some discussion and buildup—a problem such as the famous one propounded by Girolamo Cardano (also known as Cardan), "to find two numbers whose sum is 10 and whose product is 40." He said of the problem, "This is obviously impossible." (Why?) But then he went on to say, "Nevertheless, let us operate." Completing the square in the same manner as he had with earlier "possible" problems, he arrived at $5 + \sqrt{-15}$ and $5 - \sqrt{-15}$ as the new "sophisticated" numbers that would satisfy the conditions.

Some historical stories and topics are useful for outside reports, independent study, and club programs. Other situations provide the setting for encouraging student "discovery." Certainly a student who has seen the sequence of squares of integers associated with Pythagorean diagrams for figurate numbers (see NCTM 1969, pp. 53–58) is on the verge of discovering for himself the formula $n^2 + 2n + 1 = (n + 1)^2$ and the fact that the sum of the n successive odd integers beginning with 1 is n^2, as well as the fact that the second differences of a quadratic function tabulated for equal increments of the dependent variable are constant. History can do more than present stimulating problems and ideas; it can also help students to perceive relationships and structure in what appears to be a tangled web of geometry, algebra, number theory, functions, finite differences, and empirical formulas.

For the teacher, a historical view helps to determine what "modern mathematics" should really be. History shows that contemporary mathematics is a mixture of much that is very old—counting, for example, and the Pythagorean theorem, which is still important—with newer concepts, such as sets, axiomatics, structure. These newer aspects may be more significant for their clarifying and unifying value than for their explicit content. If a perception of structure in an old system makes a segment of mathematics easier to comprehend and extend, if new symbolism and terminology help to integrate and systematize a body of knowledge that was growing unwieldy, then these new perceptions of structure and symbolism make the teaching, learning, and using of mathematics easier. Often (not always) "modern mathematics" may be merely a modern perception of several old topics. The important thing is neither to throw out all that is old nor to add whatever is new but to develop and pass on to our students new syntheses of old ideas and systems

159

as well as to introduce new concepts and systems that are appropriate. Insight into the development (history) of ideas can serve to improve both the curriculum maker's choices and the teacher's power to communicate insights and stimulate interest.

REFERENCES

AAAS, Cooperative Committee on the Teaching of Science and Mathematics. "Preparation of High School Science Teachers," *Science,* CXXXI (April 8, 1960), 1024–29.

Bunt, Lucas N. H. "The Training of a Mathematics Teacher in the Netherlands," *AMM,* LXX (June–July 1963), 660–64.

Conant, Levi Leonard. *The Number Concept, Its Origin and Development.* New York: Macmillan Co., 1931.

Dodd, W. A. "A New Qualification for Mathematics Teachers in the United Kingdom." *MT,* LVI (May 1963), 311–13.

Freitag, Herta T., and Freitag, Arthur H. "Using the History of Mathematics in Teaching on the Secondary School Level," *MT,* L (March 1957), 220–24.

Hadamard, Jacques Salomon. *Essay on the Psychology of Invention in the Mathematical Field.* 1945. New York: Dover Publications, 1954.

Jones, Phillip S. "The Binary System," *MT,* XLVI (December 1953), 575–77.

———. "The History of Mathematics as a Teaching Tool," *MT,* L (January 1957), 59–64.

Kemeny, John G. "Report to the International Congress of Mathematicians," *MT,* LVI (February 1963), 66–78.

MAA. Commission on the Training and Utilization of Advanced Students of Mathematics. "Report on the Training of Teachers of Mathematics," *AMM,* XLII (May 1935).

———. Committee on the Undergraduate Program in Mathematics. "Recommendations for the Training of Teaching of Mathematics," *MT,* LIII (December 1960), 632–38, 643.

National Council of Teachers of Mathematics (NCTM). *Historical Topics for the Mathematics Classroom.* Thirty-first Yearbook of the NCTM. Washington, D.C.: NCTM, 1969.

Newman, James Roy, ed. *The World of Mathematics.* 4 vols. New York: Simon & Schuster, 1956; paper, 1962.

"On the Mathematics Curriculum of the High School," *MT,* LV (March 1962), 191–95.

Schumaker, John A. "Trends in the Education of Secondary School Mathematics Teachers," *MT,* LIV (October 1961), 413–22.

Seidenberg, A. "The Ritual Origin of Geometry," *Archive for History of Exact Sciences,* I (1961/62), 488–527.

———. "The Ritual Origin of Counting," *Archive for History of Exact Sciences,* II (1962), 1–40.

Sullivan, John William Navin. *The History of Mathematics in Europe.* London: Oxford University Press, 1925.

Teaching Mathematics in Secondary Schools. (Ministry of Education Pamphlet No. 36.) Her Majesty's Stationery Office, 1958.

Vogel, Kurt. *Vorgiechische Mathematik.* Hannover: Hermann Shroedel Verlag, 1958.

Whitehead, Alfred North. *Science and the Modern World.* New York: Macmillan Co., 1925; paper, Macmillan Co., Free Press, 1967.

Whitehead, Alfred North, and Russell, Bertrand. *Principia mathematica* (2d ed.). 3 vols. London and New York: Cambridge University Press, 1925–27.

6

From *A History of Mathematics Education in the United States and Canada,* NCTM's Thirty-second Yearbook (1970)

A careful look at the list of all seventy-five NCTM Yearbooks on pages 5–8 reveals that NCTM published two different yearbooks in 1970. Mary M. Lindquist, NCTM president from 1992 to 1994, recommended that we include the first chapter from the first volume of 1970, the Thirty-second Yearbook titled *A History of Mathematics Education in the United States and Canada.* This relatively brief chapter by Phillip S. Jones and Arthur F. Coxford Jr., **"The Goals of History: Issues and Forces,"** provides a thorough introduction to this still widely used resource. Anyone who would like an across-the-spectrum view of the history of mathematics education in the U.S. and Canada will appreciate this explication of the 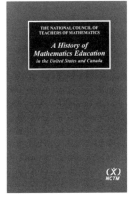 forces, issues, and historical periods that provided the focus for that yearbook.

It should be noted that NCTM past president Glenda Lappan (whose own selection from 1983 is provided in chapter 11) suggested the possible inclusion of a different chapter from this 1970 volume. That chapter, "Mathematics Education on the Defense: 1920–1945," provides an interesting read of policy issues related to mathematics teaching and curriculum during World War II. The introductory chapter we have included here may well entice the reader to seek out the Thirty-second Yearbook and review that later chapter, as well as all the others it contains.

The Goals of History: Issues and Forces

Phillip S. Jones
Arthur F. Coxford Jr.

1. Our goals

THERE are many kinds of history directed toward many goals. The chronology of events may serve for some as an end in itself, and for others as the basis of forecasts or of the search for causes and effects. Patriots or chauvinists may seek with history to revive feelings of nationalism, to justify past actions, to create or enhance a national image. The history of a special field may seek to clarify its boundaries, content, or methods and to increase understanding of the present nature and values of the discipline. Our chief goal in this history of mathematics education is to direct the thoughts of mathematicians, educators, and even the public at large to the issues of today, by showing that although the content and methods of mathematics education have changed significantly over the years, strong continuing forces imply that further change is imminent and important. The continuing development of new mathematics, new uses of mathematics, new pedagogical devices, and changing goals for a changing society all demonstrate the need for continued change in mathematics education.

However, a thoughtful view of history shows that many apparently new ideas are actually old and that they merit a revival only after the lessons of the past have been studied. The old approaches and materials should be examined in the light of both old experience and a new situation. For example, Colburn's "inductive approach" to arithmetic and algebra has many elements in common with the "discovery teaching" of today; and attention to numeration systems and geometry in the elementary school has been urged on many occasions since the beginning of the nineteenth century. History suggests that past innovations have been adopted in response to changing needs and philosophies, although rarely with a speed that could be called revolutionary, and that they have often failed to be as magically effective as their proponents expected. It is hoped that steadier progress will result from a survey of the causes and effectiveness of past proposals and projects. For this reason our plan is to stress issues, and the forces leading to action on these issues, throughout this book.

2. Our themes: forces and issues

We would like this book to be not a mere catalog of facts about mathematics education in North America, but a description of forces and issues related to mathematics education based upon such a catalog of facts. We regard *issues* as questions with reference to which there has been or may be some debate. Many issues exist in mathematics education today. Many of these are the same as, or slight modifications of, issues that have occurred frequently in our history. The emergence both of these issues and of proposed answers to them is due to a variety of *forces*—in mathematics itself, in educational theory and psychology, and in our changing culture. Our ultimate goal, then, is to provide a better understanding of the issues of today and the forces which gave rise to them, and to suggest answers for them. We believe that this may be achieved through a tracing over the past centuries of emerging issues and changing forces.

We believe that the major issues and forces relating to mathematics education can be considered under two headings: *curriculum* and *instruction*. The former, of course, includes both the content and the grade placement of materials taught. The latter is chiefly concerned with the method of teaching but also includes provisions for individual differences. This means that a complete separation of these two headings, curriculum and instruction, is not possible, since questions on individual differences are certain to involve questions of mental maturity which are related to the grade placement of materials in the curriculum.

Although we shall treat the evolution of mathematics education chiefly under these two headings, two other categories will frequently appear as the sources of issues and forces. These are the changing views of *mathematics* itself and of the newly developing fields of *educational philosophy* and *psychology*. During the period of our story mathematics first developed in association with studies of the physical world, becoming a major tool in the scientific and technological revolution of the seventeenth, eighteenth, and early nineteenth centuries. Then, in the nineteenth and twentieth centuries, it became a collection of man-made, axiomatic structures rooted in the interests and intuitions of its builders but not in the physical world. Connections of mathematics with the physical world often developed later out of the work of persons other than mathematicians. As this view of mathematics developed, rigor, abstraction, and generalization increasingly characterized the work of the professional mathematician.

Educational psychology developed in somewhat the reverse manner. From the musings of educational philosophers it moved first into laboratories in which behavior was studied and then out to share in the redesigning of textbooks and the development of teaching procedures to embody the ever-changing views of the learning process and problem solving.

Further, although the authors believe in minimizing the distinctions between elementary and secondary education and in stressing continuity and articulation rather than differences in grade placement or course content, we have found it convenient to divide our discussion of forces and issues emerging through history in accord with the elementary and

166

secondary levels of education. The extent and manner to which this distinction is being blurred today is a current issue, but we are writing history. We cannot deny that in the past one could differentiate rather sharply both the curriculum and instructional procedures according to grade level. For this reason we think that under today's circumstances our audience may be larger and more understanding if we heed this separation but try to point out both the issues involved in eliminating it and the forces tending to drive us in this direction. Along with this, we shall point out the issues and methods that are common to instruction at both levels.

These considerations have led us to design a book in six parts:

Part 1: Mathematics in the Evolving Schools

Part 2: Forces and Issues Related to Curriculum and Instruction, K–6

Part 3: Forces and Issues Related to Curriculum and Instruction, 7–12

Part 4: The Education of Teachers of Mathematics

Part 5: School Mathematics in Canada

Part 6: Epilogue: Present-Day Forces and Issues

3. Progress by periods

In Part One we will present the facts of the history of mathematics education, stopping occasionally to indicate themes that will be further developed in later parts. Our aim has been to clarify issues and identify forces which may both produce issues and help resolve them.

We have divided the story into five periods. Although time is a continuum which can be uniquely and sharply separated by specifying a single point, educational and cultural forces and issues are always overlapping and intermingling throughout all periods of time. However, it seemed sensible to us to begin with only a brief reference to the period prior to the first settlement in the United States and Canada (1607) and to extend the early history of mathematics education across the period of the beginnings of schools to the time of Warren Colburn, whose first book was published in 1821. We have selected Warren Colburn as representative of the beginning of a new concern for pedagogy. Although Colburn's arithmetic (Colburn 1821) and algebra (Colburn 1825) written on "inductive principles" are neither the first mathematical books written or published in the Western Hemisphere nor the first to give thought to pedagogy, they represent a new and more general concern for pedagogical processes in this country.

For our second period we have chosen 1821 to 1894. The latter year dates the beginning of three new forces in mathematics education in North America. This was the year of the founding of the American Mathematical Society (AMS), which was derived from the New York Mathematical Society initiated in 1888 (Smith and Ginsburg 1934, pp. 105–6). The AMS itself has probably not been a significant factor in secondary and elementary education. However, certain early members of the society, notably Professor E. H. Moore, were much concerned with the problems of secondary school mathematics and gave time and attention to them, as well as giving this area of investigation the prestige that came from their interest in innovation in the teaching of mathematics. The year 1894 also saw the publication of the *Report of the Committee of Ten on Secondary School Studies* (NEA 1894), the committee having been appointed two years earlier by the National Education Association (NEA). The appointment of this committee typifies two forces in American mathematics education: (1) the concern of persons with a major initial interest in education as a whole for the specialized subject-matter field of mathematics and (2) the influence of national committee reports as stimulators of reform.

The founding of the AMS and the role of the NEA in appointing the Committee of Ten represent another significant force that affected education during the latter part of the nineteenth century and the first part of the twentieth. This force was the rapid development of many professional organizations, both those associated with the organization and administration of the schools and those associated with scholarly research. The *Report of the Committee of Ten* will be discussed later, but it is interesting to note how many still-current issues received its attention: the introduction of more geometry of an intuitive sort into the elementary schools; the earlier beginning of algebra; the incorporation of solid geometry with plane; and the development of what later was called a "double track" program after the first year of algebra, with one syllabus for those planning to continue into college and another for those not so bent. This last provides an example of an issue and a related force. The issue is this: For what students and for what goals should the school program be defined? The force is the pressure on the secondary schools to plan curricula and guide students with college entrance requirements and examinations in mind.

We have chosen to terminate our third period with 1920, the date of the founding of the National Council of Teachers of Mathematics (NCTM) and of the issuing of the preliminary report of the National Committee on Mathematical Requirements, *The Reorganization of the First Courses in Secondary School Mathematics* (National Committee on Mathematical Requirements 1920). This committee had been appointed by the Mathematical Association of America (MAA) in 1916, prior to the entry of the United States into World War I. The major concern of the MAA, itself only recently founded (1915), was for the undergraduate mathematics curriculum in the colleges. Its sponsorship of this committee emphasizes the effect of college requirements and ideas on the secondary school curriculum and also the continuing concern of many mathematicians at the college level for improvement in the schools. A number of persons with close relations to the secondary schools—both persons training teachers and persons teaching in the schools— were sought out and included on the committee. The committee was spoken of with great approval by the first president of the NCTM, C. M. Austin, in his initial note to the members of that organization at the time it took over the *Mathematics Teacher* as its official journal in 1921 (Austin 1921, p. 1). This period, 1894–1920, might then be labeled "The Growth of College and University Influence," or perhaps "First Steps toward Revision."

Our fourth period, 1920–45, introduces the effect of two strong forces outside the control of school or college personnel: economic depression and war. We have chosen 1945 as the terminal year of this period because it marked the end of World War II and the appearance of two of the reports of the Commission on Post-War Plans of the NCTM (1944, 1945). This committee was the first of a number of committees, commissions, and studies to be set up as a result of concern for the inadequacies in mathematics education that were brought out during the war.

The reports of the Commission on Post-War Plans may be regarded as one of the links between the issues of the prewar period and the reforms that began to get under way about 1952. Prior to the war mathematics educators had struggled to answer many criticisms. There were two major types: (1) The mathematical (arithmetical) competence of high school graduates was viewed as inadequate for many occupations such as business, elementary school teaching, and the armed services. (2) Much of secondary school mathematics, even extending to long division, was regarded as of questionable value for general education by many educational philosophers, psychologists, and guidance specialists. The Commission on Post-War Plans clearly exposed fallacies in this view. Other educators began to stress that the war had not only demanded mathematical competencies at many levels but had also led to the development of many new applications with postwar, peacetime significance. Both the applications of mathematics and much of the actual mathematics were new. This led to the view that to be prepared for unforeseen needs one must know and understand the structure of mathematics, not merely its facts and operations.

Two major issues of the prewar period persisted after the war: (1) Does school mathematics instruction serve the needs of individual students? (2) Does school mathematics instruction contribute effectively to the solution of the problems of our society? Before the war many people would have answered both questions with a no. The answer remained the same; but the reasons behind it, the remedies called for, and even the persons giving the answer changed during the war and the years immediately following it when proposals for extended experimentation and reform were developed.

This led to our definition of the years from 1945 to the present as our fifth and last period, a period of experimentation and reform which some termed a "revolution." Naturally, changes so extreme as to be called revolutionary brought forth some protests and reaction.

4. Summary and preview

In Part One, then, we will seek to provide the factual framework from which emerge the *issues* that have called for concern and change and the *forces* that have given rise to these issues and given direction to the changes.

In our introductory remarks we have passingly focused on such issues as these:

1. *What should be the goals of mathematics education?* Providing for personal and vocational practical needs? Providing "mental discipline"? Training in logic and problem solving? Preparing for further study?

2. *How can mathematical education in both content and instruction be adapted to the varied needs, capacities, and interests of students?* By "double tracks"? Self-instruction?

These are perhaps the *only* issues in the sense that perhaps all others are in some way subsidiary to them. However, we have also noted, implicitly at least, these additional issues:

3. *What mathematics should be taught?* "Facts" and operations—or structure?

4. *What students should take mathematics?* Everyone? All college preparatory students? Only would-be scientists and engineers?

Some issues which will arise in later chapters are:

5. *What persons or groups should direct mathematics education?* Professional mathematicians? If so, should they be the pure or the applied mathematicians? Educators? Psychologists? Mathematics educators? The general public?

6. *How can one provide for the experimentation needed to guarantee continued tested progress in both curriculum and instruction?*

7. *What levels of rigor are sound and desirable at different stages of a student's development?*

8. *What is the role of applications and mathematical models in motivating and clarifying instruction?*

9. *How do we teach so that students perceive the excitement and beauty of mathematics as well as its facts and theorems?*

10. *Can we teach so that students will "discover" and be more creative?* If we can, should we?

Forces that have been mentioned in our introductory remarks include:

1. *Practical needs*—of explorers, soldiers, navigators.

2. *The research and beliefs of psychologists and philosophers*—with reference to the goals and methods of instruction.

3. *College entrance examinations and requirements.*

4. *The presence in the schools of increasing numbers of students with increasingly varied interests and abilities, but all needing to be retained and educated.*

This list is incomplete, and some of these forces are subsidiary to others in conception or importance. However, it is hoped that this introductory list may make the reader more alert to perceive our theme and goal: the presentation of the history of mathematics education as the story of changing issues and forces as they affect practices which, in turn, provide a series of approximations to changing educational goals. We hope that such perceptions may stimulate, prepare for, and help determine the best directions for future changes.

REFERENCES

Austin, C. M. "The National Council of Teachers of Mathematics." *MT* 14 (January 1921): 1–4.

Colburn, Warren. *An Arithmetic on the Plan of Pestalozzi, with Some Improvements.* Boston: 1821. Later editions had various titles, including *First Lessons in Arithmetic on the Plan of Pestalozzi* (1822) and *Intellectual Arithmetic upon the Inductive Method of Instruction* (1826).

———. *An Introduction to Algebra upon the Inductive Method of Instruction.* Boston: Cummings, Hilliard, & Co., 1825.

Commission on Post-War Plans of the NCTM. "First Report of the Commission on Post-War Plans." *MT* 37 (May 1944): 225–32.

———. "Second Report of the Commission on Post-War Plans." *MT* 38 (May 1945): 195–221.

National Committee on Mathematical Requirements of the MAA. *The Reorganization of the First Courses in Secondary School Mathematics.* USBE Secondary School Circular no. 5. Washington, D.C.: GPO, 1920.

NEA. *Report of the Committee of Ten on Secondary School Studies with the Reports of the Conferences Arranged by the Committees.* New York: American Book Company, 1894.

Smith, David Eugene, and Jekuthiel Ginsburg. *A History of Mathematics in America before 1900.* MAA Carus Mathematical Monographs, no. 5. Chicago: Open Court Publishing Co., 1934.

7

From *The Teaching of Secondary School Mathematics,* NCTM's Thirty-third Yearbook (1970)

John Dossey served as NCTM president from 1986 to 1988. He recommended that we include a chapter from the second of two NCTM yearbooks published in 1970. That volume, *The Teaching of Secondary School Mathematics,* has served as a valuable resource for mathematics education faculty members having the responsibility for a secondary mathematics methods course.

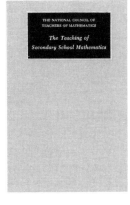

The selection that we have included here is **chapter 7, "Concepts,"** by Kenneth B. Henderson. It provides a thorough consideration of concepts and conceptual understanding in secondary mathematics, particularly those related to algebra and geometry. For just one way to think about the depth of this chapter, consider its first sentence: "One can argue that concepts are the basis of formal education." Readers with an interest in the psychology of learning, and particularly in teaching concepts, will find this chapter to be an important and valued read.

Concepts

Kenneth B. Henderson

O
NE can argue that concepts are the basis of formal education. Since formal education proceeds mainly through language and is highly concentrated, concepts loom large in importance. Principles—generalizations and rules of procedures—are formulated by means of concepts. Attitudes, too, are grounded on concepts. The time mathematics teachers devote to teaching concepts is some evidence of their importance. Hence it is proper that attention in this book on the teaching of mathematics be given to the teaching of mathematical concepts.

When one observes the teaching of concepts, he becomes aware that various teaching strategies are employed. Consider the teaching of the concept of a trapezoid. One teacher might teach this concept by drawing several representations of trapezoids and telling the students that a trapezoid is a quadrilateral having two sides parallel.

A second teacher might begin by pointing out that to some mathematicians *trapezoid* means a quadrilateral having just one pair of sides parallel and to other mathematicians *trapezoid* means a quadrilateral having at least one pair of sides parallel. He then draws two figures, one meeting the conditions of the first conception and the other the conditions of the second conception. He contrasts the two figures and then stipulates that for the present geometry course the first meaning of *trapezoid* will be used.

A third teacher might begin by drawing some geometric figures on the chalkboard, labeling some trapezoids and some not-trapezoids and in each case telling why he labeled them as he did.

From these three strategies, one can ask several questions. One question is this: Under what conditions is each strategy effective, and under what conditions is each strategy ineffective? Another question: Are there other strategies, and if so under what conditions is each effective? Finally, one can ask a question of theory: How can strategies be determined? It is to these questions that this chapter will be directed.

Organization of the Chapter

The rest of the chapter is divided into four main sections. The first section is an explication of a concept as a theoretical construct in a theory of pedagogy. It is felt that an explication is necessary because *concept*, like so many other terms in pedagogical theory, is used ambiguously and loosely. Since what is presented in the other sections depends on what is meant by a concept, it seems appropriate to make this concept as clear as possible. Then one is better able to decide whether differences of opinion, if such exist, are semantic or factual.

The second section offers a taxonomy of concepts. It will be argued in the third section that how a concept can be taught is determined, in part, by the nature of the concept. (It is also determined, in part, by the nature of the students.) Hence a logical prelude to the third section is one that seeks to present a functional classification, one whose implications for teaching are fairly forthright.

The third and fourth sections offer a model for teaching mathematical concepts. The model, like the discussion in the previous sections, is believed to be general enough to pertain to the teaching of concepts in any subject. That it is a teaching model rather than a learning model is considered an advantage. Attempts to deduce effective teaching models from learning models have not been particularly successful. Bugelski, an educational psychologist, has pointed this out (1964), and he supports his contention by quoting similar opinions by other learning theorists.

Explication of Uses of "Concept"

The term *concept* is so loosely used that the only safe inference one can draw from a use of the term is that some kind of knowledge is referred to. Van Engen (1953, p. 69) pointed out this confusion when he wrote on the teaching of concepts of mathematics. The situation is little different today. When one reads the objective, "to teach the concept of a rational number," he may find one teacher considering the objective has been attained when he teaches the students what a rational number is and they are able to discriminate between rational numbers and other numbers. Another teacher may feel that this is not enough—that to attain the objective he should go beyond the definition and teach properties of rational numbers other than definitional ones. We hear the statement, "Tom doesn't have a concept of the distributive postulate of multiplication over addition." What inferences can we draw? That Tom does not know this postulate, that is, that he cannot state it in some language? That Tom can state the postulate but cannot apply it? And when someone speaks of "having a concept of how to factor a trinomial," does he mean anything more than knowing how to factor a trinomial?

Some learning theorists simply use the term *concept* without defining it; others indicate what they mean by the term. In the first case, a concept appears to be the set of associations a person has with the term that designates the concept. This point of view emphasizes the individualistic nature of a concept, but in so doing it loses theoretical power; it is not useful in a theory of teaching mathematics. Ordinarily, the mathematics teacher is concerned about a restricted set of associations which are supposed to be invariant as to concept-holders; he is not concerned with the emotional and attitudinal connotations of the designating expression.

Some theoreticians appear to identify a concept and an abstraction. A child has his attention directed repeatedly to different sets of two objects. When he becomes aware of the common property of *twoness* amidst the differences, he has made the abstraction. It is the abstracted property which is the concept of two. Travers (1963, p. 133) appears to characterize this theoretical point of view by his statement, "The process of identifying the attributes that characterize a particular category is referred to as the process of 'concept attainment.'" He continues, "Attaching a word to a concept so that it is conventionally labeled is not an essential part of the concept-attainment process." This is consistent with his concept of concept attainment. It behooves one, however, to remember that neither of these statements is a contingent (factual) statement; the logic is that of analytic (tautological) statements. In a different theory, the second statement might not follow. For example, to psychologists identified by Pikas as of a theoretical orientation such that "a concept comes into existence through the conditioning of one word-label to several stimuli," (1966, p. 232), the existence of a term to designate the concept is a necessary condition for a concept; no name, no concept.

Hunt, Marin, and Stone offer the following: "A *concept* is a decision rule which, when applied to the description of an object, specifies whether or not a name can be applied" (1966, p. 10). This definition would seem to imply that the existence of a designating expression or a name, not necessarily the conventional name, is necessary for a concept.

Each of the two theoretical positions, namely, a concept as an abstraction where no name is involved and a concept as a set of conditions for the use of a name or designating expression, is sensible. A concept as an abstraction without name serves to explain informal concept attainment; it also directs laboratory experiments where the subject learns to classify objects by practicing with examples and nonexamples. But in the formal education of the secondary school concepts are infrequently taught in this way. It is more efficient, particularly in mathematics, to teach one concept by relying on other concepts, already taught. As Carroll observes, "There is a gap between the findings of psychologists on the conditions under which very simple 'concepts' are learned in the psychological laboratory and the experiences of teachers in teaching the 'for real' concepts that are contained in the curricula of the schools" (1966, p. 179). Johnson and Stratton (1966) confirm this judgment and point out that in formal education verbal explanatory teaching prevails partly because the conventional symbol designating the concept is usually not antecedently meaningless. Hence the theoretical construct advanced by Hunt, Marin, and Stone seems more appealing.

The two theoretical positions can be made compatible by conceiving of two kinds of concepts, nonverbal and verbal, corresponding respectively to the two theoretical positions. It is generally agreed that it is possible for a person to have a concept without verbalizing it. When a concept is assigned a name and attains sufficient currency to be communicated, it passes into the verbal category. It is the verbal concepts that are part of the subject matter of the school. We now turn to an explication of such concepts in the context of formal (institutionalized) education.

A proposal for a concept of a verbal concept in the context of formal education

In attempting to decide what a verbal concept is to be for the present discussion, that is, what is to be denoted by *verbal concept*, it is profitable to observe what teachers do when they say they are teaching concepts. When one does this, he is struck by the high correlation between having the objective of teaching a concept of *t* and teaching how to use *t*. In other words, about the same behavior is manifested by a teacher who is teaching a particular concept as is manifested when the teacher teaches how to use the term that designates the concept. As an instance, a teacher who teaches the students the concept of a hexagon does about the same things as he would do if he were teaching the students how to use the term *hexagon*. This is the case whether he teaches by exposition or by discovery. If he defines *hexagon*, this is a clear case of teaching how to use the term. If he tells the students what a hexagon is, this amounts to indicating the necessary and sufficient conditions for the use of *hexagon*. And if the teacher draws figures on the chalkboard calling certain ones *hexagons* and others *not-hexagons*, hoping that the students will induce (discover) the set of distinctive properties of a hexagon, again he is thereby indicating how *hexagon* can be used—what objects can be named by the term. The argument is similar whether the teacher uses a nonconventional appellation like "figures like that" while pointing to some paradigm or the students are allowed to invent their own name. Hence it does not seem ill-advised to consider teaching a student a verbal concept and teaching him how to use a term designating the concept as denoting the same pedagogical activity.

It seems consistent with the foregoing to regard a verbal concept as an ordered pair, one component being a designatory expression, a name, and the other being one or more rules for using the designatory expression. (Some may wish to say "meanings" rather than "rules for using." But *meaning* is not a well-defined term; one has only to consult an unabridged dictionary to realize its ambiguity.) Hence one component is in the metalanguage and the other in either the object language or the metalanguage, depending on its use. An elaboration of this theoretical position is given in Henderson ([a] 1967).

A verbal concept as an ordered pair as described above has some kinship with Church's position, "Every concept of a thing is a sense of some name of it in some (conceivable) language" (1956, p. 7), and also with that of Hunt, Marin, and Stone, namely, "A *concept* is a decision rule which, when applied to the description of an object, specifies whether or not a name can be applied" (1966, p. 10). Moreover, as we shall see, it is useful in pedagogical theory.

Hereafter, instead of speaking of verbal concepts, we shall speak more simply of concepts.

Three uses of terms can be identified. The teacher can talk about the properties or characteristics of objects named by the term if such objects exist. For example, he can use *function* to talk about the properties or characteristics functions have in common or conditions under which an object would be a function or the term properly applied. We shall speak of this as employing the conventional connotation of the term.

178

There is also the subjective connotation of a term, that is, the association both substantive and emotional a person has with a term. The subjective connotation will vary from person to person. For example, to many mathematics students, *operation* connotes doing something and getting a result even though this characteristic may not be part of a conventional connotation of the term that would appear in a definition. Because of the preciseness of mathematics, mathematical terms have fewer subjective connotations than those of subjects like social studies, English, and driver training.

The teacher can use *function* to identify particular relations that are functions; he can give examples of a function; he can also identify objects that are not functions. We shall speak of this as employing the denotation of the term. In using either the connotation or denotation of a term, the teacher is operating in the object language, that is, he is talking about the properties of functions or identifying particular functions.

Finally, the teacher can use the metalanguage and talk about the term itself rather than about its referents. For example, he can define the term *function*. Or he can give expressions that purport to be synonymous to *function*. We shall speak of this as employing the implication of the term.

These three uses of a term—the connotative, denotative, and implicative—become, as we shall see later, the three ways of teaching a concept. Some concepts can be taught by all three ways, others by only two. To support this contention, we turn to an analysis of concepts.

A Functional Taxonomy of Concepts

Since it is to be argued that the nature of a concept should, in part, determine the moves in teaching the concept, it would appear that a taxonomy of concepts should be offered.

The taxonomy of Bruner, Goodnow, and Austin

Before the work of Bruner, Goodnow, and Austin (1956), few psychologists used a taxonomy any more analytical than a partition of the set of concepts into such subsets as familiar or unfamiliar, easy or difficult, concrete or abstract, and precise or vague. Bruner and his coworkers considered it worthwhile to conceive of conjunctive, disjunctive, and relational concepts. We are more aware of conjunctive concepts and more familiar with them than with disjunctive and relational. Consider the concept of a parallelogram as a quadrilateral whose opposite sides are parallel. For something to be a parallelogram, two conditions are necessary: (1) it must be a quadrilateral and (2) the opposite sides must be parallel. The conjunction of these two conditions is sufficient. This concept of a parallelogram is a conjunctive concept. As pointed out above, to Bruner and his coworkers a concept is associated with a category. They identify a conjunctive category as "one defined by the *joint presence* of the appropriate value of several attributes" (p. 41). Hence a conjunctive concept is one that is determined by a conjunctive category. The specific determiner of a conjunctive concept is the explicit or implicit force of "and."

By a similar analysis a disjunctive concept is one determined by a value of at least one of two or more attributes. The concept of a nonnegative rational number as zero or a positive rational number is an example of a disjunctive concept. The specific determiner of a disjunctive concept is the explicit or implicit force of "or."

Bruner, Goodnow, and Austin stipulate, *"The relational concept or category* is one defined by a specifiable relationship between defining attributes" (p. 43). An example is the concept of *larger than* in the sense that something is larger than something else. (The term *relational concept* is not an insightful one since both conjunctive and disjunctive concepts are determined by "a specifiable relationship between [among] defining attributes.")

Bruner's taxonomy has directed psychologists' and teachers' attention to factors other than psychological when considering how concepts are learned. One now finds educational psychologists less disposed to write about the teaching and learning of concepts as if all concepts are alike. But the taxonomy is not analytical enough to be of great help to teachers. Nor is the research based on this model and emanating from the psychological laboratories particularly useful to classroom teachers. They do not teach verbal concepts in the inefficient way psychologists maneuver their subjects into learning concepts designated by nonsense symbols, for example, *dak, jon, ruz,* or *vek.* This is not to disparage such research, but only to argue its general irrelevance to formal education and the need for a different taxonomy.

Denotative contrasted with attributive concepts

Consider the concept of a true statement and that of truth. If we say that both "true statement" and "truth" refer, not only are the referents different but they differ in quality. "Truth" purports to refer to a property—the property that characterizes all true statements and serves to distinguish them from other statements. Truth is what we attribute to true statements. Hence we might say that truth represents a level of abstraction above that of true statements. We shall characterize the concept of a true statement as being denotative or a denotative concept; we shall characterize the concept of truth as being attributive or an attributive concept.

Every concept is either denotative, nondenotative, or attributive. Conceiving of a concept as a set selector, a denotative concept. is a concept whose referent set is not empty; a nondenotative concept is a concept whose referent set is empty. Thus "true statement" and "pi" are denotative concepts; there are true statements and there is a number π.

"Even prime number greater than 2" and "fifty-first state of the United States" are nondenotative concepts; there is no even prime number greater than 2, and in 1970 there is no fifty-first state of the United States.

Ordinarily teachers do not teach nondenotative concepts, but one can conceive of such concepts. And many teachers can remember uses of a nondenotative concept, for example, that 9 is an extraneous root of $\sqrt{x} = 3$. According to the way *root of an equation* is defined, "extraneous root" is as self-contradictory as "round square." Neither term denotes. The same can be said for "negative mantissa." There are no negative mantissas.

An attributive concept is not a set selector. Truth, elegance, skewness, consistency, and independence ordinarily do not determine sets and hence are attributive. In contrast, true propositions, elegant proofs, skew lines, consistent propositions, and independent axioms denote sets and hence are denotative concepts.

It should be noted that a concept can be either denotative or attributive. Consider the concept of *congruence*. If congruence is conceived as a relation, that is, a set of ordered pairs determined by "*x* is congruent to *y*," the concept is denotative. If it is conceived as the property of all pairs of objects *x* and *y* such that *x* is congruent to *y*, it is attributive. Hence it is worth making explicit that being denotative or attributive is not an inherent property of a concept; it is a result of how the concept is used by a teacher or author or a group of individuals.

It follows from this classification that a teacher might teach a denotative concept by exhibiting, naming, or talking about members of the referent set. He cannot use this move in teaching an attributive concept.

Conceiving of relations such as parallelism, equivalence, and similarity as sets of ordered pairs has the pedagogical advantage shared by all denotative concepts; the teacher can give examples and discuss the characteristics of members in the referent set of the concept. If these relations are conceived of as properties, hence attributive concepts, they are less readily taught and more difficult to learn.

Many attributive concepts, particularly those designated by terms that end in "-ness," "-ity," or "-tion" (such as categoricalness, distributivity, and implication), ordinarily are taught indirectly by teaching their denotative correlates. For example, convexity is inferred from the concept of convex sets, which probably would be taught deliberately; involution is somehow abstracted from the concept of an involute. Hence, in the discussion that follows, more attention will be given to denotative concepts than to attributive concepts.

Concrete contrasted with abstract concepts

Every denotative concept is either concrete or abstract. A concrete concept is a denotative concept the members of whose referent set are material objects, that is, objects that have values of such observable properties as location in space and time, weight, volume, color, among others. The concepts of a protractor, slide rule, and compass are concrete concepts. So are those of variable and formula, but the denotation (members of the referent set) of "variable" and "formula" are symbols—marks on a surface. The concepts of Descartes and of the Pythagoreans are denotative, as is the concept of any particular living mathematician.

An abstract concept is a denotative concept that is not concrete. Hence an abstract concept denotes a set whose members do not have observable properties; they are inferred entities or theoretical constructs. Examples of abstract concepts are those of 2, $\sqrt{2}$, ratio, real number, and function. For sophisticated mathematics students, teachers properly regard such geometric concepts as circle, polygon, line, and point as abstract concepts; a circle, polygon, line, or point can be represented (pictured), but the representation is not the inferred entity. Most of the concepts a mathematics teacher teaches are abstract concepts.

One of the stable findings of research on concept learning is that concrete concepts are easier to learn than abstract concepts. Pikas states, "Concepts of things are formed more easily than concepts of forms, colors, and numbers" (p. 229).

There are no concrete, attributive concepts. This follows from the definitions of *attributive concept* and *concrete concept*. Although some persons—perhaps most persons—would regard an attributive concept, for example, triangularity, as abstract, such a conception in the present taxonomy is not particularly helpful. Since there are no concrete, attributive concepts, the set of attributive concepts is identical to the set of abstract, attributive concepts. Hence we apply Ockham's razor and speak only of attributive concepts.

That this classification is functional is evident from a consideration of the difference in how the concepts of (a set of) two objects, a concrete concept, and two, an abstract concept, are taught to primary pupils. The former is easier to teach; the teacher can exhibit sets of two objects. The latter is hard to teach, in part, just because the number two cannot be exhibited. A teacher who wishes to draw a distinction between a number and a numeral finds he can ostensively exhibit numerals; "numeral" is a concrete concept. But he can only name numbers; "number" is an abstract concept. Lest one draw an invalid inference from these examples, it needs to be pointed out that concrete concepts are not necessarily more easily taught and more readily learned than abstract concepts. The concept of 500 million, an abstract concept, may be easier to acquire than a concrete correlate, the concept of 500 million people.

Singular contrasted with general concepts

Consider the concept of Euclid and that of two. Each of these denotative concepts has a referent set having just one member. Contrast these concepts with those of micrometer and integer. Each of these denotative concepts has a referent set having more than one member. We shall speak of the first two concepts as singular and the second two concepts as general.

Every concept is either singular or general. A denotative concept is singular if the referent set is a singular set; it is general if the referent set is a nonsingular set. An attributive concept is singular if the term designating the concept can be applied to just one object; it is general if it can be applied to more than one object.

Of course, all proper nouns and all uniquely designating expressions determine singular concepts. Such nouns and expressions are constants. Ordinarily, common nouns and designating expressions that do not denote a unique referent determine general concepts. Thus "linear function" is a general concept; there are many linear functions. But "the linear function determined over the reals by $y = 3x - 1$" is a singular concept; there is just one such function. Similarly, whereas "positive rational number" is a general concept, "+10," "+2$\sqrt{3}$" and "e" are singular concepts.

Whether an attributive concept is considered singular or general depends on whether the plural of the name of the concept makes sense. Is there just one consistency or are there consistencies? One freedom or several freedoms? One kind of problem solving or several kinds of problem solving? However one answers these questions can make a difference in how the concept is taught.

To be sure, other categories of concepts can be appended, for example, familiar or unfamiliar, precise or vague, conjunctive or disjunctive, and simple or complex. How far the classification is extended by any person probably depends on whether the extension is functional, that is, whether the extension can be correlated with possible or actual differences in how teachers teach the various kinds of concepts.

To summarize the taxonomy presented here, we use the tree diagram seen in figure 7.1.

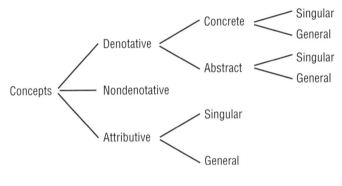

Figure 7.1

We now proceed to the moves that are available to a teacher when he undertakes to teach a concept, pointing out how the kind of concept permits or precludes certain of these moves. The set of moves to be discussed in the next section is an adaptation of those discussed in Henderson ([b] 1967).

Moves in Teaching a Concept

Smith and his coworkers (1967) made detailed analyses of tapes of classroom teaching in English, history, mathematics, science, social studies, and sociology in grades 8 through 12. To facilitate their analyses, they conceived of a *move,* a useful concept that will be used hereafter. A *move* is a sequence of verbal behaviors by a teacher and students toward attaining some objective. One common objective is teaching concepts. In identifying the moves in teaching a concept to be discussed later, Smith's findings were used. Also employed was the taxonomy discussed in the previous section. If one accepts this taxonomy, certain moves are logically possible for certain kinds of concepts and logically impossible for other kinds. The names given the various moves in the present exposition differ from the names Smith uses, but the difference is largely a matter of taste.

In the first place, in teaching a concept the teacher, as explained above, uses either the object language or the metalanguage. As one might expect, Smith found the former more frequently used than the latter. We turn first to moves in the object language, classifying them in terms of the connotative and denotative uses of the term designating the concept.

Connotative moves

A connotative move is one that is based on the connotation of the term designating the concept. If the concept is a denotative concept, a connotative move is one in which a person—teacher, student, or textbook author—states or talks about the characteristics or properties of the objects in the referent set determined by the concept. If the concept is an attributive concept, the person states or talks about the conditions that are relevant for the term to be applied to something. We shall identify eight connotative moves. For each, some examples will be given. To indicate the concept being taught in each example, the term designating the concept is set in SMALL CAPITALS.

A move usually consists of several verbal exchanges, unless, of course, it is a move in a textbook. To conserve space, only the essential sentence of each move will be given.

Single characteristic. If the concept is denotative, one characteristic or property of the referent or referents (objects in the referent set of the concept—in other words, in the denotation of the designating expression) is stated, discussed, or asked for. If the concept is attributive, a relevant condition for the applicability of the designating expression is stated or discussed.

The characteristic may be necessary, as in the case of

"A REGULAR HEXAGON is equilateral"

and

"A proof that has no gaps in the reasoning has RIGOR,"

or not necessary, as in the case of

"CIRCLES are used in design."

One can imagine the following as the first sentence in a single-characteristic move:

"Give us a property of EVEN NUMBERS."

If the student gives a property (characteristic) of an even number, the move could very well terminate and consist of two sentences. Probably such a move would appear in the context of a review of the concept of an even number after other moves had been used to teach the concept.

Sufficient condition. For denotative concepts, having a property or conjunction of properties is stated or discussed as being sufficient for an object to be in the concept's referent set, in other words, to be denoted by the designating expression. For attributive concepts, a condition (which might be a conjunction of conditions) is asserted or discussed as being sufficient for applying the designating expression.

As an example, a textbook might say,

"If a number satisfies an equation, it is a ROOT of the equation"

or

"Two terms are LIKE TERMS provided that they have the same variables to the same powers."

A conjunction of conditions is illustrated by

"If the sides are respectively congruent and the angles are also respectively congruent, the figures are CONGRUENT (figures)."

If a teacher asks,

"How do you tell whether a number is a PRIME NUMBER?"

and expects and gets the statement of a sufficient condition, his question might initiate a sufficient-condition move.

A sufficient-condition move is possible for every concept, and because of its simplicity should be helpful for just about every student. Logically, it is a more powerful move than that of giving a single characteristic.

Necessary condition. For denotative concepts, having a property is stated or discussed as being necessary for an object to be in the concept's referent set, that is, to be denoted by the expression designating the concept. For attributive concepts, a condition is asserted as being necessary for applying the designating expression.

As an example, a teacher says,

"Remember, if two equations do not have the same solution set they are not EQUIVALENT (equations),"

thereby indicating that having the same solution set is necessary for the equations to be equivalent equations.

Classification. In a classification move, a superset of the referent set is identified, discussed, or elicited by a question. The following are examples of such a move:

"A TRAPEZOID is a quadrilateral."

"AREA is a denominate number."

"Every PRISM is a polyhedron."

If a student asks, "What is a micrometer?" and the teacher replies, "It is a measuring instrument that yields very precise measurements," the student's concept of a micrometer is enhanced. He can now make valid inferences about a micrometer by virtue of the classification.

At first thought, it may seem that a classification move is hard to distinguish from a single-characteristic move. When a teacher says,

"A RHOMBUS is a parallelogram,"

is he not giving a characteristic of a rhombus, namely, that of being a parallelogram? Whereas there may be no practical difference between these two moves, there is a theoretical difference.

185

A classification move uses the "is" of set inclusion (a GRAPH is a set of points) or the "is" of set membership (PI is a member). What follows "is" in either case is a predicate noun. When "is" is used in a single-characteristic move, it is the "is" of predication and what follows "is" is a predicate adjective.

Smith (1967, p. 301) reports that the classification move was the one most frequently used by the mathematics teachers in his sample and that it was used more frequently by mathematics teachers than by teachers of other subjects. Perhaps this can be explained by the deductive nature of mathematics and the large number of denotative concepts in mathematics. Both of these characteristics make such a move easy to utilize. One can further conjecture that mathematics teachers are disposed to portray relationships. The set-membership and set-inclusion relations are effective in helping students organize their knowledge; moreover, they are easily comprehended.

It seems apparent that for a classification move to be effective the student must have a concept of whatever is used as the superset. To say, for example,

"SINE is a periodic function,"

is ineffective in teaching a concept of sine if the student does not know what a periodic function is.

Identification. This move, like the classification move, is applicable to denotative concepts. In an identification move, the referent of the term designating the concept is uniquely determined—that is, necessary and sufficient contributions are stated. In each of the following examples the object denoted by the designating expression is uniquely identified.

"A TRAPEZOID is a quadrilateral having just one pair of sides parallel."

"A polynomial having just one term is a MONOMIAL."

"A function is a LINEAR FUNCTION if and only if it is determined by $y = ax + b$ where $a \neq 0$."

Theoretically, an identification move teaches a precise concept. Moreover, it is a powerful move. The "is" involved is the "is" of identity. Two sets are designated, one of which is the referent set of the concept, and it is asserted that the two sets are identical. If A is B and B is A is taken as the form of an identification move, then four inferences are possible:

If something is an A, it is a B.

If something is a B, it is an A (converse).

If something is not an A, it is not a B.

If something is not a B, it is not an A (contrapositive).

An identification move provides a lot of information. However, such a move, just because of its nature, is one of the more difficult moves for students to comprehend, particularly students who have a paucity of concepts, do not see relations readily, and have short attention spans. Its very elegance for such students is a point in its disfavor. Yet printed in a textbook or written on the chalkboard so that students can reread it and ponder it, its effectiveness may be enhanced.

Some persons may wish to call an identification move a definitional move and the identity statement a definition. The reason for not doing this is given when moves in the metalanguage are discussed.

Analysis. In an analysis move, one or more subsets of the referent set of the concept are named, described, or elicited by a question. A teacher might say,

"A circle, parabola, ellipse, and hyperbola are CONIC SECTIONS."

If students know what the relations of equality, congruence, and similarity are, the teacher might use these in teaching the concept of an equivalence relation by stating,

"Equality, congruence, and similarity are EQUIVALENCE RELATIONS."

In reviewing the concept of a polynomial, the teacher might give as an exercise,

"Name three kinds of POLYNOMIALS,"

expecting the students to list monomials, binomials, and trinomials but accepting first-, second-, and third-degree as a correct answer.

An analysis move may be regarded as the converse of a classification move. As was pointed out in discussing the latter, students must have a concept of whatever is used as the superset. Similarly, in an analysis move, students must have a concept of each of the subsets mentioned for the move to be meaningful. It would be to no avail to attempt to teach the concept of a higher-degree equation by saying,

"Cubics and quartics are HIGHER-DEGREE EQUATIONS"

if the students do not know what a cubic equation and a quartic equation are.

Analogy. For a denotative concept, an analogy move is one in which the referent of the term designating the concept is said to be like something else, or in which the way in which the referent is like the other thing is discussed. For an attributive concept, for example, corresponding to, the property is said to be like some other property, and the basis of the analogy may be discussed. In either case, the analogy may be educed by skillful questioning.

One may distinguish between two kinds of analogy moves when denotative concepts are taught. In one kind the teacher uses the designating expression itself—as, for example, in a geometry class:

"ALTERNATE EXTERIOR ANGLES are like alternate interior angles in that each angle in the pair is on the opposite side of the transversal."

In this case, "alternate exterior angles" functions as a universally quantified variable, that is, for each x such that x is an alternate exterior angle. Of course, for this particular move to be effective the student must have the concepts of alternate interior angle and transversal because the concept of an alternate exterior angle is developed by means of these concepts.

In the other kind of analogy move the comparison is made between two or more members or subsets of the concept's referent set as, for example,

"Consider the equations $2x + 6 = 1$ and $x = -2.5$. These are different equations, but the solution set of each is the same set."

This was the initial move used by a teacher in teaching the concept of equivalent equations. It was followed by other moves to sharpen the concept.

This latter kind of analogy move has the psychological advantage of beginning with particulars and, if the teacher is astute, particulars with which most of the students are familiar. After the particulars are compared, the induction is made to the set of which the particulars are representative members. Theoretically, such a move should be effective with slow-learning students who have trouble handling abstractions. The geometry teacher who taught the concept of alternate exterior angles might have drawn figure 7.2 and then had the students compare the angle-pair (1, 2) and the angle-pair (3, 4). He might then have focused on the other pair of alternate exterior angles and had a similar comparison made. Considering this representation as a paradigm, he might then have employed the first kind of analogy move as stated above to achieve the generality he desired.

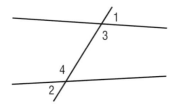

Figure 7.2

It follows from the definition of *singular* and *general concepts* that the analogy move of comparing members of the referent set is restricted to general concepts. No comparison can be made if there is only one member in the set.

Smith and his coworkers hypothesized that the analogy move where the comparison is made between members or subsets of the concept's referent set (they speak of this move as "instance comparison") ought to occur in case the teacher is teaching by discovery. "However," they report, "there were no clear cases of this kind of 'discovery learning' in the recorded discourse obtained for this study" (p. 74).

Differentiation. For a denotative concept, a differentiation move is one in which the referent of the term designating the concept is said to be different from something else or in which the way in which the referent is different is discussed. For an attributive concept of a property, for example, contraposition, the property is said to be different from some other property, and the basis may be discussed. In either case, the difference(s) may be established by pointed questioning.

As in the case of an analogy move, one can identify two kinds of differentiation moves. In one kind, the teacher uses the designating expression itself—the following, for example, as a follow-up to the analogy move in teaching the concept of alternate exterior angles:

> "But ALTERNATE EXTERIOR ANGLES differ from alternate interior angles in that they are outside the lines intersected by the transversal."

Another example:

> "Remember, a LINEAR FUNCTION is not the same as a linear equation."

One can speculate that the move was used because the teacher had learned from past experience that some students confuse linear equations and linear functions, and he sought to forestall the misconception.

In the other kind of a differentiation move, the contrast is made between two or more members or subsets of the concept's referent set. For example:

> "The POLYGONS I have drawn on the board differ in the number of sides and angles they have."

On occasion, an analogy move and a differentiation move are combined, as in the case of this example:

> "Triangles, quadrilaterals, pentagons, hexagons, and octagons differ in the number of sides and angles they have, but they are all POLYGONS."

The same theoretical advantage claimed for an analogy move can also be claimed for a differentiation move when each of these involves a comparison or contrast among members or subsets of the referent set of the concept. By starting with particulars that are familiar to students, the reasonable induction to the entire referent set which establishes the concept can more readily be followed and understood.

From the tapes Smith and his coworkers analyzed, they identified three kinds of differentiation moves. In one kind, something is stated or discussed as being the opposite of something else. In a second kind, a referent is stated as not being the same as something else or not like something else. The third kind is a more complete move in that the nature of the difference is discussed. Theoretically, the latter should be a more effective move than either of the others.

Denotative moves

We now turn to a set of moves coordinate with connotative moves, namely, denotative moves. A denotative move is one in which a person—teacher, student, or textbook author—names members or nonmembers of the referent set of the concept. Or a person designates objects and asks another person to determine whether or not a given object is a member of the referent set. Hereafter we shall speak of an example, meaning a member of the referent set of a particular concept. Hence denotative moves are exemplification moves; one exemplifies the concept of an operation by giving examples of an operation. (Some writers use the terms "exemplar" and "nonexemplar" as we have used the terms "example" and "nonexample." Others prefer the terms "instance" and "noninstance," respectively. In the interest of avoiding ambiguity and equivocation, it is advantageous to speak of an "example" of a concept but an "instance" of a generalization.)

189

It follows from the taxonomy of concepts presented in this chapter that only denotative concepts allow denotative moves. One cannot give an example of skewness, which is an attributive concept, but he can give examples of skew lines, a denotative concept. The reader should be aware that these distinctions are theoretical; the preceding statements in this paragraph are analytic rather than factual statements.

In practical situations, that is, actual teaching of verbal concepts, the denotative moves are almost entirely restricted to general concepts. This will become evident in the subsequent exposition.

Six kinds of denotative moves can be identified. Each of these will now be described and exemplified.

Example. As might be expected, in this kind of move one or more examples are given. Or the student is asked to give an example. In teaching the concept of a prime number, a teacher might say,

"Some PRIME NUMBERS are 2, 3, 5, 7, 11, and 13."

In teaching the concept of a parabola, the teacher among other moves might draw a representation of a parabola and classify it ostensively by the statement,

"This is a PARABOLA."

If a concept is singular, theoretically only one example can be given. But an example move for a singular, abstract concept is an odd move. What does it mean to give an example of π? And even for some singular, concrete concepts such a move is odd. What does it mean to give an example of the man for whom Boolean algebra is named? The very locution "an example" seems to imply a general concept.

If a concept is general, giving an example makes sense; the teacher has a choice of examples and will choose those with which the students are familiar. As might be expected, Smith and his coworkers found the example move (they speak of an "instantial move" as we have of a "denotative move" and of an "instance" as we have of an "example") one of the most frequently used moves.

It is interesting to note that the concept of a set in a textbook where "set" is used as an undefined term is taught by repeated example moves. The logical reason for this is apparent; one cannot use a connotative move without ultimate circularity.

Exclusive use of the example move cannot result in a precise concept. However, many students seem to jump to the right set of necessary and sufficient conditions, and the concept they attain is satisfactory for practical purposes. Not knowing how to explain this phenomenon of discovery, we say the students have "intuition."

Before leaving this move, it may be well to sharpen the theoretical distinction between an example move and an analysis move, which seem much alike. An example move uses the set-membership relation; an analysis move uses the set-inclusion relation. In an example move, we name members of the referent set; in an analysis move, we name subsets. Syntactically, a proper noun is found in an example move; only common nouns are found in an analysis move. Yet to many teachers, saying

"A triangle is a POLYGON"

is giving an example of a polygon. Whether the theoretical distinction between an example move and an analysis move is of practical value remains to be seen. Both moves are based on the partitioning of a set.

Nonexample. It is found that teachers name objects that are not in the referent set of a concept as well as members of the referent set; in other words, nonexamples as well as examples. A nonexample move is one in which an object that is not a member of the concept's referent set is named. For example, the teacher teaching the concept of prime number who used the example move above might say,

"But 4, 6, 8, 9, and 10 are not PRIME NUMBERS."

The nonexample move is often used when experience has shown that students make errors based on not knowing certain necessary conditions for the concept. Thus a teacher might say,

"However, $(\forall x, y)$ $[xy = yx]$ is *not* an OPEN SENTENCE even though it has variables in it"

because he has found some students jump to the conclusion that if a sentence has a variable in it, it is an open sentence.

Counterexample. Students often have misconceptions. A student asserts that a linear function is a function whose graph is a straight line. But the concept of a linear function, as taught by the textbook used in the course, restricts a linear function to a function whose graph is a subset of a straight line not parallel to either axis. The teacher can cite any constant function, for example, that determined by $y = 2$, and ask the student if it is a linear function. Since it is not, according to the way "linear function" was defined in the textbook, the teacher has posed a counterexample, one that satisfies the conditions the student laid down but cannot be called a "linear function."

In the present context, a counterexample move is a move in which an object is named or otherwise designated which falsifies a generalization purporting to characterize the members of the concept's referent set. As an example, consider the following dialogue:

Student.   A PARALLELOGRAM is a quadrilateral having one pair of congruent, parallel sides.

Teacher.   Then the figure I am drawing (fig. 7.3) is a parallelogram?

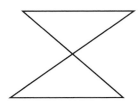

Figure 7.3

Student.   Well, no.

And the following:

Teacher.   How can we tell whether a number is a PRIME NUMBER?

Student.   A prime number is not divisible by any other number.

Teacher.   Is 7 a prime number?

Student.   Yes.

Teacher.   But 7 is divisible by a lot of numbers: 1, 7, 2, 5, and 14 among others.

In the first example, the figure drawn satisfied the conditions given by the student, but he did not want to call it a parallelogram. In the second, 7 is a number the student is willing to denote by "prime number"; but it does not satisfy the condition he asserted, namely, not being divisible by any other number.

Of necessity, a counterexample move cannot be the first move in a sequence of moves. It is effective as a rejoinder to an identification move, a sufficient-condition move, or a classification move in which the asserted generalization is false. As such, it helps the student become aware of an inconsistency in his conceptualization. Psychologically, it is effective because the student makes the discovery for himself. One might assume that the student will then be ready for a clarification of the concept. This could be done by other moves.

Counterexamples in the case of concrete concepts can be demonstrated ostensively—displayed, exhibited, or pointed to. Counterexamples in the case of abstract concepts have to be named or designated by the use of language.

Specification. In a specification move each of the members of the concept's referent set is named or otherwise designated. For example,

"The UNITS MOST FREQUENTLY USED IN MEASURING LENGTH are the inch, foot, yard, and mile"

and

"The two TRUTH VALUES are 'true' and 'false.'"

A specification move is based on determining a set by specifying (listing) its members. It is a move that, when possible, results in as precise a concept as one taught by an identification move or by a definition. Theoretically, a specification move is possible for a singular, abstract concept, but it is hard to imagine such a move and none could be found in the time available. Such a move can be imagined for a singular, concrete concept. It would be ostensive in the form

"This [designating uniquely in some way] is the DECLARATION OF INDEPENDENCE," and the person's concept of the Declaration of Independence would be clarified by noting its observable characteristics. For all practical purposes, the specification move is used to teach general concepts whose referent sets have a small number of members.

Exemplification accompanied by justification. In such a move, an example is given together with a reason why it is an example:

"'+2' is a CONSTANT because it names just one number."

"$3y^2 = y$ is a QUADRATIC EQUATION because it can be transformed into an instance of $ax^2 + bx + c = 0$ where $a \neq 0$."

Theoretically, giving an example and telling why it is an example should be an effective move because it implies a sufficient condition which most students can infer. Perhaps this is why it is used frequently in textbooks.

Nonexemplification accompanied by justification. In such a move, a nonexample is given together with a reason why it is not an example:

Teacher.	Is 3 a COMMON FACTOR of $3a$, $6a^2$, and $10ab$?
Student.	No.
Teacher.	Why not?
Student.	Because it is not a factor of $10ab$.

This move often follows the previous move. The teacher who gave $3y^2 = y$ as an example of a quadratic equation and then substantiated the assertion might conceivably follow by

"But $3y = y$ is not a QUADRATIC EQUATION because it cannot be transformed into an instance of $ax^2 + bx + c = 0$ where $a \neq 0$."

As in the case of giving an example and telling why it is an example, giving a nonexample and telling why it is not an example should be an effective move. It implies a necessary condition which most students can infer. When coupled with giving an example and a justification, sufficient and necessary conditions can be inferred. A cursory inspection of textbooks revealed that fewer nonexamples than examples are given when teaching concepts.

Moves in the metalanguage

Suppose a student asks, "What does the word *rhombus* mean?" It seems reasonable to regard this as an opening sentence in a conceptual venture and being equivalent to "What is a rhombus?" Now suppose the teacher replies,

"*Rhombus* means an equilateral parallelogram" or

"*Rhombus* means the same as *equilateral parallelogram*."

The moves are moves in the metalanguage (language about language) because what is considered is the word "rhombus," not the object rhombus. In the first case the relation "means" is used, and *a* meaning of "rhombus" is stated; only one term, namely, *rhombus*, is in the metalanguage. In the second case the relation "means the same as" is used, and a synonymous term is given; two terms are in the metalanguage, namely, *rhombus* and *equilateral parallelogram*.

A move in the metalanguage is a bit of verbal discourse in which a term designating a concept is the subject discussed; usually meaning is ascribed to the term either by stipulating what the term is to mean or, if the term has some currency, reporting what it means. Smith and his coauthors speak of moves in the metalanguage as "usage moves" or "metadistinctions" (p. 82).

In a previous analysis (Henderson [b] 1967) the set of moves in the metalanguage was divided into stipulation moves and synonymous-expression moves. On reflection, a classification in terms of a different principle seems of both more theoretical significance and more practical value. It is of theoretical significance because the moves are related to definition; it is of practical value because the two kinds of moves (definitions) have a different logic. They are substantiated in a different way by a teacher, and different rejoinders are made in case of a difference of opinion.

We shall say that there are two kinds of metalinguistic moves: stipulated definitions and reported definitions. This is based on a particular theory of definition which will now be explained.

A theory of definition

Definition is an operation performed on symbols; it is assigning a meaning to a symbol. As Copi (1961, p. 99) puts it,

> It should be remarked that definitions are always of symbols, for only symbols have meanings for definitions to explain. We can define the word "chair," since it has a meaning; but although we can sit on it, paint it, and burn it, or describe it, we cannot *define* a chair itself, for a chair is an article of furniture, not a symbol which has a meaning for us to explain.

It follows from this theoretical conception of definition that to define we have to talk about the *definiendum* (the symbol being defined), and this requires that the definiendum be in the metalanguage. Hence

"By *arithmetic mean of two or more numbers* we shall mean the average of the numbers" is a definition, with *arithmetic mean of two or more numbers* the definiendum. But

"The arithmetic mean of two or more numbers is the average of the numbers" is *not* a definition (although many persons regard it as a definition because they have a different theory of definition). It is an identity statement in the object language, and according to the taxonomy of moves discussed above it would be classified as an identification move.

Definitions can be classified according to the purpose of defining. A person may wish to introduce a new term to serve as an abbreviation of a longer term. To do this, he has to assign a meaning to the term; he has to stipulate how the term is used. Or a term may be used ambiguously—for example, *circle* as designating either a set of points in a plane equidistant from a point in the plane or the union of the set of points described above and the set interior to these points—and a teacher wishes to restrict its meaning. He can do this by stipulating what the term is to mean in his exposition. Definitions used for these purposes will be called *stipulative definitions*. Stipulative definitions sometimes are called *verbal* or *nominal definitions*.

A second purpose of defining is manifested when a person wants to report how a term is, in fact, used. For example, a textbook might state,

"An expression of equality between two ratios is called a *proportion*"

194

to report how a large number of mathematicians use the term proportion. Definitions used to report a conventional meaning of a term will be called *reported definitions.*

It is important to note the difference in the logic of stipulated and reported definitions. A stipulated definition is a proposal for the use of a term, as in

"Let us use $'f^{-1}'$ to stand for the inverse of f,"

or it is a resolution concerning what a symbol is to mean, as in

"We shall abbreviate 'a mapping of A into B' to '$f: A \to B$.'"

Hence stipulated definitions are not truth-functional; it makes no sense to say "true" or "false" to a stipulated definition. Such a definition is not right or wrong. Logically, stipulated definitions cannot be items in a true-false test.

A reported definition, on the other hand, is a factual statement. For example,

"A parallelogram having a right angle is called a *rectangle*"

either is or is not a correct report of how, in fact, mathematicians use *rectangle*. In contrast to stipulated definitions, reported definitions are truth-functional. If the report is correct, the reported definition is true; if not, it is false. If definitions are included in true-false tests, they must logically be regarded as reported definitions.

Stipulated-definition move

In accordance with the foregoing theory of definition, which is accepted by most logicians, we shall divide the set of moves for teaching a concept by the concept's designating expression into stipulative-definition moves and reported-definition moves. This classification is based on the use of the definition rather than its form; hence it is in contrast to the classification of moves in the object language, which is based on the form of an utterance rather than on its use.

A stipulated-definition move, obviously, is one in which a stipulated definition is used. When mathematics is taught rather than developed, few stipulated-definition moves are used to teach concepts. Even a move like

"By *area* we mean the nonnegative number assigned to an enclosed surface by a measure function,"

though stipulative in form, probably is used to teach a conventional concept of area. The same can be said for

"The ratio of the greatest possible error of a measurement to the measurement itself is called the *relative error of the measurement*,"

which is not clearly stipulative in form, though it is in a possible form for a stipulative definition.

About the only use of a stipulative-definition move in teaching concepts would be for a teacher to encourage his students to invent a name for an idea. He could then use a stipulative definition to reinforce the concept. No examples of such a use were found.

Reported-definition move

A reported-definition move is one in which a reported definition is used. The following are probably reported definitions:

195

"The intersection of the coordinate axes is called *the origin*."

"A prism is called *a right prism* if and only if a lateral edge is perpendicular to a base."

Reported-definition moves are found chiefly in textbooks; teachers tend to use the object language. Both of the moves above can readily be changed to identification moves by talking about the origin and a right prism instead of talking about "the origin" and "a right prism." More generally, every reported definition can be readily transformed into an identity statement, and conversely.

A reported-definition move has great applicability and results in a precise concept. It is an efficient move, and effective when followed by denotative moves. But it has the same disadvantages an identification move has. Its very efficiency makes it hard for slow-learning students. Moreover, when such a definition is used for a textbook, some students memorize it without understanding it and hence without being able to use it. Yet because of the nature of mathematics and the precision of its concepts, definitional (metalinguistic) moves are used more frequently in mathematics than in any other subject.

Strategies in Teaching Concepts

It is rarely the case that a teacher uses only one move in teaching a concept. Usually he employs several moves in a sequence. Following Smith, we shall call a sequence of moves a *strategy*.

From the analysis of the tapes, Smith reports that strategies composed of what we have called connotative moves and those composed of both connotative and denotative moves are most common (p. 84). An example of the former is a sufficient-condition move preceded or followed by one or more single-characteristic moves. An example of the latter is an identification move followed by one or more example moves. The reverse sequence is also used; a teacher will give one or more examples and then make the concept precise by an identification move.

Ginther (1964) made a study of the use of definitions to teach concepts in first-year algebra textbooks and geometry textbooks. In this study an identification move and a sufficient-condition move as well as a stipulated-definition move and a reported-definition move were considered definitions. His findings confirm those of Smith as to the kinds of strategies used. In the sample of twenty-three textbooks Ginther used, connotative moves were the most common in both the algebra and geometry textbooks, comprising 66 percent of the definitions in algebra and 89 percent of the definitions in geometry textbooks. Denotative moves comprised 29 percent of the moves in the algebra textbooks and 10 percent of the moves in the geometry textbooks.

Although one can find common strategies by surveys similar to those cited above, it is another matter to ascertain under which conditions the various strategies commonly employed are effective. One might conjecture that some are more effective than others for slow learners; for certain kinds of concepts (e.g., conjunctive vs. disjunctive concepts, or concrete vs. abstract concepts); for changing a concept already held by a student; and for concepts of great importance, which should be thoroughly mastered (in contrast to those of less importance, for which somewhat less precision is essential).

Each teacher probably has his own ideas about the relative efficacy of various strategies for teaching verbal concepts. Two experimental studies sought to secure some dependable evidence. Rollins (1966) chose three strategies appropriate for teaching mathematical concepts by guided discovery. For each concept taught, three strategies were defined. One consisted of a sequence of example moves. A second consisted of a sequence of pairs of example and non-example moves. The components of each pair (example followed by nonexample) were selected essentially by using an identification move. For example, if a sector of a circle is to be conceived of as the region in a circle bounded by an arc of the circle and two radii, a representation of a sector would be given and the student told that the region represents a sector. The examples sought to show that such properties as placement of the sector in the circle, the size of the sector, and the angle determined by the radii were irrelevant. A region of a circle other than a sector would be given and the student told that the region did not represent a sector. The nonexamples sought to identify the necessary conditions: being a region of a circle and being bounded by an arc of the circle and two radii. Examples of two pairs of moves in this strategy are shown in figure 7.4, the examples at the left and the nonexamples at the right.

The third strategy consisted of the same example and nonexample moves, but not in pairs; all the example moves were used first.

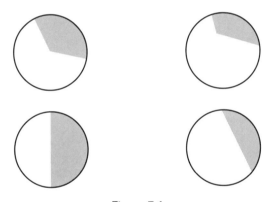

Figure 7.4

Rollins used a programmed text and a group of eighth-grade students. The California Short-Form Test of Mental Maturity, 1957, was used to partition the group into high-, average-, and low-ability groups. It was found that the students learned concepts by each of the three guided-discovery strategies. The students of high ability learned the most and those of low ability the least, as could be expected, but it was not concluded that any one of these strategies was more effective than another for teaching verbal concepts, regardless of the students' ability level.

Rector (1968) selected four strategies: (1) characterization, (2) characterization-exemplification, (3) exemplification-characterization, and (4) exemplification-characterization-exemplification. A characterization strategy is one in which only characterizing (connotative) moves are used. An exemplification strategy is one in which only exemplification (denotative) moves are used. By experimentation, he sought to determine the relative efficacy of these strategies in teaching mathematical concepts to low-ability and high-ability students. Moreover, he defined operationally three levels of understanding. These he designated as (1) knowledge and comprehension, (2) application, and (3) analysis, synthesis, and evaluation. They are listed in increasing order of understanding, that is, a student who can analyze, synthesize, and evaluate, using a concept, is regarded as having greater understanding of the concept than a student who can merely apply the concept. A student who can apply a concept is regarded as having greater understanding of the concept than a student who only has a knowledge or comprehension of the concept.

Rector's subjects were college freshmen. The eleven concepts taught were selected from probability theory and were initially unfamiliar to the students. Instruction was by a programmed text to control the moves and strategies.

Rector found concepts were attained by both ability groups by each of the strategies. Moreover, each strategy enabled the students to attain each of the three levels of competence. When a comparison of the relative effectiveness of the four strategies was made, it was found that the characterization strategy was the most effective for attaining the level of knowledge and comprehension. However, there was no evidence to support the hypotheses that any one of the four strategies was more effective than another in attaining the two higher levels of competence, (2) application and (3) analysis, synthesis, and evaluation.

One can conclude from these findings that a teacher is largely free to choose among the various possible strategies in teaching a concept in mathematics. Employment of a variety of strategies may well serve to sustain interest and provide practice in the different kinds of thinking involved.

REFERENCES

Bruner, Jerome S., Jacqueline J. Goodnow, and George A. Austin. *A Study of Thinking*. New York: John Wiley & Sons, 1956.

Bugelski, Bergen R. *The Psychology of Learning Applied to Teaching*. Indianapolis: Bobbs-Merrill Co., 1964.

Carroll, John B. "Words, Meanings, and Concepts." *Harvard Educational Review* 34 (1966): 178–202.

Church, Alonzo, *Introduction to Mathematical Logic*. Princeton, N.J.: Princeton University Press, 1956.

Copi, Irving M. *An Introduction to Logic*. 2d ed. New York: Macmillan Co., 1961.

Ginther, John L. "A Study of Definitions in High School Mathematics Textbooks." Ph.D. dissertation, University of Illinois, 1964.

Ginther, John L., and Kenneth B. Henderson. "Strategies for Teaching Concepts by Using Definitions." *Mathematics Teacher* 59 (May 1966): 455–57.

Henderson, Kenneth B. (a). "A Logical Model for Conceptualizing and Other Related Activities." In *Psychological Concepts in Education*. Chicago: Rand McNally & Co., 1967.

———— (b). "A Model for Teaching Mathematical Concepts." *Mathematics Teacher* 60 (October 1967): 573–77.

Henderson, Kenneth B., and James H. Rollins. "A Comparison of Three Stratagems for Teaching Mathematical Concepts and Generalizations by Guided Discovery." *Arithmetic Teacher* 14 (November 1967): 583–88.

Hunt, Earl B., Janet Marin, and Philip J. Stone. *Experiments in Induction.* New York: Academic Press, 1966.

Johnson, Donald M., and R. Paul Stratton. "Evaluation of Five Methods of Teaching Concepts." *Journal of Educational Psychology* 57 (1966): 48–53.

Pikas, Anatol. *Abstraction and Concept Formation.* Cambridge, Mass.: Harvard University Press, 1966.

Rector, Robert. "The Relative Effectiveness of Four Strategies for Teaching Mathematical Concepts." Ph.D. dissertation, University of Illinois, 1968.

Rollins, James H. "A Comparison of Three Stratagems for Teaching Mathematical Concepts and Generalizations by Guided Discovery." Ph.D. dissertation, University of Illinois, 1966.

Smith, B. Othanel, et al. *A Study of the Strategies of Teaching.* Urbana, Ill.: University of Illinois Press, 1967.

Travers, Robert M. W. *Essentials of Learning: An Overview for Students of Education.* New York: Macmillan Co., 1963.

Van Engen, Henry. "The Formation of Concepts." In *The Learning of Mathematics: Its Theory and Practice.* Twenty-first Yearbook of the National Council of Teachers of Mathematics. Washington, D.C.: The Council, 1953.

8

From *The Slow Learner in Mathematics*, NCTM's Thirty-fifth Yearbook (1972)

Max Sobel served as NCTM president from 1980 to 1982; he also served on the editorial panel for NCTM's Thirty-fifth Yearbook, a volume he later used as the text for a graduate course he taught. Without question, the title of this 1972 yearbook—*The Slow Learner in Mathematics*—would not be acceptable today, but the focus of many of the chapters still has great appeal to special educators and to any elementary or middle-grades teachers seeking instructional strategies for students in need of intervention.

We have included here **chapter 5, "Adjustment of Instruction (Elementary School),"** by Charlotte W. Junge. This chapter presents strategies for diagnosing pupil performance and organizing the classroom for learning, as well as a host of teaching strategies sampling many of the expectations in the elementary curriculum. Most of these expectations remain important to those charged today with implementing the Common Core State Standards. As with other chapters from the Thirty-fifth Yearbook, this selection is an important read for special education teachers, interventionists, and elementary and middle-grades classroom teachers.

Adjustment of Instruction (Elementary School)

Charlotte W. Junge

O NE of the most important facts revealed by educational assessment in mathematics instruction is the wide range of achievement levels in any class in our schools. A recent study (see table 5.1) shows the diverse scores made on a test given to sixth-grade students selected at random from the whole country (11*, p. 402).

Table 5.1

Variability in Scores on Sections of the California Arithmetic Achievement Test of 100 Children in Grade 6.1 with IQs Ranging from 90 to 100

	Raw Scores			Grade Scores		
	Lowest	Highest	Mean	Lowest	Highest	Mean
I. Reasoning						
A. Meanings	2	13	8.6	2.7	8.0+	5.7
B. Signs and Symbols	5	14	11.8	3.4	7.4	6.2
C. Problems	1	12	7.0	2.6	8.0+	5.4
Totals on A, B, C	16	35	27.4	3.9	7.7	5.9
II. Fundamentals						
D. Addition	4	15	10.0	3.6	8.4	6.0
E. Subtraction	1	15	9.6	3.4	8.0+	5.9
F. Multiplication	3	15	7.5	4.5	8.0+	5.8
G. Division	3	15	8.7	4.4	7.9	6.1
Totals on D, E, F, G	21	52	35.7	4.8	7.4	6.0

Note: Reprinted by permission of the publisher from *Discovering Meanings in Elementary School Mathematics*, 5th ed., by Foster E. Grossnickle, Leo J. Brueckner, and John Reckzeh (New York: Holt, Rinehart & Winston, 1968), table 18.2, p. 402. © 1947, 1953, 1959, 1963, and 1968 by Holt, Rinehart & Winston, Inc.

---

*The numbers in parentheses in this chapter are those of the entries in the References section at its end.

Note that on test C the grade scores range from 2.6 to 8.0+. On test E they range from 3.4 to 8.0+. The ranges are wider on tests of single skills than for the general abilities included in the totals.

Although this table represents but one set of data, its scores are typical of achievement scores in mathematics, and it portrays the difficulty of trying to teach the same mathematics lesson to all pupils. The able learner is not challenged to the full extent of his ability, and the slow learner meets steady frustration from tasks not paced to his level of achievement. It is not possible to maintain adequate levels of achievement for each child by giving standard prescriptions.

However, the most common practice among teachers of mathematics is to have all children work with the same instructional materials and to assign the same exercises to all. A survey of the intraclass grouping practices of 1,392 teachers in grades K–6 showed that only 33 percent of these teachers grouped pupils for arithmetic instruction. The majority did not believe grouping for mathematics instruction was as important as grouping for reading (2, p. 310.)

The fact that instruction in mathematics is still largely total-class instruction may be the result of failure to understand the proper use of new teaching materials, the concern for excellence and the maintenance of high standards, a lack of suitable instructional materials commensurate with the range of abilities within the class, or the fact that current curricula are described in terms of class or grade-level achievement.

The Nuffield Project (1), which is being developed in England, is a notable exception to this trend. The project directors seek to identify clearly defined developmental stages in a child's growth and to develop a curriculum around these stages. This project suggests a new approach to the organization of learning experience in mathematics, one in which instruction is individualized and children do not necessarily move together as a group. Booklets from the Nuffield Project describe learning activities and tell exactly how children can work individually and in small groups in making mathematical discoveries and in recording their findings.

Diagnosing Pupil Performance

The slow learner, while resembling the average and the above average student in general physical development, chronological age, and interests common to his age group, may not learn intellectual things at the same rate as other children, owing either to lack of potentiality or to personal and emotional factors that interfere with the ability to achieve. Grossnickle writes that "the methods by which slow learners master the concepts and skills of mathematics are not unique or strikingly different from those used by children of greater learning ability. Slow learners, however, cannot learn skills as rapidly as children of higher ability" (11, p. 421). Smith refers to research by W. M. Cruickshank indicating that retarded learners (1) are reasonably like other children in areas of computation, although they are more careless than average children and use more "primitive" habits, such as making marks and counting on their fingers, (2) have greater difficulty in identifying and understanding which process should be used in problem solving, (3) lack skill in separating irrelevant facts from the significant dimensions of a problem, and (4) have greater difficulty with the reading and language peculiar to arithmetic (26, p. 163).

One of the first responsibilities teachers have in adjusting instruction for the slow learner is to develop and use procedure for diagnosing learning difficulties and determining readiness for new learning. These procedures should focus not only on the types of errors made by each child but also on the processes used by the student in solving mathematical problems.

There are basically four ways to proceed in diagnosing pupil performance:

1. *Observe the child at work.* Note his patterns of study, his attitude toward his work, his interest or lack of interest in it. Observe his habits of work. Does he approach his as signed tasks with some plan, or does he seem to simply "try anything once"? Is he overly dependent on classmates for help? Conversely, are classmates "helping" by doing his work for him rather than helping him think through his problem? Observations can be facilitated by the use of checklists and attitude scales; but there is no substitute for the good teacher's careful, clinical observation in determining difficulties met by children in learning mathematics.

2. *Interview the pupil.* Have the pupil "think aloud" and tell the steps he has used in solving a problem. This is particularly helpful in locating errors in thinking, as well as in computation.

    Develop short diagnostic exercises in which the child works without paper or pencil and responds orally to the teacher's questions. This technique is particularly valuable in determining readiness for new learning. For example, the teacher can use the previous knowledge of the child in diagnosing his readiness to discover such relationships as

$$8 + 6 = 10 + 4$$

Ask the child to give "other names" for 6, such as

$$6 = 2 + 4$$

Ask what must be added to 8 to make 10:

$$8 + 2 = 10$$

Ask the child to study this exercise:

$$8 + 6 = 8 + (2 + 4)$$

$$(8 + 2) + 4 = \square$$

Encourage the child to take these steps mentally, using manipulative materials when they facilitate his thinking.

Provide opportunities for the children to estimate answers and to explain their thinking in arriving at the estimates. For example:

$$
\begin{array}{r}
42 \\
-26 \\
\hline
\end{array}
$$

Slow child: More than 10; I count 26, 36, then [counting on fingers] 37, 38, 39, 40, 41, 42. It's 6 more than 10.
Slow child: About 20; if 42 was 46, it would be 20. So it's a little less than 20.
Slow child: About 20; I think 20 from 40.
Bright child: Think 20 from 42 is 22; take away 6.
Bright child: Think $(46 - 26) - 4$.

Tape recordings can be made of these interviews and the child's responses can be carefully studied to determine errors in thinking, faulty logic, and inaccuracies in computation. Portable video tape recorders are valuable tools in diagnosis. Videotaped interviews permit the teacher to observe student reaction while studying the mental processes used by the child. Recordings of class activities permit the teacher to assess the success of her work with children, and are helpful in planning follow-up lessons.

3. *Analyze written work.* Study the students' written assignments (diagrams, examples, and problem solutions) to determine what kinds of errors are made and how often they are made, as well as to discover any errors being made by the entire class. Charts of the results of this analysis will be of assistance in arranging individual practice and small-group work.

Low achievement in mathematics is frequently accompanied by low achievement in reading comprehension and in language skill (8; 25). In fact, low achievement in mathematics may be due to difficulty in reading and comprehending the precise vocabulary and abstract symbolism of modern programs. (See the discussion that follows under "Instruction in Reading Mathematics.") Mathematics shorthand (If $\triangle ABC \cong \triangle XYZ$, then $\overline{AB} \cong \overline{XY}$) can create serious problems for slow learners not yet ready for abstract symbolism.

Diagnosis should include an analysis of the child's ability to read and comprehend directions, verbal problems, and mathematical symbolism.

4. *Use tests as clinical tools.* A well-coordinated testing program consisting of standardized achievement and diagnostic tests, teacher-made tests, and diagnostic assignments is a necessary part of a successful teaching program for slow learners. Available evidence (25) indicates that slow learners experience relatively less difficulty with computation than with those aspects of mathematics involving reasoning; however, assessment should be made of computation as well as reasoning and problem solving.

Standardized diagnostic and achievement test results provide a perspective of the class as a whole. An item analysis of errors and a determination of the nature of the errors on these tests are basic to planning appropriate learning experiences. Schacht (25) found, for example, that inability to read and comprehend diagrams is a common source of errors that low achievers make in problem solving, common fractions, and measurement. Failure to understand the decimal numeration system was also found to be a common difficulty in the four operations with whole numbers, common fractions, and decimal fractions.

Diagnostic assignments prepared by teachers and designed for easy scoring have special value in helping children appraise their progress and in reinforcing learning. The following are illustrations of kinds of items that might be used:

a) Here are some ways to name the number 12:

$$8 + 4 \quad 3 + 9 \quad 15 - 3 \quad 3 \times 4$$

Can you think of still other ways?

b) Write two facts shown by this number line.

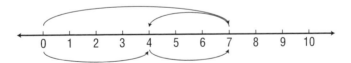

c) Write the missing numerals.

$$
\begin{array}{r}
1\,4\,9 \\
\times\ 6 \\
\hline
\end{array}
$$

□□ (6×9)

□□□ (6×40)

□□□□ (6×100)

□□□□ (6×100) + (6×40) + (6×9)

207

d) Here are sets of multiples for 2, 3, and 4:

$$A = \{2, 4, 6, 8, 10, 12, 14, 16, \ldots\}$$
$$B = \{3, 6, 9, 12, 15, 18, \ldots\}$$
$$C = \{4, 8, 12, 16, 20, \ldots\}$$

List four more multiples in each set. Name the multiples in set *A* that are in set *B*. Which of these multiples are also in set *C*? What is the least common multiple of 2, 3, 4?

e) Finish this division example:

$$35\overline{)6475}$$

$$\begin{array}{r} 1 \\ 35\overline{)6475} \\ \underline{35} \\ 29 \end{array}$$

f) Write an equation or inequality for these English sentences:

(1) I am thinking of a number. Multiply it by 2, and the product is 56.

(2) I am thinking of a number. Multiply it by 4, and the product is less than 32.

(3) I am thinking of a number. Divide it by 3, and the quotient is 29.

Diagnostic assignments may be used as periodic tests, as inventories prior to introduction of more complex ideas, or as single items embedded in each lesson to assess understanding and guide review and reteaching.

Diagnosis of learning difficulties is not an easy affair; but it makes instruction easier in the end. Constant, careful diagnosis eases the work of the teacher and improves learning for children.

Educational Programs for Slow Learners

Basically, there are three ways of providing a setting in which children have an opportunity to realize their full potential. The first is through special forms of classroom organization, the second is through curriculum adjustments, and the third requires special instructional adjustments by the classroom teacher. The first two of these ways are discussed below as subtopics of this section. The third way, special instructional adjustment, the writer considers important enough to merit a separate section—"Strategies for Teaching Slow Learners"—which constitutes the latter part of this chapter.

Organizing the classroom for learning

Homogeneous grouping. Grouping on the basis of achievement and/or intelligence is one approach used frequently to provide for individual differences. This may result in one of these arrangements:

1. Special classes are organized for slow learners in mathematics. Here both content and methods of teaching are adjusted and class size is small, with work highly individualized.

2. Remedial classes enable slow learners to be removed from their regular class group for a specified number of periods a week. These remedial classes function best at late third-grade level, for a faulty foundation in fundamental understandings and skills blocks growth in the application of these understandings and skills to the more complex operations of the middle school years. The goal of remedial instruction at this point is to return the child to his regular classroom as soon as achievement warrants.

3. All mathematics classes are scheduled at the same hour and pupils are assigned to sections appropriate to their learning level. This arrangement provides each teacher with a small group of children having like abilities in mathematics and permits her to devote full class time to their needs. At the end of the class period the children return to their respective homerooms. Reading groups are frequently scheduled in a similar manner.

Mathematics laboratories. These are special classrooms equipped with individual study centers, or carrels, and a wide variety of manipulative and visual materials as well as printed materials at various levels of difficulty. Pupils are referred to the laboratory by the homeroom teacher for a few periods each week. Close cooperation between the laboratory teacher and the homeroom teacher is essential in maintaining a consistent pattern of instruction for each pupil.

Nongraded team teaching. This organization provides for small-group and individualized study as well as large-group instruction. One member of the teaching team should be a special teacher with preparation in mathematics as well as elementary education. This teacher guides instruction in mathematics and assists other members of the team in planning learning activities for small groups and individuals.

Computer-assisted instruction. A new development in classroom organization is the use of computers in diagnosing difficulties and prescribing appropriate instructional activities for individual pupils. Computer-assisted, individualized instruction is being tested in a variety of school situations and at various grade levels as a means of helping teachers with the arduous tasks of diagnosis, remediation, and drill. Notable among these experiments are those under the direction of Patrick Suppes at Stanford University.

Curriculum adjustments for slow learners

Lerch and Kelly state that the program of study for slow learners "should be a mathematics program in the true sense, and not primarily a remedial arithmetic program" (16, p. 232).

Remedial instruction has only immediate usefulness and temporary value. It is useful for minor problems but does not strike deep at the roots of any problem. The slow learner is in need of a program of study that will give insight into the operations and relationships of mathematics. The goals of instruction should be immediate, tangible, and practical. The curriculum should allow for flexibility and the application of mathematical processes in social and vocational situations. Specifically the curriculum should meet these conditions:

1. It should be well structured and systematic as well as paced to the individual's level of maturity. Pupils whose development is below average need a longer period of time than the average student to master new content. The timetable within a sequence of learning activities should enable pupils to proceed at a pace that will motivate and challenge but not frustrate.

2. It should establish a minimal program which includes new approaches to computation, new treatment of traditional topics (measurement, common fractions, graphs, and problem solving), and selected new topics (Paschal indicates that disadvantaged children appear to think in spatial terms and that instruction for these children might begin with a unit on geometry [20, p. 6]).

3. It should make wide use of visual and manipulative materials, progressing toward abstract representation but at a more deliberate pace than for average and above-average pupils.

4. It should give special help on the vocabulary of mathematics. Precise terminology should not be required until the teacher is sure understanding is developed.

5. It should include instruction on how to locate information, how to use the textbook, how to study, how to remember, and how to check computation.

6. It should relate mathematics to other curricular areas, particularly to science, social studies, and art.

Strategies for Teaching Slow Learners

Attitudes toward mathematics are of great importance among all students but are particularly important among those who experience difficulty in learning. Negative attitudes of fear, dislike, frustration, and outright rejection are often firmly fixed among these students. They do not respond to long-range educational goals. Their goals are immediate, practical, and self-centered. For this reason it is necessary to provide such a student with problems related to his experiences and to the real world in which he lives.

Familiar experiences

1. *The school lunchroom* can provide many opportunities for applying basic addition and subtraction understandings. A menu and price list, such as shown in table 5.2, can be placed on the chalkboard and toy money used for the transactions.

Every day give each child 25¢ to spend. In the beginning. have the children tell orally the selections they would make from the menu in addition to the regular lunch. The computation should be done by the individual child and checked by the class. After a few days ask the children to buy certain extras for a given amount and compute the change from a quarter. (Buy some fruit, milk, and cookies; buy cookies and ice cream; buy a roll and milk, etc.) Finally "story problems" may be written by the children, placed in a card file, and used for additional practice in problem solving.

Table 5.2

Menu	Price
Regular lunch	20¢
Milk (one glass)	5¢
Bread (one slice)	1¢
Rolls (each)	2¢
Cookies (each)	2¢
Ice cream (one cup)	5¢
Orange	3¢
Apple	3¢

2. Readiness for multiplication and division can be built by *marching activities* similar to those used in the gymnasium. For example, select eight children and have them march single file, in pairs, and four abreast. Count the marchers by ones, twos, and fours. Using felt-board cutouts or magnetic disks, picture the groups as in figure 5.1. Children can observe that an increase in the size of the marching groups shortens the length of the marching line.

Fig. 5.1

Repeat this activity, using groups of six, ten, or twelve children. Each time illustrate the marching groups with objects or diagrams (arrays). Finally ask the children to use objects or diagrams to show all the ways a group of eighteen children can march so that the smaller groupings are the same in size (ones, twos, threes, sixes, nines).

Later these diagrams can be associated with appropriate multiplication and division facts, as in figure 5.2.

4 twos are 8 $4 \times 2 = 8$
2 fours are 8 $2 \times 4 = 8$
In 8 there are 4 sets of two.
In 8 there are 2 sets of four.
$8 \div 2 = 4$ and $8 \div 4 = 2$

Fig. 5.2

3.   Current events can often provide the motivation for introducing new concepts or applying concepts already learned. One such activity in a sixth-grade classroom grew out of *the launch of Apollo 13*. At the time of launch the teacher carefully noted (from the TV screen) the initial launch data with respect to altitude, rate (feet per second), and distance down range from the launch site. These data were placed in two tables, one showing altitude and speed, the other showing altitude and distance down range in the first ten minutes of flight.

A large grid was placed on the chalkboard; and the children, with teacher assistance, made two graphs, using the information in the tables. The point at which Apollo 13 began its "pitch program" became readily apparent on the one graph. On the other, the children noted the vertical rise of Apollo 13 in early flight and the flattened trajectory as it moved at increasing velocity into earth orbit. Through this activity words and figures took on new meaning for these sixth-grade children. Incidentally, much computation was involved in changing feet per second to nautical miles per hour!

4.   In another sixth-grade classroom, problems related to the spring *"paint up, clean up" activities* in the community were used to create interest in mathematics and to provide practice in problem solving. Many examples of the use of ratio were found in mixing liquid fertilizers, determining the amount of grass seed needed for a particular plot of ground, and mixing paint for prospective home repairs. The girls in the class found and solved problems using jam and jelly recipes that called for a certain ratio of juice or pulp to sugar.

5.   At holiday time a project related to *mailing parcel-post packages* provides opportunities for developing skill and understanding in measuring. Have packages of different shapes for children to measure. Apply the postal regulation that limits the weight to 70 pounds and

requires that the package not exceed 100 inches in length and girth combined. (Girth is the distance around the widest part of the package.) Figure 5.3 gives one example.

Length 18˝
Girth 42˝
60˝

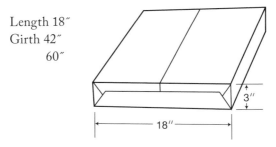

18″

3″

Fig. 5.3

6. Geometry is implicit in many of the out-of-school activities of children. In one classroom a study of right triangles was introduced in the following way:

Teacher: "On your way to school this morning, how many cut across a vacant lot?"

Almost every pupil had crossed a lot somewhere en route. The others recalled crossing a vacant lot when going to a nearby store, or to church, or to a friend's house.

Teacher: "Why did you cut across the lot?"

Almost in a chorus they answered, "Because it's shorter that way!"

Teacher: "How do you know it's shorter? Could you prove it?"

The pupils couldn't prove it exactly, but they knew it must be shorter because everybody went that way whenever he could.

They were all anxious to know how to prove it arithmetically, so the teacher asked several children to go to the chalkboard and draw a diagram of themselves crossing a lot. They were directed to "be sure to name the intersecting streets."

When the drawings were finished, the pupils observed that all the figures had the same shape even though they were different in size—they all were triangles. Other pupils went to the board and measured the corner angles in these triangles. They discovered that all were ninety-degree angles, or right angles.

On the playground the children proved by actual measurement that by cutting across the lot they did save distance, and how much they saved. (The same type of demonstration can be done in a gymnasium.) Then back in the classroom, where the original triangles were still on the board, the children identified the base and altitude of each. (In an earlier study of triangles in general the children had become familiar with base and altitude.)

Teacher: "Now in the right triangle we give a name to the third line also. We call it the hypotenuse. See if you can tell me three facts that you know about the hypotenuse from what we have said and talked about so far. These facts refer to length and position."

All agreed that (1) it is the longest of all three lines; (2) it is a shorter line than the base and altitude added together; and (3) it is the side opposite the right angle.

7. Slow learners will develop basic understandings in geometry much more readily if frequent reference is made to the use of geometry in daily life.

   a) A fourth-grade teacher skilled in the use of the camera took many colored photographs of buildings, bridges, streets, cars, airplanes, plants, and animals (fish, turtles, etc.) and had the children study them to find different shapes. The basic shapes—circle, square, rectangle, triangle, and pentagon—were identified again and again. Illustrations of point, ray, line segment, and line were also easily identified in the photographs.

   b) Requiring young children to bring a toy or object from home to illustrate a geometric shape being studied adds meaning to their understanding of space and shape. Permit the child to describe the object he brings, naming the shape it represents.

   c) In learning to understand the concept of volume, slow learners will profit from filling various familiar containers with water or sand and comparing the sizes of the containers by finding which hold the same amount or which of two containers will hold more. To help them find volume more precisely, it may be best to use cubes or rectangular prisms that hold an exact number of cubes of equal size (13, p. 427).

   d) Often a child who has no difficulty in finding the perimeter or the area of a rectangle or a square will fail to see a relationship between what he has done and finding the amount of fencing needed for a yard or the amount of tile needed for a bathroom wall. "Matching activities" similar to the following may be used to build understanding of such relationships:

   Match each item in column I with either item *a* or item *b* in column II.

	I		II
	What must I find when I—		I must find:
	1. frame a picture?		*a.* perimeter
	2. build a fence?		*b.* area
	3. tile a bathroom wall?		
	4. cover a floor with linoleum?		
	5. measure glass for a window?		
	6. put tape around a table cloth?		

Patterns and relationships

The slow-learning pupil has acquired some understanding and some skill. He needs the kind of instruction that enables him to collect in a unified way the things he has learned and helps him extend these learnings into new situations by careful emphasis on pattern and relationships. Here are several illustrations:

1.   When pupils are involved in using two-place numerals in adding and subtracting, help them see relationships to previously learned concepts. For example,

$$4 + 3 = 7$$

so 4 tens and 3 tens are 7 tens, and

$$40 + 30 = 70$$

Illustrate with manipulative materials (bundles of tens and ones, dimes and pennies, tens blocks, Dienes multibase blocks, the abacus, etc.) and record the results with numerals. One teacher (10, p. 231) found that using a color code for place value helped the slow learner at this point. If money is used as a visual aid, the cents column in addition or subtraction is recorded with blue chalk or crayon, the dimes are recorded with red, and so on. This same color reference is used with the abacus; blue beads mark units, red beads mark tens, and so on. Once place value is understood, the color code is replaced and an addition grid or table is developed using multiples of ten, as in figure 5.4.

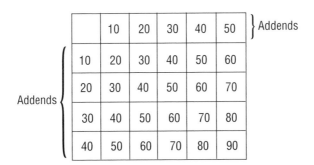

Fig. 5.4

2.   In developing number patterns with young children it is helpful to proceed as follows:

a)   Use large wooden beads (as in fig. 5.5) or kindergarten blocks arranged in a pattern dependent on color. Show the pattern once, repeat it in part, and have the children tell what should be added next.

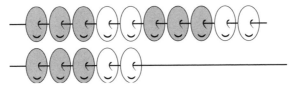

Fig. 5.5

215

b) Repeat the activity, this time varying the pattern in terms of shape (fig. 5.6).

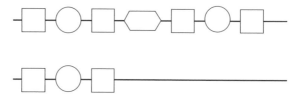

Fig. 5.6

c) Follow this with a pattern involving both shape and color (fig. 5.7).

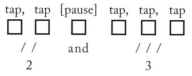

Fig. 5.7

d) Using rhythm sticks, tap out a pattern; ask the children to listen carefully and repeat the pattern by tapping on their desks. Permit the children to make up their own patterns to use with their classmates.

e) Ask the children to listen to a pattern tapped on a small drum and to record the pattern by using, first, markers, then tallies, and finally numerals. For example,

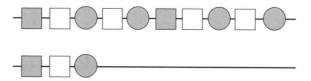

tap, tap [pause] tap, tap, tap

/ /   and   / / /

2   3

f) Build patterns using numerals. Begin with a number chart, erasing or removing numerals in a pattern. Inquire what numeral should be removed next.

1, ___, 3, ___, 5, ___, 7, ___, 9, 10, 11, etc.
2, ___, 4, ___, 6, ___, 8, ___, 10, 11, 12, etc.
10, ___, 8, ___, 6, 5, 4, 3, 2, 1

g) Start a number pattern on the chalkboard; ask the children to write the next three (or five) numerals in the pattern. These may be built to develop increasingly complex relationships.

2, 5, 8, 11, ___, ___, ___

2, 4, 8, 16, ___, ___, ___

3, 4, 6, 9, 13, ___, ___, ___

3.   When children are developing skill in basic multiplication facts, the following types of charts and diagrams can be used to build understanding of the commutative property, of the relationships between multiplication and division, and of patterns that exist in different multiplication tables.

a) Use arrays to visualize the commutative property. These may be pictured easily on graph paper. (See fig. 5.8.)

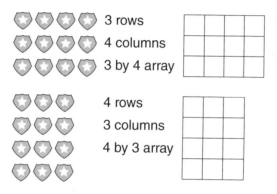

Fig. 5.8

b) Use number lines to visualize relationships and patterns.

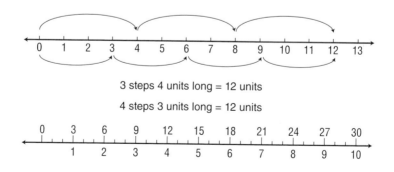

This product line can be read "2 threes are 6; 3 threes are 9," and so forth, or "In 12 there are 4 threes; in 15 there are 5 threes."

217

c) Multiplication tables may be built first on an array pattern.

$$
\left.\begin{array}{l}
0\ 0\ 0\ 0 \leftarrow 4 \\
0\ 0\ 0\ 0 \\
0\ 0\ 0\ 0 \\
0\ 0\ 0\ 0
\end{array}\right\} \leftarrow 8 \left.\begin{array}{l} \\ \end{array}\right\} \leftarrow 12 \left.\begin{array}{l} \\ \\ \end{array}\right\} \leftarrow 16
$$

This array is read:

$$
\begin{array}{ll}
1 \text{ four } = \ \ 4 & 1 \times 4 \\
2 \text{ fours } = \ \ 8 & 2 \times 4 \\
3 \text{ fours } = \ 12 & 3 \times 4 \\
4 \text{ fours } = \ 16 & 4 \times 4
\end{array}
$$

d) The familiar multiplication table can be studied for patterns, too. In the table here shown, for example, notice that all products

$$
\begin{array}{ll}
4 \times 1 = \ \ 4 & 4 \times 6 = 24 \\
4 \times 2 = \ \ 8 & 4 \times 7 = 28 \\
4 \times 3 = 12 & 4 \times 8 = 32 \\
4 \times 4 = 16 & 4 \times 9 = 36 \\
4 \times 5 = 20 & 4 \times 10 = 40
\end{array}
$$

are even numbers. Notice also that the ones digits form a repeating pattern of 4, 8, 2, 6, 0; 4, 8, 2, 6, 0. Comparing this table with that where 8 is a factor reveals similarities.

$$
\begin{array}{ll}
8 \times 1 = \ \ 8 & 8 \times 6 = 48 \\
8 \times 2 = 16 & 8 \times 7 = 56 \\
8 \times 3 = 24 & 8 \times 8 = 64 \\
8 \times 4 = 32 & 8 \times 9 = 72 \\
8 \times 5 = 10 & 8 \times 10 = 80
\end{array}
$$

4. Practice should be more than a drill. Activities given for practice, which may be developed first as small-group or class activities and later used for individual study, should stress the seeing of patterns, the verbalizing of relationships, the search for clues that will help children use known learnings to solve new problems:

$$
\begin{array}{lll}
6 \times 30 = 180, & \text{so} & 12 \times 30 = \underline{\quad} \\
7 \times 90 = 630, & \text{so} & 14 \times 90 = \underline{\quad} \\
20 \times 5 = 100, & \text{so} & 20 \times 10 = \underline{\quad}
\end{array}
$$

$$
\begin{array}{lll}
32 \div 4 = \ \ 8, & \text{so} & 64 \div 4 = \underline{\quad} \\
& \text{and} & 64 \div 8 = \underline{\quad} \\
80 \div 8 = 10, & \text{so} & 160 \div 8 = \underline{\quad} \\
& \text{and} & 160 \div 10 = \underline{\quad}
\end{array}
$$

5. At upper-grade levels, visual materials and diagrams prove useful in extending measurement to division of rational numbers. Here are two teaching strategies:

a) Start with a verbal problem: "Tom has 12 inches of paper tape. He wants to make as many ½-inch labels as he can for his rock collection. How many labels can he make?"

$$12" \div \tfrac{1}{2}" = \square$$

Provide each child with 12 inches of adding-machine tape and a ruler. Mark the tape in inches and half inches. Summarize:

> In 1 inch there are 2 half inches.
> In 2 inches there are 4 half inches.
> In 3 inches there are 6 half inches.

So in 12 inches there are 2 × 12, or 24, half inches. Tom can make 24 labels.

b) Have the students use graph paper marked in 1-inch areas to solve this equation:

$$15\tfrac{1}{2}" \div \tfrac{1}{2}" = \square$$

Ask them to measure 15½ units on a grid like the one shown in reduced size in figure 5.9. Select 1½ inches as the unit of measure. Say, "Count to find the number of 1½ inches in the grid."

"There are 10 pieces (sized 1½ inches) and ½ inch over."

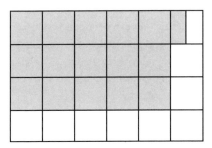

Fig. 5.9

Next have them count the half inches in 15½ inches. Say, "How many pieces 1½ inches in size can be measured on 31 half inches?" By folding the 1½-inch measure, help the children *see* that the ½ inch over is $^1/_3$ of 1½. So

$$\frac{31}{2} \div \frac{3}{2} = 10\frac{1}{3}$$

Repeat this procedure many times, using other examples solved with graph paper and/or number lines to build understanding of measuring with like units as background for division with a fraction divisor.

219

Instructional activities that build step by step are important in teaching slow learners. A central idea or concept needs to be analyzed carefully and developed in such a way that the pupil is led from one level of learning to the next. Drill will be a necessary part of these activities, but it should be delayed until the pupil has some insight into the processes on which he must drill. *A slow learner is not helped by mere repetition. An enlarged view, carefully developed, is better for him than endless review and repetition.* The opening of a larger view is illustrated in the preceding paragraphs that describe a variety of visual and symbolic approaches to building understanding of the multiplication of whole numbers. It does a slow learner little good to be supplied with one explanation or one way of solving a problem. He needs to experience the understanding by development through as many sensory modalities as are useful (seeing, hearing, touching, writing) and with a variety of applications to his own experiences. The steps in learning need to be so carefully scaled that his progress toward mastery is like walking a ramp rather than climbing a stair!

Instruction in reading mathematics

Direct instruction in vocabulary, reading, and the use of verbal and symbolic cues in problem solving is basic to teaching slow learners in mathematics.

Learning the language of mathematics, like learning any language, develops as a result of interaction of the learner with his environment. Many slow learners are retarded in educational development because contact with their social and physical environment was limited in their preschool years. A deficiency in verbal development in these early years has an adverse effect on the pupil's later ability to abstract ideas, to use symbolic cues, and to solve verbal problems.

Reading in mathematics requires (1) careful attention to detail, (2) a questioning, critical, reflective frame of mind, and (3) a familiarity with mathematical language and symbolism as well as familiarity with the "usual," more general, vocabulary and symbolism. In the study of mathematics, children must read and interpret words that carry not only a usually understood meaning but also a special meaning. Words such as *set, power, point, product, root, rational, irrational,* require special interpretation, since many pupils already have nonmathematical referents for them.

Similarly, the precise symbolism and terminology in today's curricula must be carefully taught to make certain that children can not only "read," or recognize, them but also understand them. Examples are *congruent, rectangle, plane, region, array, set of points, rotation.* Even a cursory examination of a child's textbook impresses the reader with extensive use of mathematical shorthand. For example, such symbols as $\therefore, =, \neq, <, >, \leq, \geq, \pm, \cup, \cap, \sim,$ $\overrightarrow{AB}, \overrightarrow{CD},$ and $\angle CDE$ are in common use in elementary textbooks, as is such punctuation as

$$A = \{1, 3, 5, 7, \cdots \},$$
$$B = \{1, 3, 5, 7\},$$
$$(4 + 6) + 3 = 4 + (6 + 3),$$

and

$$346_{\text{eight}} = [3 \times (8 \times 8) + (4 \times 8) + 6]$$

These words, signs, and symbols seldom appear in isolation: they are usually integral parts of grammatically correct mathematical sentences. Contemporary programs compound reading problems for children who have difficulty with handling abstract ideas and with reading per se. Consequently, systematic instruction in reading mathematics must be a part of any program of study for slow learners.

Lyda and Duncan (17) recommend that all teachers consider the feasibility of incorporating direct study of quantitative vocabulary as an important aspect of the class period in arithmetic. Their study with second-grade children showed that direct study of quantitative vocabulary contributed significantly to growth in problem solving.

In describing a Miami, Florida, project on teaching culturally disadvantaged children, Paschal advocates an audiolingual method. This, says Paschal, "is basically a method of sustained practice in the use of the language in the relationship of teacher-speaker, student-hearer, student-speaker situations" (21, p. 370). The teacher in this project always used the correct term in presenting the idea but did not insist that the child use the correct term at the initial learning level. Some activities used by this teacher follow:

1. The teacher displayed models of circles, triangles, rectangles, and squares and asked the children to identify them. If necessary, the teacher supplied the name of each shape and the children touched the models, ran their fingers around them, and found other objects in the room shaped like them.

2. The children placed their hands behind them, and the teacher placed an object in the hands of each, who told what shape he had by feeling the object carefully.

3. Different geometric shapes were placed around the room before the children arrived. When they had gathered, the teacher held up an object shaped like a triangle and had the children identify it. Then they were asked to look around the room and find other objects shaped like the triangle. This game continued with other geometric shapes.

As many sense modalities as possible should be used in helping slow learners form quantitative concepts. Cuisenaire rods, magnetic disks, felt pieces, and attribute blocks are excellent at this level. As objects are manipulated the pupil tells what he sees, what relationships he observes, and the teacher supplies more precise terminology where it is needed. The pace of instruction in building vocabulary and mathematical readiness will of necessity be slow. Stable understandings will not develop if instruction is rushed and symbolic representation introduced too soon.

At upper grade levels when children have gained some facility in reading, such activities as the following can be used in building vocabulary:

1. Words can be matched with definitions, objects, or pictures.

2. Geometric figures, parts of drawings, and steps in computation can be labeled.

3. The meaning of a phrase, a concept, or a word can be demonstrated by using concrete materials or diagrams. For example, the meaning of *average* can be demonstrated by the use of books, coins, or blocks. A mirror may be used to demonstrate symmetry, and congruence can be shown by placing one geometric shape upon another.

4. A picture dictionary of mathematical terms and meanings can be constructed, as a project for an individual pupil or for a small group.

5. Bulletin-board displays can be designed in which terms or definitions are connected by yarn to an appropriate diagram, picture, object, or example. This same approach can be developed on an electric board and used for individual practice.

When reading from a mathematics textbook slow learners should be given the same kind of direct vocabulary instruction as is commonly used in reading classes. New or difficult words or phrases should be identified and placed on the chalkboard or on charts for discussion and recognition. Guided silent and oral reading should precede individual solution of verbal problems. Problems should be read silently, discussed orally by the group, and solved individually by each child.

Children who have difficulty in reading and solving verbal problems need to develop a *way* of reading mathematics:

1. A first reading to get an overall picture
2. A more careful, slower reading for details
3. Scanning to locate specific data
4. Occasional reading between the lines to note implied data necessary for solution
5. Thinking about relationships in the problem before attempting solution

The mathematics textbook should be used, on occasion, in the reading class, and attention centered on reading for specific purposes rather than on solving problems. One skill at a time might be selected for this instruction. For example, the student might read—

1. to locate specific information;
2. to select appropriate data (to build an awareness of extraneous data);
3. to perceive relationships;
4. to interpret maps, scales, charts, graphs, diagrams, etc.;
5. to determine the question(s) asked;
6. to determine the operation(s) to use;
7. to get the main thought and express it in the student's own words.

Games, puzzles, and recreations

Games, puzzles, and recreations are appealing to almost all children. Paschal, in discussing a program for disadvantaged children, advocates the use of games in maintaining class attention and in giving practice in listening and speaking (21, p. 372). He indicates that the games used for these children should be sharply defined and structured, and the rules definite and easily understood. Furthermore, he recommends games that are person-centered and concerned with direct action and visible results.

Games may be used for practice, such as the various forms of bingo that are in common use for practicing on multiplication and division facts.

Another game that children find interesting is called "The Top." In the game as illustrated in figure 5.10, you begin with the number 13.

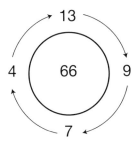

Fig. 5.10

You may choose whether to add, subtract, multiply, or divide the numbers. Follow the directions of the arrows. Calculate until you reach 66. As the game starts, the children shout, "The top is spinning!" In turn, each child carries out one operation. The first child may choose to subtract, saying,

"13 – 9 = 4"

The next child may elect to add, saying,

"4 + 7 = 11"

The third child may multiply,

"11 × 4 = 44"

The next player may subtract,

"44 – 13 = 31"

The child who reaches 66 wins the game.

"Can You Match Me" is a game that two or more pupils can play on the chalkboard or with figures placed on a feltboard or even with small objects, which can be grouped and regrouped to show the operation called for. One player writes a mathematical expression which can be matched in several ways by other players. For example:

Given:

$$\frac{3+17}{4}$$

Matching equivalent expressions:

$$20 \div 4 \qquad \frac{23-3}{4} \qquad (2 \times 10) \div 4 \qquad 5 = (3 + 17) \div 4$$

"T Square" is a game that can be used to practice basic facts in any of the four operations. The first player calls out five numbers, which are written on the left side of the T square. He then call something like "plus 6," "times 7," or "divide by 4," which the players write at the top of the "T" before immediately writing the correct answers on the right side of the "T." The first child to complete the T square with correct solutions wins the game.

+6	×7	÷4
9	4	28
7	6	36
8	8	16
6	5	24
5	3	32

A puzzle similar to a magic square is placed on the chalkboard (see fig. 5.11). The children copy it on graph paper and fill in the missing numerals. Similar puzzles can be made simply by changing the numeral in square A or by altering the directions.

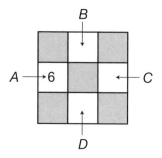

1. B is 6 times as large as A.
2. C is 4 less than B.
3. D is 8 more than A.

(If you work the puzzle correctly, the total of A, B, C, and D is 88.)

Fig. 5.11

224

Similar games can be found in teacher's guides to textbooks, the *Arithmetic Teacher*, and special books devoted to games and puzzles. A number of the references listed at the end of this chapter include descriptions of games.

Supplementary assignment sheets

Another successful way of adapting instruction to a range of learning abilities is to vary assignments in terms of depth and scope. Supplementary assignment sheets may be prepared in advance in such a way that they can be reused later. These assignments may be enclosed in clear plastic envelopes obtainable at most school and office supply stores, or they may be laminated. In either case the child marks directly on the plastic, using a marking pencil or pen suitable for transparencies. When corrected, the plastic may be wiped clear and reused by other children. Yet another procedure is to place the assignment sheets in file folders and have the child answer on separate sheets or answer strips. Keys for the exercises may be provided, and each child may check his own assignment, noting where he needs further study. For slow-learning children, the assignments should be limited to a few simple items with easily understood directions. Here are a few examples:

STRANGER IN THE FAMILY

In each column find the number sentence that does not belong in the family. Mark it out.

$$4 \times 6 = 24 \qquad 4 + 7 = 11$$
$$24 \div 4 = 6 \qquad 11 - 4 = 7$$
$$3 \times 8 = 24 \qquad 11 - 7 = 4$$
$$6 \times 4 = 24 \qquad 11 = 5 + 6$$
$$24 \div 6 = 4 \qquad 11 = 7 + 4$$

MAKING CHANGE

How many different ways can you make change for 11¢ which Bill and Joe can share?

Bill	Joe
10¢	1¢

Using the Hundred Chart

Make a chart like the one shown in figure 5.12.
Draw a ring around multiples of 2.
Put a slash mark on the multiples of 3.
Draw a triangle around multiples of 5.

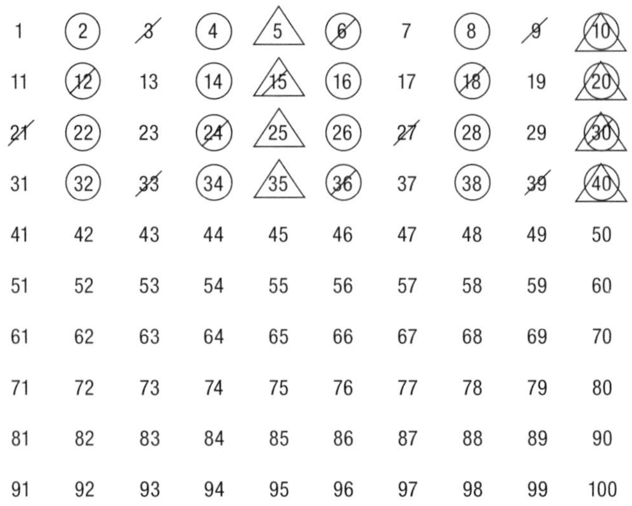

Fig. 5.12

Wallpaper Arithmetic

[Provide each child with a set of directions and several sheets of graph paper.]

Here is a "3" pattern (fig. 5.13). What would a "2" pattern look like? What would a "9" pattern look like? Would a "5" pattern be attractive?

1	2	3	4	5	6	7	8	9	10
11	12	13	14	15	16	17	18	19	20
21	22	23	24	25	26	27	28	29	30
31	32								
41									
51									
61									
71									
81									
91									

Fig. 5.13

Matching Verbal Problems with Equations

[Prepare sets of verbal problems and equations. Children read the problems and select the appropriate equation for each problem.]

Bill and Tom have 12 marbles.

Jane needs 2 more pennies to buy a 10¢ candy bar.

At 6¢ each, how many candy bars can Tom get for 30¢?

If one can of soup serves 3 children, how many cans are needed for 12 children?

$$2 + N = 10¢$$
$$6 \times N = 30¢$$
$$\triangle + \square = 12$$
$$N \times 3 = 12$$

Practice in Problem Solving

Be careful. These problems have numbers that are not needed. Estimate each answer. Then find the exact answer. Check by comparing.

1. Bob had $1.50. He wanted to buy six-ounce packages of cookies at 29¢ a package. Did he have enough money?

2. Jim says that five 18-inch towel racks at $1.39 each should cost about $5 + $2, or $7. Is he right? What is the exact cost?

3. Jack bought five packs of notebook paper for 47¢. There were 50 sheets of paper in each pack. He estimated the cost of one pack to be $1/_5$ of 50¢, or _____. Was his estimate a good one?

4. Find the cost of a game at 49¢, a 100-piece jigsaw puzzle at 28¢, and a pencil at 9¢.

5. Jane can get 27-inch-wide gingham for 49¢ a yard and 36-inch-wide gingham for 59¢ a yard. She needs $1/_2$ yard of gingham. She has a quarter. Which kind of gingham should she buy?

Practice for Estimating Quotients

Which statements are true? Which are false?

$3 \times 6 < 19$	$3 \times 60 < 190$	$6 \times 90 < 520$
$30 \times 6 < 190$	$6 \times 6 < 52$	$6 \times 900 < 5,200$

Make these sentences true.

$$22 = (\square \times 4) + 2 \qquad 220 = (\square \times 40) + 20$$
$$43 = (\square \times 6) + 1 \qquad 430 = (\square \times 60) + 10$$
$$19 = (\square \times 3) + \triangle \qquad 190 = (\square \times 6) + \triangle$$
$$52 = (\square \times 6) + \triangle \qquad 520 = (\square \times 60) + \triangle$$

227

Find the quotients. What is the remainder in each example?

$$3\overline{)19} \qquad 30\overline{)190} \qquad 60\overline{)190}$$

$$6\overline{)52} \qquad 60\overline{)520} \qquad 900\overline{)5,200}$$

$$7\overline{)26} \qquad 7\overline{)570} \qquad 70\overline{)640}$$

$$9\overline{)81} \qquad 90\overline{)635} \qquad 80\overline{)254}$$

Written materials

A variety of written materials is necessary in any classroom where instruction is adjusted to the needs and abilities of the learner. Not only is a suitable textbook or textbook series necessary, but many different textbooks should be available in sets of six or eight. These textbooks may be individualized by removing the binding and mounting the separate practice units in file folders. Practice activities on problem solving, basic division facts, addition of fractions, and so forth, may be grouped according to learning difficulty and stapled into folders. The children work through the material in each folder, writing answers on separate answer sheets. Self-scoring devices may be included in the folder for ready use by the student. Old textbooks, or new textbooks other than the basic text, can be cannibalized and used in this way to provide needed extra practice.

Programmed textbooks and workbooks are available from many publishing houses. A variety of these should be available for individual use in each classroom. They should be purchased in small sets and handled in such a way that they, too, are reusable. Most publishers have such materials coordinated with their textbook series.

A wealth of good material is available from government agencies. For example, the U.S. Office of Education has "Space Oriented Mathematics for Early Grades," which is a set of workbook materials by Edwina Deans; a circular by Patricia Spross entitled *Elementary Arithmetic and Learning Aids;* and a bulletin, edited by Lauren Woodby, entitled *The Low Achiever in Mathematics.* Materials of the workbook type can be obtained also from the Treasury Department and the Forest Service of the Department of Agriculture. These are well done and are interesting to older students.

There are a growing number of supplementary books in mathematics written for elementary school children. Some of these are factual in character; others would be classed as fiction but are sound mathematically and carry high interest for elementary school children. Certainly a portion of the school's allotment for the library should be spent for such publications. Recent issues of the *Arithmetic Teacher* carry reviews as well as listings of these books as they become available.

Manipulative materials

Slow learners in mathematics should make frequent use of manipulative and visual materials. Many of these may be made by the teacher and the children; others can be collected and adapted to classroom use; still others may be purchased from school supply houses and textbook publishing houses.

Children's toys can be good mathematics materials in many instances! These should be used as teaching materials at all steps in learning—to aid discovery, to provide practice, to test understanding. They should be used by the children and by the teacher for demonstration. Various construction toys such as Tinkertoys, building blocks, link blocks, D-sticks, and parquetry tiles are useful in teaching geometry. Dominoes and various card games not only capture interest but provide practice in counting and computation. A trip through the toy department of any large department store will reveal, to the perceptive teacher, many toys and games that can serve the cause of mathematics! The writer discovered that the cardboard coin holders used by coin collectors can be used as arrays in multiplication and to develop concepts of percent. For the teacher who needs templates, they are excellent for tracing circles of various sizes!

It is impossible to list or describe the hundreds of visuals available for teachers. The *Arithmetic Teacher* often includes descriptions of teacher-made instructional devices. Books dealing with the teaching of elementary school mathematics are veritable gold mines of ideas for teachers seeking suggestions for making and using manipulative materials.

Slow-learning children need to resort constantly to objective materials, pictures, diagrams, and dramatization to help them "see" relationships. The written record and verbalization should accompany the visualization to make sure that the bridge is built between the operation and the algorithm.

Within-class grouping

Intraclass, or within-class, grouping of students is one of the more successful ways of adjusting instruction to the needs of the slow learner. This type of grouping should be flexible and temporary, with the size and the number of groups varying with the intent of the lesson.

For example, in a class working on division with a one-digit divisor, one group of children might be solving "$248 \div 8$" by using bundles of tens and ones. Their approach might be to expand 248 into 24 tens and 8 ones and to divide the 24 tens into 8 groups of equal size (or to separate the 24 tens into 8 groups of 3 tens each) and then to divide (or separate) the 8 ones. The written record could read:

$$
\begin{aligned}
24 \text{ tens} \ \div \ 8 \ &= \ 3 \text{ tens} \ \text{ or } 30 \\
8 \text{ ones} \ \div \ 8 \ &= \ 1 \text{ one} \ \text{ or } \underline{1} \\
248 \ \div \ 8 \ = \ 30 + 1 \ &= \ 31
\end{aligned}
$$

229

A second group could be solving "248 ÷ 8" by subtracting, using the "ladder" algorism:

$$
\begin{array}{r|l}
8 & 248 \\
  & \underline{80} \qquad 10 \times 8 = 80 \\
  & 168 \\
  & \underline{160} \qquad 20 \times 8 = 160 \\
  & 8 \\
  & 8 \qquad\; \underline{1 \times 8 = 8} \\
\end{array}
$$

$$31$$

A more advanced group of children could be using a standard algorism, first estimating the answer.

$$
\begin{array}{r}
1 \\
30 \qquad \text{Estimate: About 30, since } 8 \times 30 = 240. \\
8\overline{)\,248} \\
\underline{240} \\
8 \\
\underline{8} \\
\end{array}
$$

Working within this general framework of grouping according to levels of thinking, the teacher provides opportunities for the slow learner to make his discoveries under guidance, to share and discuss his findings with other children, to move slowly by well-graduated steps from objective materials to pictorial representation, and finally to move to the written record using numerals.

Grouping by maturity levels develops very naturally from a type of teaching that encourages discovery. In fact, the class sharing and discussion involved in group discovery may be one of the best ways to meet individual differences.

Grouping need not always be used for discovery lessons; it can be used effectively to arrange practice activities for small groups of children with similar needs.

1. Pupil teams can be formed for practice on basic facts and for remedial help on any of the operations with whole numbers and fractions. Usually one member of the team should be competent and able in the content being practiced. Assignment cards can be prepared in advance for each team and appropriate concrete aids set aside for use by those children who need to check their solutions with objective materials.

    Occasionally a pupil team may be constituted of two children who both need help. Recently a sixth-grade teacher who was unable to motivate an underachiever to master basic multiplication facts assigned him the responsibility of helping a fourth-grade child who needed similar help. As the older boy assisted the younger one, he found it necessary to assume a teacher's role, to prepare his materials in advance each day, and when challenged by his young student, to learn the basic facts himself! At the end of one week the sixth-grade boy had mastered the facts he had not known, become interested in his participation, and developed more self-esteem.

2. A listening post equipped with a tape recorder and several sets of earphones makes it easy to administer special practice activities. Before class time the teacher reads practice activities or directions for individual work onto the audio tape. During class time, while the teacher works with part of the class, the children who need extra practice gather around the tape recorder, put on the headsets, and carry out the previously recorded activities.

Tapes may be developed with a pause, or "silent spot," to allow time for the pupil to produce the correct response. Sound effects such as those from rhythm-band instruments might be incorporated to help children perceive the meanings: (Beep, beep, beep, pause, beep, beep: 3 + 2 = 5.) The children may reproduce the pattern with counters, with pictures, or with the appropriate algorithm.

Tape may be indexed, stored, and used for group testing and practice in the same manner. However, many teachers prefer to create practice materials appropriate to the immediate needs of their students.

Help for teachers

Teacher need help themselves if they are to succeed in teaching slow learners. The strategies described above call for careful, time-consuming preparation. As a final strategy, therefore, administrators should provide clerical assistance for preparing materials and recording test data. They should also, perhaps, provide teacher aides and remedial tutorial help for disadvantaged children.

Summary

If mathematics instruction at the elementary school level is to be successful, then teachers must take cognizance of the variations in the groups taught, and assign expectations to individual pupils instead of groups of pupils. Readiness for learning must be determined, learning difficulties diagnosed, and the classroom organization and environment made conducive to learning. Instructional materials commensurate with the needs of the children should be provided and teaching strategies employed that will help each pupil perform learning tasks in which he can be successful.

There is no single, easy solution in the education of the slow learner. His problem is complex. Teaching must take this into account.

REFERENCES

1. Biggs, Edith E. *Mathematics in the Primary School.* New York: British Information Service, 1965.

2. Brewer, Emery. "A Survey of Arithmetic Intraclass Grouping Practices." *Arithmetic Teacher* 13 (April 1966): 310–14.

3. Buffie, Edward G., Ronald C. Welch, and Donald D. Paige. "From Diagnosis to Treatment." In *Mathematics: Strategies of Teaching,* Modern Elementary Methods Series. pp. 182–214. Englewood Cliffs, N.J.: Prentice-Hall, 1968.

4. Chandler, Arnold M. "Mathematics and the Low Achiever." *Arithmetic Teacher* 17 (March 1970): 196–98.

5. Davidson, Patricia S. "An Annotated Bibliography of Suggested Manipulative Devices." *Arithmetic Teacher* 15 (October 1968): 509–12, 514–24.

6. Davies, Roberta. "Low Achiever Lesson in Primes." *Arithmetic Teacher* 16 (November 1969): 529–32.

7. Gibb, E. Glenadine. "Through the Years: Individualizing Instruction in Mathematics." *Arithmetic Teacher* 17 (May 1970): 396–402.

8. Gilmary, Sister. "Transfer Effects of Reading Remediation to Arithmetic Computation When Intelligence Is Controlled and All Other School Factors Are Eliminated." *Arithmetic Teacher* 14 (January 1967): 17–20.

9. Golden, Sarah R. "Fostering Enthusiasm through Child-created Games." *Arithmetic Teacher* 17 (February 1970): 111–15.

10. Green, Roberta. "A Color-coded Method of Teaching Basic Arithmetic Concepts and Procedures." *Arithmetic Teacher* 17 (March 1970): 231–33.

11. Grossnickle, Foster E.; Leo J. Brueckner; and John Reckzeh. *Discovering Meanings in Elementary School Mathematics.* 5th ed. New York: Holt, Rinehart & Winston, 1968.

12. Hammitt, Helen. "Evaluating and Reteaching Slow Learners." *Arithmetic Teacher* 14 (January 1967): 40–41.

13. Higgins, Jon L. "Sugar-Cube Mathematics." *Arithmetic Teacher* 16 (October 1969): 427–31.

14. Kaplan, Jerome D. "An Example of a Mathematics Instructional Program for Disadvantaged Children." *Arithmetic Teacher* 17 (April 1970): 332–34.

15. Keiffer, Mildred C. "The Development of Teaching Materials for Low-achieving Pupils in Seventh and Eighth Grade Mathematics." *Arithmetic Teacher* 15 (November 1968): 599–604.

16. Lerch, Harold H., and Francis J. Kelly. "A Mathematical Program for Slow Learners at Junior High Level." *Arithmetic Teacher* 13 (March 1966): 232–36.

17. Lyda, W. J., and Frances M. Duncan. "Quantitative Vocabulary and Problem Solving." *Arithmetic Teacher* 14 (April 1967): 289–91.

18. National Council of Teachers of Mathematics. *Instruction in Arithmetic,* Twenty-fifth Yearbook. Washington, D.C.: The Council, 1960.

19. National Society for the Study of Education. *Individualized Instruction,* Sixty-first Yearbook, pt. 1. Chicago: University of Chicago Press, 1962.

20. Paschal, Billy J. "Geometry for the Disadvantaged." *Arithmetic Teacher* 14 (January 1967): 4–6.

21. ———. "Teaching the Culturally Disadvantaged Child." *Arithmetic Teacher* 13 (May 1966): 369–74.

22. Reed, Mary Katherine Stevens. "Vocabulary Load of Certain State Adopted Mathematics Textbooks, Grades One through Three." Ed.D. dissertation, University of Southern California, n.d.

23. Richards, Pauline L. "Tinkertoy Geometry." *Arithmetic Teacher* 14 (October 1967): 468–69. Reprinted in *Readings in Geometry from the "Arithmetic Teacher,"* pp. 60–61. Washington, D.C.: National Council of Teachers of Mathematics, 1970.

24. Ridding, L. W. "Investigations of Personality Measures Associated with Over and Under Achievers." *British Journal of Educational Psychology* 37 (November 1967): 397–98.

25. Schacht, Elmer James. *A Study of the Mathematical Errors of Low Achievers in Elementary School Mathematics.* Ed.D. dissertation, Wayne State University, 1966. Ann Arbor, Mich.: University Microfilms (no. 67–10, 488).

26. Smith, Robert M. *Clinical Teaching: Methods of Instruction for the Retarded.* New York: McGraw-Hill Book Co., 1968.

27. Spitzer, Herbert F. *Enrichment of Arithmetic.* St. Louis: McGraw-Hill Book Co., Webster Division, 1964.

28. ———. "Providing for Individual Differences." In *Teaching Elementary School Mathematics,* pp. 314–30. Boston: Houghton Mifflin Co., 1967.

29. Stenzel, Jane G. "Math for the Low, Slow, and Fidgety." *Arithmetic Teacher* 15 (January 1968): 30–34.

30. Turner, Ethel M. *Teaching Aids for Elementary Mathematics.* New York: Holt, Rinehart & Winston, 1966.

9

From *Measurement in School Mathematics*, NCTM's Thirty-eighth Yearbook (1976)

Estimation, as approximation, can be thought of as the mental use of units of measurement when applied in contexts and problems that involve measurement. Shirley Frye, who served as NCTM president from 1988 to 1990, recommended a selection from the Thirty-eighth Yearbook, *Measurement in School Mathematics:* **chapter 5, "Estimation as Part of Learning to Measure,"** by George W. Bright.

This chapter will appeal to all those interested in the importance of estimation in measurement, regardless of context and grade or instructional level. In it, Bright presents the foundations of estimation, as well as some very strong sections on preparing to teach estimation and on estimation in the classroom. The concluding section on sample estimation programs provides some thoughtful considerations for teachers, math leaders, curriculum specialists, and anyone else teaching this topic.

Estimation as Part of Learning to Measure

George W. Bright

MEASUREMENT can be taught more successfully if estimating is one kind of instructional activity. Since activities involving estimation not only fit into many different classroom situations but also encompass a wide variety of behaviors, frequent use of estimating is both possible and desirable. As they gain experience, students are likely to estimate more accurately and are therefore likely to enjoy estimation activities.

Probably the most common use of estimation is illustrated by this type of exercise: *Estimate the width of this page to the nearest centimeter.* This activity focuses on an identified attribute of a specified object, which is in full view, in order to obtain an estimate in terms of a known unit, here communicated by the name of the unit, *centimeter.* Since attention is called more to the measure (number) than to the unit of measure, other kinds of activities should be used to balance this one-sided approach.

Throughout this essay, attention will be given only to estimating measurements that are made up of both a number and a unit. In particular, the estimation of number or numerousness is intentionally ignored. This kind of estimation is certainly a part of measurement and is important in mathematics instruction, but a body of literature already exists on this subject (see Bakst [1937] and Payne and Seber [1959]). The purposes of this essay are—

1. to examine the relationship between estimation and measurement;

2. to provide examples of ways of teaching estimation;

3. to show how estimation activities can be sequenced to become a useful part of mathematics instruction at all levels.

The author would like to thank John G. Harvey of the University of Wisconsin–Madison for his thoughtful comments, criticisms, and suggestions.

Foundations of Estimation

Estimation of measurements is the use of units of measure in a strictly mental way, without the aid of measurement tools. Estimation provides a means of applying measurement directly to the real world. Skill in estimating can improve verbal communication, for when several units of measure are used to express an idea, mental comparisons of the relative sizes of these units must be made. But before amplification for these statements is provided, several terms need to be defined.

Definitions

A *measurable attribute* (hereafter simply *attribute*) of an object is a characteristic that can be quantified by comparing it to some standard unit. For example, the mass of a rock is a measurable attribute, but the hardness of that rock is not, since the standard hardness scale is not based on a single unit. Typical attributes are length, area, volume, mass, temperature, time, and electric current.

Measuring is the process of comparing an attribute of a physical object to some unit selected to quantify that attribute. The comparison may be direct, such as when a measuring tape is held around one's waist; or it may be indirect, such as when the width of a river is measured through the use of similar triangles. Measuring also takes place when the volume of a box is computed after its length, width, and height are measured or when the circumference of a ball is measured by wrapping a string around the ball and then laying the string next to a ruler.

A *measurement* or a *measure* is the result of measuring. Both are descriptions of an attribute of an object, but there is one major difference between them. A measurement is made up of both a number and a unit of measure; for example, the circumference of the teacher's head is 59 cm. A measure is a number alone and can be derived only when the unit of measure is understood or is available for examination; for instance, in terms of the width of the ring finger, the width of the hand across the knuckles is about 4.

Estimating is the process of arriving at a measurement or measure without the aid of measuring tools. It is a mental process, though there are often visual or manipulative aspects to it. It requires that several ideas be firmly in mind: (1) the unit of measure to be used, (2) the size of that unit relative to familiar objects or to other units of measure for the same attribute, (3) other measurements in that unit, and (4) a commitment to perform the estimating so that the product is as close to the actual measurement as possible. Estimating is guessing, but the guessing must be educated. Wild guessing is not true estimating, except insofar as such behavior represents the first stages of the development of estimating skill.

An *estimate* is the product of estimating. It is the description of an attribute in terms of either a number and a unit or a number alone.

Two distinctions among terms must be clearly delineated. First, measuring and estimating are processes, whereas measurements, measures, and estimates are products of these processes. Second, physical tools are required for measuring but not for estimating.

Measurement and estimation

The mathematical view of measurement is somewhat different from the view of the preceding section. In the mathematical sense, measurement is a correspondence between an attribute of an object and a positive real number. The correspondence or assignment represents a comparison between some unit and the object in question. That is, measurement for a specified attribute is a function from a set of objects to the set of real numbers. The determination of the specific function for a given characteristic depends on the attribute and the unit selected to measure that attribute, although a simple change of variables will translate the values for a function defined for one unit to the values defined for any comparable unit. Unfortunately, the pedagogical consequences of such translations are not nearly as trivial as the mathematics of the situation might suggest.

The function is clearly many to one, for many objects can and do have the same measure, at least within the tolerances inherent in physical measurement. Mathematically, this means that it is not possible to reverse the assignment of numbers to objects. In the context of the real world, however, the assignment can, in a real sense, be viewed in the reverse way. That is, a measurement picks out the set of objects that would have that measurement. For example, in the typical measurement activity

measure the length of the table to the nearest centimeter,

the stress is on the mathematical assignment of a measure to an object. If a standard unit like the centimeter is to have meaning for students, however, there ought to be opportunities to use the assignment in the reverse direction. The following exercise might accomplish this goal:

Use your measuring tape as a searching device to find in the room something that is 45 cm long.

The student is asked to locate one object that is assigned the value 45 by the measure function for length based on the standard unit of 1 cm. This activity helps to establish a mental picture of the size of a centimeter, or in this example, the size of 45 centimeters. The better the student's mental picture of 45 cm, the more efficient will be the search. At least part of the reward for a better mental picture is a more rapid completion of the exercise. The example incorporates a safety valve in the form of the instruction "use your measuring tape," for if students cannot visualize a length of 45 cm, the measuring tape provides a physical model that can be compared to objects in the room.

Kinds of estimation

There are eight basic kinds of estimation. Each of the paths in figure 5.1 represents one of these. It is assumed that the unit has been selected for each kind.

Although further subdivisions could be made in some of the categories, the scheme of figure 5.1 is fine enough for purposes of comparison. Four kinds of estimation, those in

239

class A, can be used to illustrate and emphasize the mathematical properties of a measure function; four others, those in class B, can be used to illustrate the reverse relationship between measurements and the objects that would be assigned those measurements.

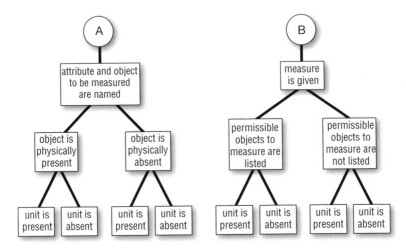

Fig. 5.1. Kinds of estimation

In making the estimates in class A, students guess the measure for a named attribute of an object; the object may be present or absent. Such activities provide practice in assigning a measure to an object. Presumably, students mentally compare the given unit with the named object and determine the appropriate measure. With some kinds of units, however, students may not compare in the expected way. For example, in estimating the area of a rectangular object, some students will estimate the respective linear dimensions and compute an estimate of the area by applying the standard area formula. Also, when the named object is present, its position relative to the observer affects the accuracy of the estimate, at least for attributes such as length and area. Accuracy will be greater when the object is viewed straight on than when it is viewed at an angle, either horizontally or vertically.

Examples of the four kinds of estimation in class A can easily be created by satisfying each of the three conditions for each path in figure 5.1:

1. Estimate in square centimeters the area of the polygonal region shown here. [Both the object and the unit are present.]

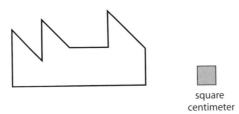

square
centimeter

2. Draw the diagonal of this page. Estimate to the nearest centimeter its length. [The object is present but the unit is not.]

3. Get a meterstick. Estimate to the nearest meter the amount of string that would be needed to make a tennis net. [The unit is present but the object is not.]

4. Estimate the volume of a garbage can in liters. [Both the object and the unit are absent.]

Students should understand that in each example they are to estimate to the nearest unit, just as they would measure to the nearest unit to check their estimates. The fourth example might be solved by mentally picturing a liter container and then mentally stacking such containers in the garbage can. Alternatively, if the garbage can is round, the height and the radius of the base could be estimated in decimeters and a volume estimate could be computed by the formula $V = \pi r^2 h$. Since the computed volume would be in cubic decimeters, the estimate in liters would have the same measure.

Since students are asked to estimate, the usual right/wrong dichotomy does not apply to their answers. Rather, there is a degree of correctness, which depends on the discrepancy between the estimate and the subsequent measurement of the object. Improving estimates should be the primary goal of these activities.

In making the estimates in class B, students either choose an object from a list of permissible objects or name an object of their choice to which a specified measure could be assigned. Such activities help students picture mentally the size of a given unit of measure. Presumably they picture the measurement in terms of the repetitions of the unit and compare this mental picture with objects (specified or unspecified) around them. Alternatively, the student may imagine a succession of objects, assign a measure to each, and compare these numbers to the specified measure. Performing the activity the first way rather than the second will probably be more useful in developing sound concepts. These kinds of estimation do not reinforce the exact mathematical concepts of measurement, but they do make units of measure more accessible to students for use in practical situations.

Examples of class B estimation may help to clarify the discussion:

1. Get a kilogram mass. Which of the following has a mass closest to 2 kg?

    football, baseball bat, hockey puck

    [Permissible objects to measure are listed and the unit is present.]

2. Which of the following would normally have a temperature closest to 5°C?

    ice cube, candle flame, Gulf of Mexico

    [Permissible objects are listed but the unit is absent.]

3. Make a model of one square meter. Name something having an area of 6 m². [Permissible objects are not listed but the unit is present.]

241

4. Name something 7 dm long. [Permissible objects are not listed and the unit is absent.]

The objects that are named for the last two exercises should be chosen so that the measure of each is as close as possible to the specified measure.

One of the three conditions determining each kind of estimation is the presence or absence of the unit of measure. When the unit is absent, students must mentally picture the unit before they can begin the process of making an estimate. Consequently, students can be expected to make less accurate estimates when the unit is absent simply because an additional mental process is involved. An error in the mental picture of a unit is likely to be reflected in the estimates made for that unit.

Estimation made with the unit present can help solidify a student's mental image of the size of that unit if such activities are interspersed with those involving estimation in the absence of the unit. Once developed, the mental picture will make students more comfortable in situations requiring measuring.

Preparation for Teaching Estimation

Purposes of estimation activities

The primary purposes of including estimation exercises in the mathematics curriculum are, first, to help students develop a mental frame of reference for the sizes of units of measure relative to each other and to real objects, and, second, to provide students with activities that will concretely illustrate basic properties of measurement. Developing a frame of reference may be the more crucial for most students, since for many people the importance of mathematics lies in its usefulness in the real world. Estimation would seem to be more useful than understanding fully the mathematical structure of measurement.

Not every problem that could be solved by estimation is in fact solved that way. Some such problems are "solved" by a habitual response to a recurring situation. For example, when a new kettle is needed, the choice must be made from among a few standard sizes. It seems doubtful that the frequent cook really pays much attention to the actual capacity of the pot. Rather, the desired size is sensed intuitively. The goal of estimation in the public school is not to create such habituated responses, but to help students—who are novices in measurement—to develop skill that will give them flexibility in dealing with a wide variety of situations.

Good estimation skills come from exposure to all eight kinds of estimation previously outlined. If a shirt pattern calls for 1 m of material 90 cm wide, one must make an on-the-spot judgment of how much material 120 cm wide to buy. This is a more sophisticated problem than merely adjusting the dimensions so that the total area is the same, for the determining criterion of a successful decision is whether the pieces of the pattern will fit onto the cloth. Successful mental juggling of this sort can be enhanced by previous practice in estimating.

Developing estimation skills

Developing skill in estimating is probably best done by having students first make an estimate and then make the measurement as a check. There are no data to indicate whether students should be introduced to a unit of measure through a guess/check procedure or should be provided with opportunities to measure with the given unit before making estimates. The latter alternative seems logically more defensible, for it seems better to give students an opportunity to develop a feel for the size of the unit before expecting them to use that unit to make estimates of the same attribute of other objects. In any event, since what is being developed is the students' skills in making good estimates, it seems important that opportunities be provided in situations that are not personally threatening. Students should be encouraged to make good estimates, but they should not be penalized even for wildly inaccurate ones. Activities should be sequenced so that all students can improve the accuracy of their estimates and can develop skills that are as good as their needs demand.

Self-checking of estimates helps students develop self-correcting skills to take into the real world. That is, students learn to make their own checks. These checks will not always take the form of measuring, for such accuracy is not always needed. Sometimes a sufficient check is provided by comparing the named object to one that has already been measured.

Self-checking also helps to distinguish the act of measuring from the abstract concept of measurement. An understanding of the abstract properties of measurement should be accompanied by an understanding of the kinds of, and reasons for, errors that appear in real-world measurement data. Self-checking provides an experiential background from which errors in measuring can be explained and accurate measuring skills can be isolated, studied, and improved. Estimation exercises help students develop an appreciation for the realistic limits of the accuracy of physical measurements.

Evaluating estimation skills

Developing skill in estimating probably cannot be put in proper perspective unless some longitudinal records are kept relating students' estimates and corresponding measurements. Such records are useful, however, only if they can be interpreted in ways that help students improve their estimates. One simple procedure is to keep for each student a list of pluses and minuses—a "+" to indicate that the estimate is larger than the corresponding measurement and a "–" to indicate that the estimate is smaller. Those students who consistently have pluses (minuses) know, or at least can be told, that they need to make an adjustment in their estimates by making them smaller (larger). Unfortunately, this kind of record gives no hint as to the amount of adjustment that is needed.

An alternative procedure is to graph the data of each student's estimates on perpendicular axes, with the measurements graphed on the horizontal axis and the estimates graphed on the vertical axis. The graphed points can be compared with plotted points on the line of the equation $y = x$. Suppose, for example, that the following estimates and measurements were made in centimeters:

Estimate	7	30	40	5	15	90	110	75	60	12	20	80	25	35	2	40
Measurement	6	26	35	5	4	87	92	65	54	14	20	64	27	29	2	39

These pairs would be graphed as shown in figure 5.2. Note first that most of the points of this student's estimates lie above the line $y = x$, and second that as the measurements get larger, the amount of overestimation (not necessarily the percent of overestimation) increases. A decrease in estimates is called for, with the amount of decrease increasing as the size of the object increases.

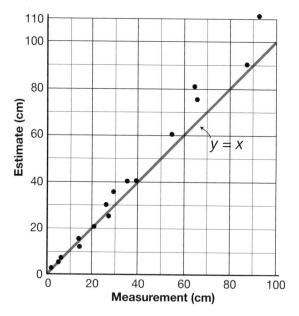

Fig. 5.2. Graph of estimates and measurements

The data could be reorganized as shown in figure 5.3 to display the percentage of discrepancy rather than the actual amount. On the average the estimates here appear to be a little more than 10 percent too large. Encouraging the student to decrease estimates by this amount for a short period of time ought to correct this bias.

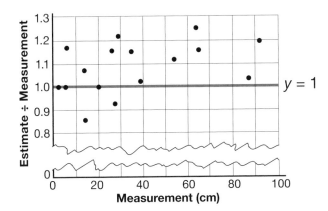

Fig. 5.3. Graph of percent of overestimate

Alternatively, students can be helped to remake their mental image of the unit being used. Overestimating suggests that the mental image of the unit is too small. This image might be corrected by having a physical model of the unit present while making a series of estimates or measurements. Newly measured objects could serve as referents for future estimates that would, it is hoped, be more accurate. In the remaking of the mental image, the focus of attention is on the size of the unit.

Error patterns in estimates may be related to the attribute being estimated. Patterns as clear as the one shown in figures 5.2 and 5.3 are most likely to occur for length measurement. For units of area and capacity, older students are likely to estimate linear dimensions and then compute estimates of the desired attribute. If they realize that the computed estimate can be balanced by overestimating and underestimating the different linear dimensions, then the error patterns may appear more random than systematic. Direct estimation of volume and mass is normally more difficult, since interferences arise from the shape of the container or object, the manner in which the estimate is made, and the material from which the object is made.

Estimation in the Classroom

Sample estimation activities

Mathematics textbooks for public school students include various kinds of activities designed to improve students' estimation skills or to help establish referents for specific measurements. Introductory exercises often highlight the fact that measurements are approximations; for example, a unit and a picture of an object are presented, and students are asked to fill in an accompanying blank (fig. 5.4). Such exercises also provide a good beginning for the studying of estimation, since it is the property of "about-ness" that is the focus of estimating. After the transition is made from arbitrary to standard units, the following exercises would be appropriate:

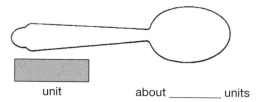

unit   about _____ units

Fig. 5.4. Sample introductory exercise

1. Without using a ruler, draw a segment that you think is 15 centimeters long. Check by measuring.

2. Guess the width of your hand to the nearest centimeter. Then find the actual measure.

Such activities help students attach meaning to names of units in the manner suggested by the Nuffield project (1967, p. 81):

> To ensure that the words [*centimeter* and *meter*] are becoming meaningful it is useful to encourage children to estimate distances before actually measuring them. The degree of accuracy of the estimates, and the use or misuse of vocabulary will help the teacher towards assessing individual progress.

Correct usage of names of units of measure tends to accompany the development of mental images of the sizes of these units. Body measurements are usually helpful in reaching this goal (fig. 5.5). Body parts can serve as referents that students can use for comparing with other objects.

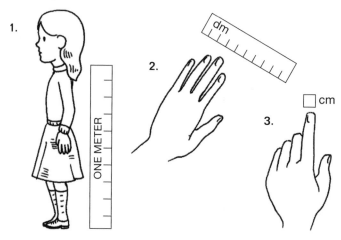

Fig. 5.5. Body referents

246

Although this technique is easily implemented in the classroom, it has a distinct disadvantage: as students grow, so do their body referents. Over an extended period of time, remeasuring must be done, and estimating procedures must be adjusted accordingly.

The use of referents can be externalized by having students select objects that do not change in size. The mass of a book might be used as a referent for one kilogram, and the width of a door might be taken as a model of one meter. Some objects are difficult to carry around; so students may experience some inconvenience in finding a handy referent for a specific task. Too, as students grow, their perception of the sizes of objects may change, since their size relative to the objects has in fact changed. This should not be a serious problem if estimation exercises are a regular part of the curriculum. Probably a combination of body measurement models and external models provides the most applicable set of referents. Referents should be selected not only from objects in the classroom but also from objects in the home. This will facilitate the transfer of estimation skills from the classroom to the outside world. Without such explicit facilitation, transfer may not occur at all.

Once frames of reference have been established for units of measure, the referents can be refreshed and extended through activities such as the following:

1. Complete this sentence correctly:

    A football field is about 1 _____ (meter, hectometer, kilometer) long.

2. Choose the best measurement for each object:

3 grams
30 grams
3 kilograms

25 milliliters
250 milliliters
250 liters

3. Choose the best unit for measuring each object:

kilometer
meter
millimeter

kilogram
gram
milligram

4. Which is closest to the area of a table top?
   1 cm^2, 1 dm^2, 1 m^2, 100 m^2, $10,000 \text{ m}^2$

5. Take a large can and punch a small hole in the bottom. Guess how many milliliters of water would run out of the hole in 60 seconds. Check yourself.

Each of these problems could be solved by having students actually go out and measure the objects directly. Normally, however, this is not desirable, for the important objective of these activities is to develop the ability to compare objects mentally. In accomplishing this objective, teachers should not always allow students to verify their answers directly, though occasional verification can reinforce students' answers and reward their progress in improving their skills.

In spite of the variety of estimation exercises found in mathematics textbooks, two major impressions stand out in currently available series. First, estimation does not seem to be an important topic, and second, the extent and proper sequencing of activities to develop estimation skills seems to be neither understood nor appreciated. In NCTM's Twenty-fourth Yearbook, Payne and Seber suggested broad guidelines for teaching estimation (1959, p. 190):

> The practical value of being able to estimate the size of physical objects is recognized by almost all people. . . . The teacher should realize that this kind of estimation is difficult and somewhat cumbersome to teach. There are several reasons for this. It is time consuming; it may be difficult to apply a measuring instrument to the quantity; and a certain degree of mathematical maturity seems desirable, if not necessary. But this does not mean that we should ignore this useful aspect of estimation. It is even more reason for giving concentrated and systematic study to this topic.

In spite of this comment made more than fifteen years ago, textbooks have continued to ignore the systematic development even of minimal estimation skills. The few activities that are included in most series are scattered more or less indiscriminately throughout the volumes. Often they are segregated in brief sections headed "Explore," "Investigate," or "Extra for Fun" and thus become attachments to, but not part of, the mainstream of the content development.

Unfortunately, such superficial treatment of estimation is often accompanied by a similarly superficial treatment of measurement. This double dose of inadequate instruction leaves a quite distorted view of measurement that is usually never corrected. For example, abstraction in the process of measurement is often begun and concluded in one motion, simultaneously with the introduction of the standard units.

At the time of the initial push to rewrite mathematics textbooks to include the metric system, numerous supplementary workbooks, filmstrips, cassettes, and packages of materials were produced. The role of estimation in the instruction of these programs is noticeably different from its corresponding role in formal textbooks. Estimation is not only used more but also sequenced more carefully.

The most frequent kind of exercise is still the estimation of an attribute of a given object, such as the length of a pencil or the perimeter of a room. The reverse process is included, however; for example, students are asked to find an object that weighs the same (has the same mass) as an object of known weight (mass). The actual mass of the named object can then be estimated or measured to determine how closely it corresponds to the mass of the given object. A variation of this activity is the classifying of objects into categories such as the following: much less than one square meter, less than one square meter, exactly one square meter, more than one square meter, much more than one square meter. Estimation skills would not need to be completely developed to succeed in this activity. Nevertheless, activities of this kind would seem to be effective in developing and maintaining mental images of the sizes of many kinds of units of measure.

Careful sequencing of these activities is needed to help students build their estimation skills. Incorporating sets of sequenced activities into curriculum materials would seem to be critical in providing adequate opportunity for students to develop estimation skills, for "the need for giving pupils repeated experience cannot be overemphasized" (Payne and Seber 1959, p. 190). Estimation is not a once-in-a-while thing. It is a dynamic process that must be developed through extensive practice over a long period of time.

Games can be used both to develop and to maintain estimation skills, provided that estimation is needed to win the game. Most commercial games deal with abstractions of measurement rather than with manipulative skills. Games to teach estimation, therefore, will most likely be homemade and adapted to take advantage of the special objects that are available in a particular classroom. Team competition games and scavenger hunts are perhaps the easiest to organize. Teams can be asked either to select one object to match each of several measurements or to select several objects to match a single measurement. After estimating, students can measure the objects, and the team with the lowest cumulative error is declared the winner.

Ways of making estimates

Not all estimation skills are the same or are developed in the same way. For example, the estimation of a length could be accomplished either by imagining copies of the unit laid end to end alongside the object or by comparing the object to some other object whose measure is known. An estimate of time might be made by comparing the interval to an internalized standard or by counting the repetitions of an external event, such as the number of songs played over the radio. Estimating area can be done either directly or by first estimating linear dimensions and then computing an area estimate. The latter procedure is probably the most common method among people who are familiar with the standard area formulas, but it is doubtful that this procedure is particularly beneficial in helping people visualize the relative sizes of units of area. A more useful procedure to accomplish this end is to ask students to estimate an area by comparing it directly to one of the standard units of area. Estimating the area of an irregular shape can often be facilitated by mentally rearranging the area to form a more standard shape. The emphasis of these activities should be on visualizing the entire area at once, comparing it to the area unit most nearly the same size, and then comparing the remainder to smaller area units. The relationships among units will be stressed in this way, and more accurate estimates will probably be obtained.

A similar approach can be taken for estimating volumes of containers. Ideally, students should develop skill in estimating the capacity of an object directly, though for regular shapes (e.g., prisms, cylinders, spheres), one could estimate one or more lengths and then compute an estimate of the volume. Grocery sacks are useful in helping students visualize volumes. The shape of these sacks makes the computation of the actual volume relatively easy, and students can readily imagine filling the sacks with liter containers, since the normal purpose of sacks is to be filled. Jars (common objects that are also normally filled) have the disadvantage of often having odd shapes; so estimating volume may become confounded by a lack of skill in conserving volume. One way to provide practice in estimating volume is to fill a set of jars with different amounts of water and have students order the jars solely on the basis of the amount of water in each. This activity should of course be preceded by a lot of practice in measuring the volumes of jars. (Teachers, too, are frequently misled by the shape of a container, although this is probably not really a manifestation of the lack of conservation of volume but instead relates to their lack of practice in estimating volumes.)

The estimation of the mass of an object requires understanding the physics of the real world, so that an accurate estimate of mass can be derived. When an object is lifted, the attribute that is felt is the weight of the object, that is, the force of gravity pulling the mass of the object toward the center of the earth. Compounding the problem is that the *way* an object is lifted can alter how heavy the object feels. For example, an object lifted close to the body will "feel" lighter than the same object lifted at arm's length, and an object lifted by one hand will "feel" different from the same object lifted with both hands. Consequently, to compare objects with respect to lighter and heavier, the way each object is held should be standardized. From this type of comparison, objects whose masses have been measured can be employed as referents that can be used to generate an estimate of the mass of a given object.

Sample estimation programs

Given the variety of estimation activities and the problems inherent in developing estimation skills, how should estimation be taught? First, and most important, familiar objects need to be measured to develop a set of referents for the unit of measure being studied. This means that some experience in direct measurement should be provided immediately after the unit is introduced. Only a few measurements would seem to be called for, though, before students are asked to estimate measurements prior to actually performing the measuring. Initial estimates may be very wild indeed, but this should not discourage the use of estimation activities in the classroom.

Second, students should be encouraged to make reasonable estimates. If estimates are nothing but wild guesses, they are of no use in any practical situation. Third, all eight kinds of estimation (see fig. 5.1) should be practiced. The net effect of incorporating all eight into a curriculum is a mutual reinforcement of skills. Although it may be possible to develop only one or just a few of these skills, skill in all eight will give students far greater flexibility in dealing with real-world problems.

Fourth, activities should provide practice in estimating both with a variety of units that measure the same attribute and with a variety of units that measure different attributes. Although there may not be much transfer of skill from one unit to another, the variety of exposure will help students develop flexibility in using estimation skills.

Fifth, estimation skill must be practiced in order to be maintained. Estimation should not be viewed as a topic that can be studied during one period of time and then never mentioned again. It should be used in a variety of situations, certainly not all of which are associated with mathematics topics. Whenever a measurement must be made—making a drawing, building something, making a map—an estimate could be made.

These five guidelines hold not only for instruction of students but also for their teachers and parents. Because of the historical lack of estimation activities in the curriculum, all three groups start with a similar degree of lack of skill. Since the groups will not progress at the same rates, the structuring and sequencing of activities should probably be somewhat different for each.

Students. The most important element of instruction in estimation for students is the provision for sufficient opportunity for the growth of their skills. Certainly as more and more units of measure are introduced, review and renewal of skills are needed. One effective way to do this is through short activities or games that provide practice in a single kind of estimation, for example, finding four objects, the sum of whose lengths is 2.5 m. Every student will not become equally proficient, but all students need to be given opportunities at each level in their own cognitive development to improve their skill. Estimation activities should parallel the measurement activities that are already part of the mathematics program, though opportunity for estimation should be provided independently if measurement is not included as part of the standard curriculum.

Estimation can be used to help teach a variety of concepts common to the mathematics curriculum of public schools. In the following examples, which are meant to be representative of the possibilities, it is assumed that each estimate is checked by measuring.

First, estimation can help solidify the concepts of number and counting as well as the concepts of more/less, larger/smaller, heavier/lighter, and so on. In making estimates of length, young students can be directed to answer the question, "How many of this unit would be as long as this object?" Answering this question requires the conceptualization of the number of units that must be put together to make the estimate. The relationship between number and unit becomes the central problem in obtaining an accurate estimate. As the size of the unit increases, the number of units needed decreases. If the concept of number is poorly understood, students cannot be expected to succeed in making such estimates.

Second, estimation can help describe the world. Amount (e.g., length, mass, temperature) is an important part of one's understanding of the world. Estimation skill can help organize this world so that patterns can be observed.

Third, computation can be practiced through estimation. Sums, products, and differences can be embedded in estimation exercises: estimate the total length of the diagonals of a rectangle; compute an estimate for the area of a rectangle; how much more does jar A hold than jar B? As decimal notation is developed to record metric measures, these computations can be performed on decimal numbers instead of whole numbers.

Teachers. Instruction for teachers can be more intense than for students. This would be reasonable even if demands of scheduling did not require it, for most teachers already possess concepts of measurement that can be used in developing estimation skills. It is probably more efficient to develop these skills through a daily workshop program of one-to-two-weeks' duration than through a weekly or monthly seminar. Extensive practice in estimating with rapid feedback provided by having the teachers measure the objects estimated seems to work well both in putting teachers at ease and in helping them develop skills. Exercises highlighting the symbolism and numerical relationships among various units can be intermingled with estimation activities.

It is important to provide instruction to teachers in a form that is readily transferrable to their own classrooms. This is perhaps easier with estimation and measurement activities than with computation activities, since many adults begin to develop their estimation skills from a conceptual base not much more secure or extensive than that of public school students. The kinds of activities useful in helping teachers develop their skills are the same kinds that are useful in helping students, though the amount of repetition required to improve skills is on the average less for teachers than for their students. All the sample exercises given in this essay are appropriate for use with teachers and can be used with many different kinds of students.

Parents. Concentration on the units of measure most frequently encountered in everyday living would seem good advice for any adult-education endeavor: meter, centimeter, kilometer, liter, kilogram, gram, degree Celsius. Estimation activities should be restricted to situations very similar to those likely to be encountered in the real world. Transfer of skill is optimized by this technique, and potential frustration is avoided. The comparison of this approach to the approach used for students should be explicitly made, so that parents understand the reasons for the differences.

Conclusion

Estimation can become a meaningful part of the public school mathematics curriculum. All that is required is for teachers to recognize its usefulness and take advantage of the variety of activities it encompasses. Students enjoy estimating; therefore, teachers need not fear using it as one of their teaching tools. By the way, how much string does it take to make a tennis net?

REFERENCES

Bakst, Aaron. *Approximate Computation.* Twelfth Yearbook of the National Council of Teachers of Mathematics. New York: Bureau of Publications, Teachers College, Columbia University, 1937.

Nuffield Mathematics Project. *Beginnings.* New York: John Wiley & Sons, 1967.

Payne, Joseph N., and Robert C. Seber. "Measurement and Approximation." In *The Growth of Mathematical Ideas, Grades K–12,* pp. 182–228. Twenty-fourth Yearbook of the National Council of Teachers of Mathematics. Washington, D.C.: The Council, 1959.

10

From *Problem Solving in School Mathematics*, NCTM's Forty-second Yearbook (1980)

J. Michael Shaughnessy, NCTM president from 2010 to 2012, recommended that we include **chapter 4, "Posing Problems Properly,"** by Thomas Butts, from the yearbook *Problem Solving in School Mathematics*, which was published in 1980. This was a very popular yearbook, because it was released in the same year as NCTM's *An Agenda for Action*, which recommended that problem solving be the focus of school mathematics in the 1980s. Today, at a time when the implementation of the Common Core State Standards is an important concern for many, this chapter of more than three decades ago remains a valued read for its focus on problem posing.

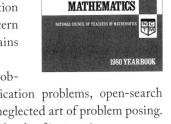

Butts begins by presenting the following types of problems: recognition exercises, algorithmic exercises, application problems, open-search problems, and problem situations. He then addresses the neglected art of problem posing. His chapter should appeal to mathematics educators at all levels of instruction.

Posing Problems Properly

Thomas Butts

OR me, and I suspect the same is true for many other people, the real joy in studying mathematics is the feeling of exhilaration that comes from solving a problem—the tougher the problem, the greater the satisfaction. But what factors initially motivate someone to *want* to solve a problem? Answers to this question can range from personal curiosity to the fear of the consequences if the solution is not handed in tomorrow, but a prime consideration has to be the manner in which the problem is posed. Examine the following problems:

Problem 1. Let $d(n)$ denote the number of positive divisors of the integer n. Prove that $d(n)$ is odd if and only if n is a square.

Problem 2. Which positive integers have an odd number of factors? (Justify your answer.)

Problem 3. Imagine n lockers, all closed, and n men. Suppose the first man goes along and opens every locker. Then the second man goes along and closes every other locker beginning with #2. The third man then goes along and changes the state of every third locker beginning with #3 (i.e., if it's open, he closes it, and vice versa). If this procedure is continued until all n men have passed by all the lockers, which lockers are then open?

These three problems are, in fact, different formulations of the same problem. (To see the equivalence of problems 2 and 3, examine a particular locker, number 12, for example. Locker 12 is touched by men 1, 2, 3, 4, 6, 12—i.e., the factors of 12. Since a locker is alternately opened and closed, lockers whose numbers possess an odd number of factors will be open in the end.)

The first version is posed in typically dry, mathematical style. The second, less ponderous, is given as a question to answer rather than as a statement to prove. The third asks this mathematical question in a very picturesque manner. I would argue that the phrasing of the third version (and, to a lesser extent, the second) would significantly motivate the potential solver to tackle the problem.

This paper, then, will be concerned with the question of how to pose (or "re-pose") a problem to maximize this source of motivation. Before carefully investigating the art of problem posing, let us first briefly discuss the different types of mathematical problems.

Types of Problems

We shall partition the set of mathematical problems into five arbitrarily titled subsets:

1. Recognition exercises
2. Algorithmic exercises
3. Application problems
4. Open-search problems
5. Problem situations

A brief summary of each category follows.

Recognition exercises

This type of exercise typically asks the solver to recognize or recall a specific fact, definition, or statement of a theorem.

Example 1: Which of the following are polynomials?

(a) $x^3 + 3x + 2$, (b) $x^3 + 3\sqrt{x} + 2$, (c) $x^3 + \sqrt{3x} + 2$, (d) $x^3 + 3/x + 2$, (e) 2

Example 2: The line segment joining a vertex of a triangle to the midpoint of the opposite side is called a ___?___ .

Example 3: If a, b, c are real numbers and $a > b$, then $ac > bc$. True or false?

Algorithmic exercises

As the adjective *algorithmic* implies, these are exercises that can be solved with a step-by-step procedure, often a numerical algorithm.

Example 4: Compute $16 + 4 \cdot (-2) - (6 \div 3)$.

Example 5: Solve $2x^2 - 3x - 5 = 0$.

Example 6: Find the center and the radius of the circle

$$x^2 - 2x + y^2 + 4y = 4.$$

Application problems

Application problems involve applying algorithms. Traditional word problems fall in this category in that their solution requires (*a*) formulating the problem symbolically and then (*b*) manipulating the symbols according to various algorithms.

Example 7: If the length and width of a rectangle are each increased by 20%, by what percent is the area increased?

Example 8: A bag of nickels, dimes, and quarters contains 435 coins worth $43.45. There are three times as many dimes as quarters. How many of each type of coin are in the bag?

A high percentage of all exercises and problems in elementary, secondary, and beginning college textbooks fall in these first three categories. The distinguishing feature of these problems is that *a strategy for solving the problem is contained within its statement*. The hurdle to overcome, then, is to translate the written word into an appropriate mathematical form so that suitable algorithms can be applied.

Open-search problems

An open-search problem is one that does not contain a strategy for solving the problem in its statement. The problem given at the beginning of this essay falls in this category. Typically such problems are phrased, "Prove that _____," "Find all _____," or "For which _____ is _____," but many other, more interesting variations are possible.

Two other open-search problems (favorites of mine) are these:

Example 9: Let $S_n = \{1, 2, 3, \ldots n\}$. For what type of integers, n, can S_n be partitioned into two subsets so that the sum of the elements in each subset is the same? For example, $S_7 = \{1, 2, 3, 4, 5, 6, 7\}$ can be partitioned into $\{1, 6, 7\}$ and $\{2, 3, 4, 5\}$.

Example 10: How many different triangles with integer sides can be drawn having a longest side (or sides) of length 5 cm? 6 cm? n cm? In each case, how many of the triangles are isosceles?

Though a higher percentage of problems in upper-level college texts are in this category, they are often unimaginatively phrased (e.g., "Prove theorem 2.3"). One of the great misconceptions involving the open-search problem is that it must be concerned with sophisticated mathematical concepts. A tragedy of some instructional methods at the elementary and secondary level is that these students are not given the opportunity to solve open-search problems that depend on the most elementary concepts. The experience of finding a pattern, so fundamental in the study of mathematics, should be offered at all levels of mathematical training. One solution to the problem given in example 9 depends only on the ability to add positive integers and to perceive that the sum of an odd number of odd integers equals an odd integer.

Problem situations

This category is best typified by Henry Pollak's exhortation: "Instead of telling students, 'Here is a problem, solve it,' tell them, 'Here is a situation, think about it.'" Included in this subset, then, are not problems per se but situations in which one of the crucial steps is to identify the problem(s) inherent in the situation whose solution will improve it. This approach, espoused by the Unified Science and Mathematics for Elementary Schools program (USMES) and others, is part of comprehensive problem solving.

Example 11: Design a parking lot. Possible problems to consider could include the following. There are many, many others.

(*a*) How large should each space be?

(*b*) At what angle should each space be placed?

(*c*) How much should be charged per car per hour if one wishes to make a 10% profit?

Clearly a gray area exists for each category, but these subsets can serve as a basis for our discussion on the art of posing problems.

Some Suggestions for Posing Problems

In this section we shall offer some specific suggestions on posing exercises and problems of the first four types: recognition, algorithmic, application, and open-search. The key step in the fifth type, problem situations, is to identify the problems, which usually fall into the application and open-search categories. Some of these suggestions will, of course, apply to several types of problems; they are placed in the category of greatest impact.

Recognition exercises

Since the main function of recognition exercises is to test the recall of definitions, basic facts and theorems, and so on, these exercises are usually posed in the *true-false, multiple-choice, fill-in-the-blank,* or *matching* format. Teachers sometimes hesitate to give such exercises for fear students will memorize without understanding. The "give an example of" problem is often effective in such situations because of its nonspecific answers. With possible prior warning, one may also give such a problem with no solution.

Example 12: Give, if possible, an example of each of the following:

(*a*) A proper fraction greater than 3/4

(*b*) A polynomial of degree 5 with four terms

(*c*) A triangle, one of whose altitudes coincides with one of its medians

(*d*) The equation of a circle that is tangent to both coordinate axes

(*e*) A differentiable function $f(x)$ that is not continuous at $x = 2$

(*f*) Two three-dimensional vectors, $\vec{A}$, $\vec{B}$, for which $\vec{A} \circ \vec{B} = |\vec{A} \times \vec{B}|$

Such exercises are sometimes more time-consuming to grade as test items, but as problems posed during class they can generate a variety of responses and often stimulate worthwhile discussion.

Algorithmic exercises

The current interest in basic skills has placed renewed emphasis on algorithmic and application problems. Computational proficiency, in its broadest sense, requires drill and practice; the challenge is to make it interesting. Two suggestions are offered in this vein.

1. *Give a sequence of algorithmic exercises with a purpose.*

Example 13: Choose two whole numbers. Find their sum and nonnegative difference. Add these results. Any observations?

Example 14: Compute:

(a) $\dfrac{1}{1\cdot 2}+\dfrac{1}{2\cdot 3}$,

(b) $\dfrac{1}{1\cdot 2}+\dfrac{1}{2\cdot 3}+\dfrac{1}{3\cdot 4}+\dfrac{1}{4\cdot 5}$,

(c) $\dfrac{1}{1\cdot 2}+\dfrac{1}{2\cdot 3}+\cdots+\dfrac{1}{99\cdot 100}$. Generalize.

Example 15: Multiply:

(a) $(1 + x)(1 - x + x^2)$,
(b) $(a + 2b)(a^2 - 2ab + 4b^2)$,
(c) $(3y - 2w^2)(9y^2 + 6yw^2 + 4w^4)$

Example 16: Find the areas of the shapes in figure 1.

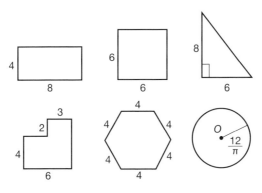

Fig. 1

Do these figures have anything in common? Do you have any observations?

Completing magic squares and magic triangles is another example illustrating this suggestion.

2. *Give the reversal of a familiar problem*

Ask the problem in the "opposite way." Some classic examples of standard problems whose reversals are also standard include the following:

(*a*) Multiply polynomials; factor polynomials.

(*b*) Given a polynomial, find its roots; given a set of roots, find a polynomial with those roots.

(*c*) Differentiation; integration.

Note that the reversal of a problem often has more than one solution. Some less familiar examples of reversal problems are these:

Example 17: Find three arithmetical problems whose solution is 13. (You can put as many boundary conditions as you wish on such a problem—e.g., specifying that at least five numbers and at least one +, ×, and ÷ sign must be used.)

Example 18: Find three solids whose surface area is 60.

Example 19: Find a cubic polynomial that has a relative minimum at (–1, 2), a relative maximum at (2, 7), and an inflection point at (1, –2).

Example 20: Write 525 as the sum of consecutive integers in as many ways as possible.

Example 21: (Cryptarithms): Although not strictly an algorithmic exercise, a cryptarithm is an excellent vehicle for testing the understanding of numerical algorithms. Three of my favorites follow:

$D = 5.$ Other letters stand for a unique digit.

(*a*)	DONALD	(*b*) HOCUS	(*c*)	ABC
	+GERALD	+POCUS		× BAC
	ROBERT	PRESTO		* * *C
				* * A
				* * * B
				* * * * * *

Application problems

Since the universal complaint against so-called word problems is their artificiality, the most obvious suggestion for improving application problems is to make them realistic. The joint MAA–NCTM *Sourcebook on Applications* gives several criteria for a "good example" in application problems, three of which are worthy of mention here:

1. Data should be realistic, both in the type of information known and the numerical values used. (A problem that asks for the length of a room given its perimeter or area would be artificial.)

2. The "unknown" in the problem should reasonably be expected to be unknown in reality. (The canonical age problem fails this test miserably.)

3. The answer to the problem should be a quantity someone might plausibly have a reason to seek.

Four examples, taken from the *Sourcebook*, illustrate these maxims:

Example 22: A six-lane track for running footraces is in the shape of a rectangle whose length is 1.5 times its width with a semicircle on each end. Each lane is to be 1 meter wide. What is the length and width of the rectangle if the inside track is to be 1500 meters long? For a 1500-meter race, the inside runner would start at the finish line. Where should the runners in the other five lanes start?

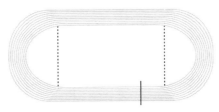

Example 23: If an 8-inch pizza serves two, how many should two 12-inch pizzas serve?

Example 24: A teacher wishes to rescale the scores on a set of test papers to "improve" everyone's grade. The maximum score is still 100, but 56 becomes a 70 on the new scale. Your score was 75. What does it become on the new scale?

Example 25: Suppose you wish to buy a new car and you feel you can afford, at most, $100 a month. If used-car loans are made for a maximum of 30 months at 14% interest, what price car should you consider?

This seems the appropriate place to mention the posing of problems containing insufficient or extraneous data. The ability to discern the data necessary to solve a problem is certainly crucial in real-world problem solving. Consequently, posing problems demanding a critical analysis of the data is to be commended.

Example 26: Two sides of a triangle have lengths 4 cm and 6 cm. Find the perimeter and area of the triangle.

Although the data are clearly insufficient to determine a unique solution, one should seek a solution of the form "any real numbers, P, A, satisfying $12 < P < 20$, $0 < A \leq 12$" rather than accepting "no solution." Problems possessing more than one solution are perfectly reasonable problems.

One could give a hokey problem with extraneous data (a person's age, shoe size, birth date, etc.), but the best application problems require the solvers to gather their own data—a type of problem more properly categorized as a problem situation.

Open-search problems

Since an open-search problem does not contain a strategy within its statement, solving such a problem requires higher-level thinking. The most important function of open-search problems is to *encourage guessing* (Pólya says, "Let us teach guessing"); writing proofs will follow. Especially with an emphasis on guessing, such problems can, and should, be used at all mathematical levels.

The fundamental axiom of posing an open-search problem is—

Pose the problem in a manner that requires the solver to guess the solution.

Though this axiom cannot always be followed, all too often a good problem is spoiled by including the answer in the statement of the problem. The problem in the first section furnishes a nice illustration; here is another:

Example 27:

Problem A. Prove that $(n - 1)! \equiv 0 \bmod 4$ if and only if n is a composite > 4.

Problem B. For which positive integers n is n a factor of the product $(n - 1)(n - 2) \ldots$ $3 \cdot 2 \cdot 1$? (For example, 5 *is not* a factor of $4 \cdot 3 \cdot 2 \cdot 1 = 24$, but 6 *is* a factor of $5 \cdot 4 \cdot 3 \cdot 2 \cdot 1 = 120$.)

In the form of Problem A, this example is typical of college texts on number theory; stated in the form of Problem B it could be posed in middle school immediately following the introduction of the concepts *factor* and *prime*. (In fact, it is almost an alternative characterization of "prime.")

Brevity is not necessarily a virtue in posing a problem. Including illustrative examples, for instance, simply enlarges the set of potential solvers.

As a rough rule of thumb, any problem whose statement includes the words "prove that," "show that," and so on, will not encourage guessing and often can be rephrased to do so.

Even problems having only one numerical answer or, perhaps, no answer at all can be posed in this way.

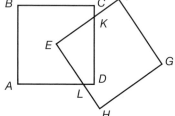

Example 28: What values are possible for the area of quadrilateral *EKDL* if *ABCD* and *EFGH* are squares of side 12 and *E* is the center of square *ABCD*?

Example 29: Find all primes that are one less than a fourth power.

"Find all" is another excellent way to pose an open-search problem that requires guessing or searching for patterns.

Here are several other suggestions for posing good open-search problems:

▶ *"Twenty Questions."* Think of some mathematical object, such as a number, geometric figure, concept, or theorem. (If it is a physical object, you can place it in a bag for added effect.) The students then attempt to identify it by asking questions, as few as possible, which you can answer only with yes or no. Discourage the question "Is it a _____?" Such an activity could be used prior to giving open-search problems as a means to encourage guessing—something that is often hard for teachers to do.

▶ *Whimsical problems.* All too often, students regard textbook problems as artificial; one remedy, cited earlier, is to make the problem as realistic as possible. Another alternative is to pose the problem in a totally unrealistic manner, making it a *whimsical* problem. Such problems often paint a vivid, sometimes ridiculous, picture in the mind. The locker problem mentioned earlier is one example—can't you just see someone walking down a row of 10,000 lockers, opening every other one? Here is another:

> *Example 30:* Two pirates were burying their booty on an island, with the Coast Guard in hot pursuit. Near the shore were two large rocks and a lone palm tree. Bluebeard A started from one of the rocks and, walking along the line at right angles to the line joining the rock to the palm tree, paced off a distance equal to that between the rock and the palm tree. Bluebeard B did a similar thing with respect to the other rock and the palm tree. They then walked toward each other and buried the treasure halfway between. Two years later the pirates returned to the island to dig up their treasure but found that the palm tree was no longer there. How can they find the treasure?

A whimsical problem can appeal to students because they do not regard it as phony and it may pique their intellectual curiosity.

Mathematical games and puzzles are still another rich source of open-search problems. They are usually posed as questions and often are whimsical problems as well.

Re-posing Problems

▶ *Given any skill or concept, it is usually possible to pose a set of nonroutine problems of varying type and difficulty involving that skill or concept.*

In this section we apply some of the foregoing suggestions to the re-posing of problems.

> *Example 31 (The straight line):* Typical problems involve finding the equation of a line given two points on the line, graphing a line given its equation, determining the equation of a line given its graph, and so on. Several nonroutine problems follow:

> *Problem A* (Recognition). Give the equations of three lines containing the point (1, 2). *Comment:* The nonspecific answers required should help those students who are merely memorizing the various formulas. Students who choose $x = 1$, $y = 2$, $y = 2x$ might be regarded as clever; they are insightful problem solvers.

Problem B (Algorithmic): Take a point, *P*, on the parabola $y = x^2$, $0 \le x \le 2$, and compute the slope of the line containing *P* and (1, 1). Do the same for six other points on the curve. What do you think the slope of the line tangent to $y = x^2$ at (1, 1) is? (See fig. 2.)

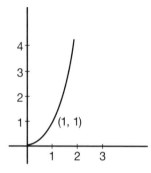

Fig. 2

Comment: This is a sequence of algorithmic problems with a purpose.

Problem C (Open search). Find all possible values for the slope of a line that contains the origin and intersects the circle $(x - 12)^2 + (y - 5)^2 = 25$.

Problem D. Write the equations of several line segments with $-2 \le y \le 2$ that will spell your first name.

▶ The problem of posing and re-posing algebraic word problems is a tough one. The solution of word problems requires the translation of English words into mathematical terms—a critically needed skill in any discipline that uses mathematics. One difficulty is that the quantity sought in a word problem requiring the solution of an equation is often known in real life. When the problem is posed realistically, however, it becomes an arithmetic problem.

Example 32:

Problem A. The Bigtown Coliseum has 20,000 seats—8,000 reserved seats and 12,000 general admission seats. Reserved seats cost $1 each, and general admission seats cost $5 each. The newest rock-group sensation, SMOOCH, is giving a concert there. If $130,000 was collected from the sale of 18,500 tickets, how many reserved seats were sold?

Comment: Surely the person selling the tickets knows how many were sold. At best that person would want to compute the amount of money received—an arithmetic problem—if that amount of money were not known.

This might be a somewhat more realistic version of the problem:

Problem B. The Bigtown . . . concert there. The promoter estimates her expenses for ushers, programs, janitorial service, and so on, at $10,000. If SMOOCH demands 40% of the price of each ticket sold plus $40,000, how many tickets must she sell to break even? To make a profit of $50,000?

Note: Attempts at humor (however feeble) in the statement of mathematical problems are usually worthwhile.

▶ In order to find interesting open-search problems for your class, you may consult one of the many problem books currently available. Offered here, then, are three examples of such problems with suggestions for re-posing them in a more challenging manner:

Example 33:

Problem A. Prove that a positive integer greater than 9, all of whose digits are identical, cannot be a perfect square.

Problem B (Better). Which squares have identical digits? (For example, can 1111 . . . 11 ever be a square?)

Problem C (Best). What is the maximum number of identical nonzero digits in which a square can end?

Comment: Problems B and C re-pose the problem as a question to encourage guessing. Version C has a positive answer (rather than "none").

Example 34:

Problem A. Show that

$$1 + \frac{1}{2} + \frac{1}{3} + \cdots = \frac{1}{n}$$

is never an integer.

Problem B. The series

$$\sum_{k=1}^{\infty} \frac{1}{k}$$

is known to diverge, and consequently the sum

$$\sum_{k=1}^{n} \frac{1}{k} = 1 + \frac{1}{2} + \frac{1}{3} + \cdots + \frac{1}{n}$$

can become arbitrarily large. If we examine a few of its partial sums, we notice that

$$\sum_{k=1}^{4} \frac{1}{k} = 2.083, \quad \sum_{k=1}^{11} \frac{1}{k} = 3.02, \quad \sum_{k=1}^{31} \frac{1}{k} = 4.0272.$$

Can

$$\sum_{k=1}^{n} \frac{1}{k}$$

ever be an integer?

Comment: Giving some numerical examples often helps clarify the problem and makes it more tantalizing to the solver.

267

Example 35:

Problem A. AEDC and *BCFG* are squares on the legs of right triangle *ABC*. Lines *EB* and *AG* are drawn. Prove that the area of triangle *AHB* equals the area of quadrilateral *CIHJ*.

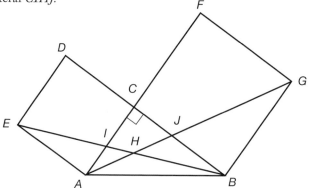

Problem B. AEDC . . . drawn. If *BC* = 8 and *AB* = 10, find the areas of triangle *AHB* and quadrilateral *CIHJ*. Can you generalize?

Comment: By quantifying the problem, we retain the guessing feature.

Summary

This article is concerned with the improvement of the oft-neglected art of problem posing. The three key words in this sentence are *neglected, art,* and *improvement.* One has only to examine the problem sections of most textbooks to see the "neglect." They often consist of lists of algorithmic exercises, unimaginative word problems, "prove that" problems, and so on. (The review problems are usually worse.) As for "art," it requires the creativity of an artist to pose a problem so that the potential solver will—

1. be motivated to solve the problem;

2. understand and retain the concept involved in the solution of the problem;

3. learn something about the art of solving problems.

The suggestions offered here will, perhaps, improve the state of the art.

The study of mathematics is solving problems. It is incumbent on teachers of mathematics at all levels, therefore, to teach the art of problem solving. The first step in this process is to pose the problem properly.

11

From *The Agenda in Action,* NCTM's Forty-fifth Yearbook (1983)

In 1980, NCTM published *An Agenda for Action* (NCTM 1980). This precursor to *Curriculum and Evaluation Standards for School Mathematics* (NCTM 1989) recommended that—

1. problem solving be the focus of school mathematics in the 1980s;

2. basic skills in mathematics be defined to encompass more than computational facility;

3. mathematics programs take full advantage of the power of calculators and computers at all grade levels;

4. stringent standards of both effectiveness and efficiency be applied to the teaching of mathematics;

5. the success of mathematics programs and student learning be evaluated by a wider range of measures than conventional testing;

6. more mathematics study be required for all students and a flexible curriculum with a greater range of options be designed to accommodate the diverse needs of the student population;

7. mathematics teachers demand of themselves and their colleagues a high level of professionalism; and

8. public support for mathematics instruction be raised to a level commensurate with the importance of mathematical understanding to individuals and society.

Three years later, in 1983, NCTM published a yearbook titled *The Agenda in Action.* From that volume, Glenda Lappan, NCTM president from 1998 to 2000, has recommended **chapter 1, "Problem Solving as a Focus: How? When? Whose Responsibility?"** by Peggy A. House, Martha L. Wallace, and Mary A. Johnson. This chapter addresses how challenges related to problem solving have become the focus of school mathematics, looking both through the lens of the classroom teacher and within mathematics teacher education. This selection has particular relevance today to the problem-solving and reasoning Standards for Mathematical Practice (National Governors Association Center for Best Practices and Council of Chief State School Officers 2010).

Problem Solving as a Focus: How? When? Whose Responsibility?

Peggy A. House
Martha L. Wallace
Mary A. Johnson

To MAKE problem solving the focus of school mathematics . . . to apply stringent standards of efficiency and effectiveness to the teaching of mathematics . . . to demand of ourselves and our colleagues a high level of professionalism—these are laudable and challenging goals. But how shall we translate them into actual classroom practice that will change the behaviors of both teachers and pupils? Therein lies the real challenge of the *Agenda for Action*. To investigate this challenge, we must first understand the complexity of teaching in the style implied by the *Agenda* and then discover ways to implement the recommendations. Let us begin this investigation by reflecting on the experience of a student teacher.

Clarifying the Problem

John was perplexed as he shared the experience of his first day of student teaching. He had been observing a junior high school mathematics class when he was approached by Del, an eighth grader, who began to ask questions about courses in high school and college mathematics. "I'm going to be an architect," Del announced, "so I'll need lots of math, won't I? I think I need trigonometry and calculus. Do you know what they are?"

As John was struggling to find a way of explaining trigonometry to an eighth grader, Del was already asking, "Can you show me some calculus problems?" John showed Del how to find the area under a curve by adding rectangles, and Del soon concluded that one could probably use a similar technique to find the volume of a mountain. As the class ended and Del was leaving, he turned to thank John for helping him and to share his philosophy of what makes a good teacher: "Teachers spend lots of time telling us things we should know. I would like to have teachers who want to make it stick!"

"I was really excited," John recalled later. "My first day out, and I had found a student who was excited about mathematics, a student who was a good problem solver! But later when I told Del's teacher about my discovery, she said, 'Del's no problem solver! He can't even add fractions yet, and he needs to know how to do that before he can solve problems!'"

Who was right, John or the teacher? Is it true that Del was not a problem solver even though he seemed much more interested and excited about mathematics than most of the students one observes? To answer these questions, we must define "a good problem solver" and decide whether we can identify a successful problem solver in such a short encounter. We must also ask, "What is problem solving?"

The common definition of a mathematical problem is a situation that involves a goal to be achieved, has obstacles to reaching that goal, and requires deliberation, since no known algorithm is available to solve it. The situation is usually quantitative or requires mathematical techniques for its solution, and it must be accepted as a problem by someone before it can be called a problem. Problem solving is the process of attacking such a problem: accepting the challenge, formulating the questions, clarifying the goal, defining and executing the plan of action, and evaluating the solution. It will involve the use of heuristics but not in a predictable manner, for if the heuristics could be prescribed in advance, they would themselves become algorithms and the problem would become an exercise.

Problem solving is a process, not a step-by-step procedure or an answer to be found; it is a journey, not a destination. Successful problem solvers can be identified by the processes or the attitudes of mind they display. Four characteristics that help identify good problem solvers are desire, enthusiasm, facility, and ability.

Recognizing a Problem Solver

A problem solver needs, first of all, the *desire* to approach the problem, accept a challenge, take a risk, find an answer, understand a question, and discover new knowledge or create a new solution. A problem does not exist until someone accepts it as a problem, and a problem solver does not exist without the desire to solve problems.

Along with this desire, a problem solver needs *enthusiasm* to proceed with the solution. Enthusiasm, as we use it here, signifies the willingness to accept a challenge or set one's own challenge. It means having the determination to investigate past the first obstacle and the perseverance to continue even when the problem looks hopeless. It means possessing the flexibility to try several methods and to look for more questions once the original problem has been solved. Enthusiasm is kindled by successful problem-solving experiences, so that the problem solver knows the delight of conquering a problem.

Problem solving also requires *facility* in using mathematics and heuristics. Facility with mathematics includes understanding fundamental concepts, relationships, and mathematical processes; it means "basic skills" broadly defined. Facility with heuristics signifies the ability to make guesses; solve simpler or related problems; construct pictures, graphs, tables, and charts; recognize and generalize patterns; make and test predictions; offer explanations; and apply results to new situations.

A successful problem solver sees the general structure of problems, separates meaningful data from irrelevant detail, thinks of several ways to approach the problem, and extends the problem and its solution into other areas. Thus, the problem solver must have a variety of techniques with which to attack a problem and the understanding needed to judge which techniques are likely to work and which will probably not. And since a problem is not solved until the mathematical aspects of the situation are worked through, the problem solver must not only know appropriate mathematical concepts but also be able to evaluate the acceptability of the solutions.

Finally, problem solving requires *ability*. Just as there seem to be naturally talented artists, musicians, and athletes, so there seem to be naturally talented problem solvers; these outstanding problem solvers, like the outstanding artists, musicians, or athletes, are sometimes thought to be born, not made. This may be true of a brilliant few in each field, but all people possess some measure of ability and this ability frequently goes undeveloped. Just as ordinary persons can learn to enjoy painting, piano playing, or tennis, so ordinary students can learn to enjoy mathematics. Just as many less-than-outstanding persons can, with training, become good artists, musicians, or athletes, so, too, can students become good problem solvers when they are given the opportunity to develop whatever talents they have. Likewise, even outstanding problem solvers can become better. Problem solving is not the domain of a select few but the right of all mathematics students.

Then what of Del? Is he a good problem solver by our definition? Probably not. True, he exhibited some measure of desire and enthusiasm; yet, according to his teacher, he still lacked certain kinds of facility, and we have very little indication of his ability. It would seem that Del is not yet a problem solver. Can teachers make Del and others like him into better problem solvers? How? And most important, how will teachers develop the skills they need to achieve the goal of teaching with a problem-solving focus?

"Making It Stick"

Del's characterization of a good teacher was one who could "make it stick." How does a teacher go about meeting Del's challenge? How, indeed, do mathematics teachers develop problem solvers with desire, enthusiasm, facility, and ability?

First, teachers must foster the desire to approach, accept, and try to solve problems. They must stimulate the curiosity that is in every student and direct that curiosity toward mathematical problems. Second, teachers must model enthusiasm for problem solving, showing students how determination and perseverance in attacking a problem often lead to the delight of solving it. Third, teachers must help students develop the facility in mathematical and heuristic skills that will enable them to solve new problems. Unlike Del's teacher, they should view a lack of facility not as a deterrent to problem solving but as an opportunity for encouraging new learning in order to make problem solving more possible. Finally, teachers must nurture their students' ability for problem solving by helping them find and develop their own strengths, recognizing their differences, and finding ways to challenge each class member in ways commensurate with the individual's ability and interests.

And how are these outcomes to be realized? Not by waiting for extensive curriculum changes, nor by adding puzzles and routine story problems to lessons, nor even by teaching heuristics to students. Some curriculum changes may be desirable, but problem solving can become the focus of mathematics without extensive curricular reform. Puzzles are fun and can be used for diversion, motivation, and enrichment, and story problems can be useful for practice and the reinforcement of procedures; yet neither is enough to make students better problem solvers. Heuristics are necessary to successful problem solving, but if they are isolated topics of instruction, then problems become exercises and the heuristics act as nongeneralizable algorithms.

In assessing the modern mathematics movement of the 1950s and 1960s, E. G. Begle once observed that mathematics educators had made progress toward solving the problem of teaching better mathematics but not of teaching mathematics better. The primary challenge of the 1980s continues to be to learn to teach mathematics better. Achieving the *Agenda* goal of making problem solving the focus of mathematics instruction demands significant changes in teacher behavior. The curriculum can be left largely unaltered, and existing textbooks can still be used (although, no doubt, used differently). But teachers cannot persist in employing the commonly observed repetitive routine of correcting the homework, going over a few examples, and assigning more homework; nor can they continue to represent mathematics as essentially computational, consisting of precise rules and algorithms and primarily justified by the importance of each topic for subsequent topics or courses.

Teaching with a Problem-Solving Focus

Implementing the problem-solving goal of the *Agenda* depends on the instructional strategies employed by each individual teacher. Consider Jane, a successful teacher with a problem-solving focus. At first glance, Jane's classroom looks like many other first-year algebra rooms. Her bulletin boards are colorful; one contains a section called "Problem of the Day" (fig. 1.1) and another called "Problem of the Week" (fig. 1.2). Jane uses a popular current textbook. Most days she requires homework from the students, and she is planning to include questions on the unit test requiring familiarity with common algorithms.

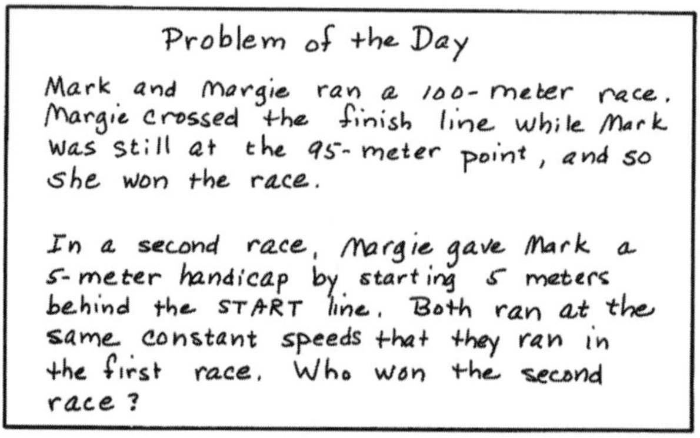

> ### Problem of the Day
>
> Mark and Margie ran a 100-meter race. Margie crossed the finish line while Mark was still at the 95-meter point, and so she won the race.
>
> In a second race, Margie gave Mark a 5-meter handicap by starting 5 meters behind the START line. Both ran at the same constant speeds that they ran in the first race. Who won the second race?

Fig. 1.1

A closer look reveals Jane's emphasis on problem solving. The bulletin boards contain not only "trick puzzles" but real problems, frequently with extraneous information or with unstated questions. Two problems are presented together, and the students are asked to find similarities and differences in the problems themselves rather than in the answers. Sometimes solutions are posted, but the challenge is to find another method of solution or to create another problem with a similar solution. Problems seem to reflect student interests, and student-produced problems are often posted. Class time is set aside to discuss solutions, solution methods, and the reasons for selecting those methods. Students are encouraged to generalize results and to investigate similar problems each time a problem is solved.

Jane uses the textbook in her teaching, but she often starts at the middle or end of a chapter and works backward. She began the present unit on linear equations by sending the students home with a variety of work problems from the end of the chapter. Since the students did not yet know the applicable algorithms, these were truly problems, not exercises as they would be later. Students were told to work as many of the problems as they were able, to classify the others, and to tell what they needed to know to solve the problems in each category. Thus they received practice in identifying common structures and in recognizing what they needed to know as well as receiving motivation for studying problems involving linear equations.

Problem of the Week

The five dart boards shown here are all squares with perimeter 48 units. In each board, the circles are congruent and tangent to each other and to the edges of the boards.

Darts are thrown without aiming, and so each dart has an equal chance of hitting anywhere on the board.

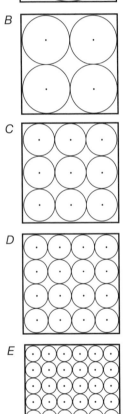

1. If a game scores one point for each dart that hits inside a circle, which board would you rather play on?

2. If you throw 100 darts, how many points do you expect to score?

3. If a game consists of 100 points, how many darts would you expect to throw?

4. If you get +1 point for each dart inside a circle and –2 points for each dart outside a circle, what score do you expect after 200 throws? How many throws do you expect to need to score 100 points?

5. If a dart hits the dot in the center of a circle, you score +100 points, but if it hits any point on the circumference of a circle, you score –10 points. Under these conditions, which board would you choose to play on?

6. In a game for three players, scoring is as follows:

   Player A: +1 for each dart inside a circle

                   0 for each dart outside a circle

   Player B: +3 for each dart outside a circle

                   0 for each dart inside a circle

   Player C: +2 for each dart inside a circle

                   –4 for each dart outside a circle

   Which player do you expect to be the winner? Does your answer depend on which dart board you choose?

7. Make up a set of scoring rules for a fair game for two persons.

Fig. 1.2

276

Today the class is conducting experiments to measure the stretch of various springs as different weights are added. The students are trying to predict results for cases they have not measured, a task involving the solution of linear equations even though they have not been given the algorithms. Later the students will play a variant of "Guess My Rule" where they already know the rule but must guess the input, given the output.

As the students become more adept at solving the equations generated through experiments, games, and textbook exercises, Jane will focus their attention on the methods they are using. Discussion will center on the different ways students are finding answers, and the class will investigate which methods are most efficient, most reliable, or easiest to use and remember. Only after this discussion will the students be told to work through the early part of the chapter where they will see the author's development of the algorithms and have a chance to practice their own.

One gets the impression while watching Jane teach that she genuinely enjoys mathematics and problem solving. She always seems excited about whatever problems her students are solving, and she usually has one of her own to work on in her spare time. One of her favorite expressions seems to be "What if . . ." and she encourages her students to ask that question often. She assigns as much small-group work as possible and promotes class discussion as the primary method of imparting information. She emphasizes the question as much as the answer and focuses on the development of the algorithms rather than on the finished product. She considers students' answers valuable in contributing to their understanding of the situation and the process. Jane's students appear secure and confident of their ability, and they ask questions freely.

Jane is not teaching problem solving. She is teaching mathematics with a problem-solving focus. She does not need a new curriculum, but she does not hesitate to rearrange the old one. And she knows how to incorporate problem solving into the curriculum.

A good teacher of problem solving possesses the same characteristics as a good problem solver. Jane, for instance, has the *desire* to find problems, generate problems, solve problems, and evaluate solutions. She shows *enthusiasm* for the entire process and for trying new approaches in focusing on problem solving. Her enthusiasm is contagious; it captures the attention of students and colleagues alike. She has the *facility* in her subject matter and instructional procedures to make her techniques work. She is well-grounded in mathematics, heuristics, and pedagogy. She knows how to communicate problem-solving strategies and how to lead her students to develop their own methods. Finally, she has the *ability* to solve problems successfully, to generalize them, and to think of several possible methods of attack. She is comfortable with her own strengths and willing to take risks.

The Role of the Teacher Educator

Jane was not born an effective teacher nor did she become one by accident. For although the *Agenda for Action* may appear to speak primarily to classroom teachers, the responsibility for implementing its goals does not end there. Teacher educators and supervisors must also share the responsibility for teaching mathematics better. Consequently, a major challenge to teacher educators and supervisors is to discover ways to help both preservice and in-service teachers become effective teachers of problem solving. This includes showing them how to add meaningful problem-solving activities to their lessons, how to decide which traditional topics can be presented in a different order or with a different emphasis, how to communicate problem-solving attitudes to their students, how to encourage their students to see the mathematics in a wide range of situations, and how to formulate their own questions and evaluate their own answers.

Teacher education programs have traditionally recognized the importance of a knowledge of mathematics, psychology, pedagogy, and aspects of general education, and these components continue to be important. However, more is needed.

First, teachers need not only the knowledge of mathematics that is represented by their ability to perform mathematical operations and procedures but also the knowledge *about* mathematics that makes it possible for them to deliver that knowledge and understanding to pupils in a meaningful way. Second, teachers must themselves become good problem solvers, but knowledge in problem-solving techniques alone will not be enough. Since teachers are teaching a process rather than a skill, they must have solved problems themselves; they must have known firsthand the satisfaction of making some progress, the frustration of encountering temporary blocks, and the delight of solving a difficult problem. They should learn to recognize mathematical aspects of seemingly nonmathematical situations and to ask questions where previously they sought only answers. Teacher education needs to provide these experiences for the teacher.

An undergraduate methods course gave Jane her start in problem solving. Problems were assigned throughout the course, and students discussed solutions and methods of solution in class. Jane learned the importance of group work and group experience in solving problems and evaluating strategies. She realized that emphasizing the process rather than the answer made students think about how they solved the problem so that the next problem seemed easier. Another activity in Jane's methods class that contributed to her success was collecting problems suitable for different age and ability levels. That set of problems formed the nucleus for the collection she uses now, but it contributed something even more valuable to her development as a teacher. In the collection process Jane discovered enough references and resources to produce usable problems for several years.

In addition to training teachers to solve problems and to find problems, teacher educators must be sure teachers know how to generate their own problems. One way to teach this skill is through assignments and discussions in methods classes or in-service programs. For example, the leader can assign group members the task of generating mathematical problems from nonmathematical items in newspapers or magazines. Then if each person is sent home with the same item, discussion can center on the number and types of problems developed; individuals can explain why they chose a particular aspect of the situation or why they presented it in a particular manner. A follow-up assignment could be to find other stories that lead to similar problems. Charles was in such a methods class that worked both with news items and with data collection and analysis of the results of laboratory experiments. He has no trouble generating questions from current events stories. Lately he has been thinking about a town he knows that holds a celebration every Friday the thirteenth. He has enough questions to keep his fifth graders busy for several days: How many such days are there this year? Does every year have a Friday the thirteenth? What is the most any one year can have? What is the longest time between two such celebrations? Is there a pattern in successive years? What happens during leap year?

Once teachers learn how to find and generate problems, they must also learn how to evaluate them in terms of their own students and the mathematics content of their courses. Again, a discussion approach is appropriate in developing evaluation skills. Sue and Brad, who teach twelfth and third grade, respectively, are in an in-service program for mathematics teachers in their district. The supervisor, a vibrant and imaginative problem solver himself, brings a variety of problems for the group each week. Sometimes everyone works on the same problem, and sometimes they group themselves according to teaching levels; but each session ends with a looking-back discussion. They consider how each problem could be used in a classroom, how it might be presented to a group or to an individual, how it could be adapted to different age or ability levels. They examine each problem to determine why and in what circumstances it is indeed a problem, and they decide with what mathematical topics it might be used. Finally, they discuss the heuristic techniques that could be exemplified by each problem and try to generalize the results or create an extension of each problem. Sue and Brad are each building a file of problems for their classes (sometimes they can each use the same problem in a different manner), and they have gained valuable skills in adapting and generalizing problems for their own use. They have also become adept at solving problems and, by watching the supervisor, have learned how to model problem-solving attitudes for their own students.

Just as discussion and group work are effective methods to use in teacher education programs, so are they effective with these teachers' own pupils. Consequently, mathematics teachers also need training in leading discussions and understanding group dynamics. Good mathematics teachers recognize the importance of creating a supportive and encouraging atmosphere, and this becomes even more important in teaching with a focus on problem solving. Students will not feel secure in attacking a problem unless they are confident that their attempts will be accepted and valued whether they discover a solution or merely refine a question. Thus, classroom atmosphere is another important topic for consideration in teacher education programs.

Even when preservice teachers have learned how to solve problems, find and generate problems, evaluate their solutions, and create a classroom atmosphere conducive to problem solving, they need, in addition, the opportunity to practice and observe these skills. Similarly, practicing teachers need to learn and share new ideas for implementing problem

279

solving. Both of these needs can be accommodated by carefully placing student teachers with cooperating teachers who are chosen for their problem-solving ability, leadership, enthusiasm, and receptiveness to new ideas. As experienced teachers learn more about teaching with a problem-solving focus, they will become better able to demonstrate this facility to their student teachers. The student teachers, in turn, will bring new ideas, problems, and strategies from their mathematics education classes, and cooperating teachers who are professionally active as leaders in their schools and districts can disseminate new information to other teachers. Thus the careful placement of student teachers and the wise planning of in-service activities could eventually facilitate the incorporation of problem solving into the mathematics teaching of many others. Only then will students begin to enjoy the full opportunity to develop their problem-solving abilities.

Conclusion

The scenario of a future in which problem solving is truly the focus of mathematics instruction is exciting and challenging. However, the degree to which the *Agenda* actually becomes *action* depends on the degree to which each mathematics teacher, teacher educator, and supervisor accepts the responsibility to create that future. There is no question that we have the facility and the ability to do this and to make it stick. The real question is, do we have the desire and the enthusiasm?

REFERENCES

National Council of Teachers of Mathematics (NCTM). *An Agenda for Action: Recommendations for School Mathematics of the 1980s*. Reston, Va.: NCTM, 1979.

———. *Curriculum and Evaluation Standards for School Mathematics*. Reston, Va.: NCTM, 1989.

National Governors Association Center for Best Practices and Council of Chief State School Officers (NGA Center and CCSSO). *Common Core State Standards for Mathematics. Common Core State Standards (College- and Career-Readiness Standards and K–12 Standards in English Language Arts and Math)*. Washington, D.C.: NGA Center and CCSSO, 2010. http://www.corestandards.org.

12

From *Learning and Teaching Geometry, K–12*, NCTM's Forty-ninth Yearbook (1987)

Mary M. Lindquist served as president of NCTM from 1992 to 1994; she was also the editor of the Council's Forty-ninth Yearbook, published in 1987. From that yearbook, she selected **chapter 17, "Geometry for Calculus Readiness,"** by Richard H. Balomenos, Joan Ferrini-Mundy, and Thomas Dick. The activities in this chapter support an expanded role for geometry in high school mathematics, one that provides readiness for calculus and develops spatial visualization. The authors note that the ability to construct pictorial representations of geometric configurations is an important component of applied problem solving involving calculus.

The chapter presents specific areas or topics of calculus that rely heavily on geometric representations. Each area is discussed using particular examples, ones that would be useful in contemporary classrooms and would most likely be accessible using technological tools. The topics here include a geometric look at the following: maximum-minimum problems, related-rate problems, volume-of-solids-of-revolution problems, and calculus problem solving in general. This selection should be of interest to teachers and teacher leaders with responsibility for courses in geometry, precalculus, and calculus.

Geometry for Calculus Readiness

Richard H. Balomenos
Joan Ferrini-Mundy
Thomas Dick

EOMETRY! Why do I need to study it?" How many times have teachers of geometry heard that question as a class is about to embark on a yearlong course in geometry? All too often the answers are dutifully given about the importance of geometry as an example of a deductive system, one of the marvels of mathematics inherited from ancient Greece. We stress the need to understand logical reasoning and its relation to axioms, theorems, and proofs, as illustrated by Euclidean geometry. In addition, many teachers view high school geometry as an isolated course having little bearing on later courses, especially college mathematics—algebra and trigonometry are seen as the courses that are central to future success in mathematics. In fact, some teachers see little relationship between the geometry experience and the development of problem-solving facility, and they let their students know it!

There is more to the answer to the question, "Why do I need to study geometry?" Evidence is growing that many of our students' difficulties in calculus are due to poor preparation in geometry. We suggest an expanded role for geometry in the high school: the study of geometry should also provide readiness for calculus and develop spatial visualization.

Geometry, Spatial Ability, and Calculus Readiness

Certain concepts of the traditional college calculus course are frequently introduced through geometric representations, and a majority of the physical problems in calculus are virtually impossible to understand without adequate visual representations. Many of the ideas needed for understanding college calculus are based on the traditional secondary school geometry course, but college calculus students are sometimes taken by surprise when somewhere in the middle of a complicated calculus problem or the explanation of a concept reference is made to similar triangles, the Pythagorean theorem, or the volume of a cylinder. It would help students if we could better prepare them for these occurrences of geometric ideas in new contexts, and this is possible in the secondary school geometry course.

In order to solve many of the applied problems traditionally included in calculus, the ability to construct a pictorial representation of some geometric configuration, based on a complicated verbal description, is essential. The typical secondary school geometry course does not impose this type of spatial demand, and as a result students greet the rather complicated figures and diagrams that they see in calculus with some consternation.

A number of the fundamental concepts in calculus, such as the definite integral, the derivative, the area between curves, the volume of solids of revolution, and the surface area of solids, frequently are introduced with heavy reliance on pictures. Despite the calculus teacher's predilection for diagrams, our research indicates that students resist the use of geometric and spatial strategies in actually solving calculus problems. The blind use of a strictly analytic formula can, of course, lead to disastrous results. For example, here is a problem that many students fail to solve correctly because of an incomplete analysis of the situation:

Find the area enclosed by the graphs of $y = x^3$ and $y = x$ between $x = -1$ and $x = 1$.

Students who evaluate the definite integral

$$\int_{-1}^{1} (x^3 - x)\,dx \quad \text{or} \quad \int_{-1}^{1} (x - x^3)\,dx$$

usually do so in the belief that the expression

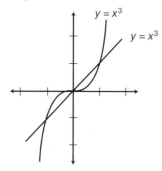

Fig. 17.1

$$\int_{a}^{b} [f(x) - g(x)]\,dx$$

will give the area enclosed by $y = f(x)$ and $y = g(x)$ between $x = a$ and $x = b$. However, this formula is valid only if $f(x) \geq g(x)$ for $a \leq x \leq b$. The student who graphs the functions involved (see fig. 17.1) is often alerted to the need of evaluating

$$\int_{-1}^{0} (x^3 - x)\,dx + \int_{0}^{1} (x - x^3)\,dx$$

as the expression giving the correct result. Without graphing, it is doubtful that a student would recognize that symmetry could also be exploited to give

$$2\int_{0}^{1} (x - x^3)\,dx$$

as an expression for this area.

Specific areas of calculus rely heavily on geometric representations. To translate ladders sliding down walls into meaningful geometric representations, or to represent sand falling onto a conical pile, students must formulate two-dimensional representations of three-dimensional dynamic situations. Following are suggestions for anticipating these areas in secondary school geometry.

Ideas for Secondary School Geometry

The secondary school geometry course should explicitly develop the capacity of students to formulate geometric representations in unfamiliar contexts and require them to use visual strategies that demand the synthesis of previously disparate geometric approaches (e.g., finding similar triangles in the cross section of an inscribed cone). We have selected four topics in calculus in which our students have particular difficulty because of the geometric and visual elements involved: maximum-minimum problems, related-rate problems, volume-of-solids-of-revolution problems, and other applied problems. In discussing each topic we attempt to highlight the ways in which these areas of calculus could be "anticipated" through specific emphases and examples that could be included in the secondary school geometry course. Calculus teachers also might find these topics well worth discussing from a geometric point of view immediately before their introduction.

A Geometric Look at Maximum-Minimum Problems

One of the more important applications of the calculus is its use in solving problems of *optimization:* under given constraints how do we maximize or minimize certain quantities? Most students appreciate this use of calculus, since they are well aware of the practical concerns of maximizing benefits for given cost or minimizing costs to realize desired benefits. However, students often view "max-min" problems as quite difficult. There are a large number of max-min problems in which the students' difficulties can be traced to inadequate geometric skills rather than inadequate calculus skills. As an example, consider the following two calculus problems:

1. Given that x lies in the closed interval $[0,3]$, find the value of x that maximizes the quantity V, where
$$V = (9 + 3x - x^2 - \frac{x^3}{3})$$

2. Find the dimensions of the right circular cone of maximum volume that can be inscribed in a sphere of radius 3 units.

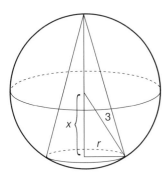

Fig. 17.2

Students generally do not find problem 1 very difficult, but many have a great deal of trouble with exercises like problem 2. The calculus used to solve problem 2 is no more difficult than that required for problem 1. Indeed, if we consider the diagram in figure 17.2 illustrating the situation described in problem 2, we see that the two problems are actually equivalent with respect to calculus. The radius r of the base of the cone satisfies the Pythagorean relationship $r^2 = 3^2 - x^2$, and the height h of the cone is $h = 3 + x$. Since the volume V of the cone is $V = \frac{1}{3}\pi r^2 h$, we next try to maximize V as a function of x.

$$V = \frac{1}{3}\pi(9 - x^2)(3 + x) = \pi(9 + 3x - x^2 - \frac{x^3}{3})$$

285

This is the same expression that appears in problem 1. Furthermore, since $0 \le x \le 3$ (cones with a height less than 3 units would certainly not give us maximum volume), the problems can be considered entirely equivalent.

Students have very little trouble applying calculus to finding the maximum of a function of one variable. Their real difficulties lie in finding the function! The crucial steps in the solution of a max-min problem are those involving the use of the constraints to rewrite the quantity to be optimized in terms of one variable. Because many of the standard max-min problems in calculus are based on geometric representations, students could benefit greatly from practice on precalculus problems that focus on the geometric strategies typically needed in the max-min exercises.

> *Calculus problem:* A sphere of radius 4 is inscribed in a right circular cone. Find the dimensions of the cone with minimum volume.

Here the crucial step is to use the fact that the inscribed sphere is of radius 4 to express the volume of the cone as a function of one variable. To focus attention on this phase of the problem, it could be rewritten for geometry students as follows.

> *Geometry problem:* A sphere of radius 4 is inscribed in a right circular cone. Express the volume of the cone as a function of its height, h.

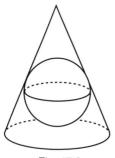

Fig. 17.3

We suggest that students be encouraged to solve problems such as this by attending carefully to their geometric components. The basic situation is illustrated in figure 17.3. Students can profit from drawing two-dimensional representations of the three-dimensional situations that arise in calculus. Geometrically, a key step in solving this problem is developing an image of the cross section. Next, students need to "disembed" the cross section and search for the relevant attributes. Recognizing that the radius $\overline{DE}$ is a useful auxiliary line segment requires both the anticipation of the use of similar triangles and the visualization of the cross section within the solid (fig. 17.4).

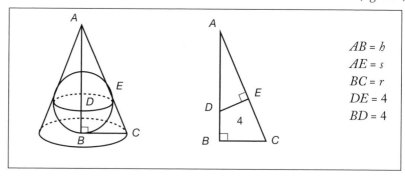

Fig. 17.4

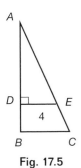

Fig. 17.5

A common student error is to visualize incorrectly and arrive at the drawing in figure 17.5. Unfortunately, ___ in this drawing is not a radius of the sphere. When the correct representation in figure 17.4 is used, $\triangle ABC \sim \triangle AED$, and hence

$$\frac{r}{4} = \frac{h}{s} = \frac{r+s}{h-4}.$$

From this it is an algebraic exercise to express r^2 as

$$\frac{16h}{h-8} \text{ and } V = \frac{1}{3}\pi\frac{16h^3}{h^2-8h}$$

A slightly different class of max-min problems is represented by the following precalculus problem; this typically appears in the calculus setting, and students assume calculus is required (fig. 17.6).

Problem: Of all triangles having two given sides, show that the triangle of maximum area is the one in which the two given sides include a right angle.

Solution: Creating a representation for this problem that suggests a method of solution requires the student to envision the various angles formed as the given sides are separated.

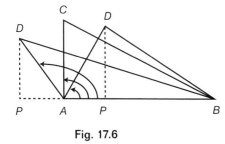

Fig. 17.6

Formulating the diagram from the verbal statement is a crucial skill in many calculus problems, as is organizing the verbal statement into what is given and what is to be proved.

Given: $AC = AD$, $AB = AB$, $\angle BAC$ is a right angle

Prove: Area of $\triangle ABC \geq$ area of $\triangle ADB$

Proof: From D we construct the perpendicular $\overline{DP}$ to line $\overleftrightarrow{AB}$. $DA > DP$, since the shortest distance from a point to a given line is the perpendicular. $DA = CA$, given. Therefore, $CA \geq DP$ and the area of $\triangle ABC \geq$ area of $\triangle ADB$, since the area of $\triangle ABC = \frac{1}{2}(AB)(CA)$ and the area of $\triangle ADB = \frac{1}{2}(AB)(DP)$.

We suggest the exercises in figure 17.7 for use with high school geometry students as a way of "anticipating" the maximum-minimum problems of calculus.

Geometry Problems Anticipating Max-Min Calculus Problems

1. Find the area of a rectangle inscribed in a semicircle as a function of its width.

2. Find the area of a rectangle of perimeter L as a function of its width.

3. Find the volume and surface area of a cube as functions of its diagonal length.

4. Prove: Of all parallelograms with given area and base, the rectangle has minimum perimeter.

5. Prove: Of all rectangles with equal area, the square has the minimum perimeter.

6. Prove: Of all triangles of a given base and altitude, the isosceles triangle has the minimum perimeter.

7. Prove: Of all triangles having a given base and a given perimeter, the isosceles triangle has maximum area.

8. Prove: Of all triangles that can be inscribed in a given circle, the equilateral triangle has both the greatest perimeter and the greatest area.

Fig. 17.7

A Geometric Look at Related-Rate Problems

In calculus, students frequently encounter problems in which one quantity is related to another and both change with time. Taking the derivative of each with respect to time yields a relation between their rates of change. Such problems are called related-rate problems. Setting up and solving related-rate problems is difficult for calculus students. It is the ability to develop a geometric representation of the physical situation from the complicated verbal description that is the real challenge. In many related-rate problems, the key to the solution is solving a geometry problem in which time is "frozen." Consider the following example:

> A train is traveling at 60 miles an hour (or 88 feet a second) on a track 50 feet below a long, level highway bridge. A car traveling at 45 miles an hour (or 66 feet a second) passes directly over the train.

The calculus problem is "How fast are the train and car separating 10 seconds later?" The geometry problem is "How far apart are the train and car 10 seconds later?" The solution in either case requires the use of the Pythagorean theorem in three dimensions. If we freeze the action at 10 seconds after the car passes directly over the train, the car will have traveled 660 feet and the train 880 feet (fig. 17.8).

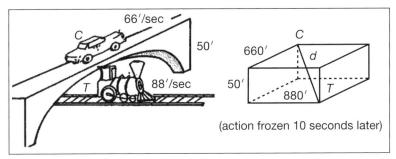

Fig. 17.8

$$d = \sqrt{660^2 + 880^2 + 50^2} = 10\sqrt{11^2 \cdot 6^2 + 11^2 \cdot 8^2 + 5^2}$$
$$= 10\sqrt{11^2(36+64)+25} = 10\sqrt{25(11^2 \cdot 4+1)}$$
$$= 50\sqrt{485} \approx 1100$$

Thus the car and train will be approximately 1100 feet apart.

The spatial demands of such a problem are considerable, and when the student is also struggling with the calculus concepts involved, the situation can be overwhelming. By focusing only on the geometry of the problem, we can encourage the student's facility in finding a visual representation, and the task of finding the appropriate right triangle in this three-dimensional representation can be dealt with separately.

The following is a calculus problem in which the properties of similar triangles are used in the solution.

Water is pouring into a conical cistern at the rate of 8 cubic feet per minute. If the height of the inverted cone is 12 feet and the radius of its circular opening is 6 feet, how fast is the water level rising when the water is 4 feet deep? (Purcell 1978, p. 160)

By "freezing the action," we convert the problem into a geometry problem: What is the volume of the water when the water is 4 feet deep? The formula for the volume of water is

$$V = \frac{1}{3}\pi r^2 h.$$

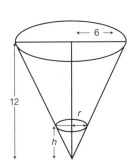

Fig. 17.9

The geometry question is to find V when $h = 4$. From similar triangles (see fig. 17.9):

$$\frac{r}{h} = \frac{6}{12}, \text{ or } r = \frac{1}{2}h \qquad V = \frac{1}{3}\pi(\frac{h}{2})^2 h = \frac{\pi h^3}{12}$$

Therefore, when $h = 4$, $V = \frac{\pi 4^3}{12} = \frac{16\pi}{3}$ cubic feet.

Notice that here again the student must visualize the cross section of the cone, label it appropriately in terms of given relationships, and choose the appropriate similar triangles.

In figure 17.10 are some geometry exercises adapted from related-rate problems as found in any standard calculus textbook; we have used texts by Swokowski (1984) and Anton (1984).

Geometry Problems from Related-Rate Problems in Calculus

1. A ladder 20 ft long leans against a vertical building. If the bottom of the ladder slides away from the building horizontally at a rate of 3 ft/sec, how far is the bottom of the ladder from the building when the top of the ladder is 8 ft from the ground? [Swokowski]

2. At 1:00 p.m., ship A is 25 mi due south of ship B. If ship A is sailing west at a rate of 16 mi/hr and ship B is sailing south at a rate of 20 mi/hr, what is the distance between the ships at 1:30 p.m.? [Swokowski]

3. A girl starts at a point A and runs east at a rate of 10 ft/sec. One minute later, another girl starts at A and runs north at a rate of 8 ft/sec. What is the distance between them 1 minute after the second girl starts? [Swokowski]

4. The ends of a water trough 8 ft long are equilateral triangles whose sides are 2 ft long. Find the volume of the water in the trough when the depth is 8 in. [Swokowski]

5. A softball diamond has the shape of a square with sides 60 ft long. If a player is running from second to third, what is her distance from home plate when she is 20 ft from third? [Swokowski]

6. The top part of a swimming pool is a rectangle of length 60 ft and width 30 ft. The depth of the pool varies uniformly from 4 ft to 9 ft through a horizontal distance of 40 ft and then is level for the remaining 20 ft. If the pool is being filled with water, find the number of gallons of water in the pool when the depth at the deep end is 4 ft. (1 gal is approximately 0.1337 ft^3). [Swokowski]

7. A stone dropped into a still pond sends out a circular ripple whose radius increases at a constant rate of 3 ft/sec. What is the area enclosed by the ripple after 10 sec? [Anton]

8. Grain pouring from a chute at the rate of $8 \text{ ft}^3/\text{min}$ forms a conical pile whose altitude is always twice its radius. What is the volume of grain when the pile is 6 ft high? [Anton]

9. A police helicopter is flying due south at 100 mi/hr, and at a constant altitude of $1/_2$ mi. Below, a car is traveling west on a highway at 75 mi/hr. At the moment the helicopter crosses over the highway the car is 2 mi east of the helicopter. What is the distance between the car and helicopter 30 minutes later? [Anton]

10. Coffee is poured at a uniform rate of $2 \text{ cm}^3/\text{sec}$ into a cup whose bowl is shaped like a truncated cone. If the upper and lower radii of the cup are 4 cm and 2 cm and the height of the cup is 6 cm, what is the volume of the coffee when the coffee is halfway up? [Anton]

Fig. 17.10

A Geometric Look at Volume-of-Solids-of-Revolution Problems

Calculus students have great difficulty with problems requiring them to find the volumes of solids of revolution. On a multiple-choice final examination problem asking students to set up the integral for a routine volume-of-solid-of-revolution problem, the average rate of correct responses over the past several semesters at this university has been 25 percent. Among college freshmen, the most commonly made errors include revolving the region about an axis other than the correct one, attempting to sketch the solid of revolution using an incorrect reflection of the region through the axis of revolution, and being unable to choose appropriately between the "disk" method and the "cylindrical shell" method because of an inadequate sketch (Mundy 1981).

Understanding the derivation of volume-of-solid-of-revolution techniques and then solving problems by applying these techniques in a meaningful way taps a varied set of spatial-geometric skills. Consider the following problem, typical of those found in all calculus textbooks:

Sketch the region R bounded by the graphs of the given equations and find the volume of the solid generated by revolving R about the indicated axis:

$$x = y^3, \ x^2 + y = 0; \text{ about the } x\text{-axis.}$$

To approach this problem successfully, the student must

- represent the information provided graphically and determine the region to be revolved;

- sketch or visualize the solid of revolution and choose the appropriate method ("disks or washers" or "cylindrical shells");

- sketch or visualize a cross section of the solid, revolve a "representative slice" to form a cross section, or place a representative shell correctly;

- express the representative volume in terms of the given relationships and finally set up an integral to compute the volume.

Only the very last of these steps actually involves a calculus concept. All the other steps require visualization and graphical representations that can be anticipated in the secondary school geometry curriculum.

Activities can be devised for geometry students that will ready them for the several phases of the volume-of-solids-of-revolution problems they will encounter later on. For example, in preparing students to sketch revolved solids, teachers can introduce exercises involving the reflection of regions in lines (see fig. 17.11).

Sketch the reflection of the given region in the given line.

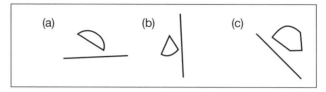

Fig. 17.11

The concept of a solid of revolution can be introduced in the context of the geometry course, ideally with the assistance of computer graphics that show the solid as it is swept out. An inexpensive substitute for such graphics are the crepe paper fold-out decorations available in card and gift shops. Wedding bells, Thanksgiving turkeys, Halloween witches' hats, and snowmen can be amazingly useful in demonstrating the concept of a solid of revolution. Asking students to sketch solids of revolution can follow the examples where the students are sketching reflections. Multiple choice problems like the one in figure 17.12 are an interesting means of approaching this phase.

Which of the following is the correct solid of revolution?

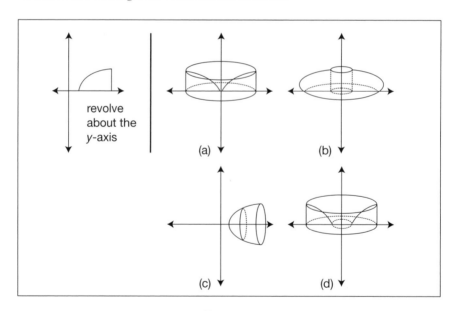

Fig. 17.12

Once students are comfortable with the concept of a solid of revolution, it is possible to combine their new understanding with their previous experience in geometry in finding volumes of solids. Examples such as the following anticipate the volume-of-solids-of-revolution problems to be encountered in calculus but can be solved by the secondary school geometry student:

Sketch and find the volume of the solid formed by revolving about the x-axis the region bounded by $y = x$, $y = (4/3)x - 2$, and the x-axis.

Solution:

volume   = "outer" volume − "inner" volume

$$= (\frac{1}{3})\pi 6^2 \cdot 6 - (\frac{1}{3})\pi 6^2 \cdot (4\frac{1}{2})$$

$$= 18\pi$$

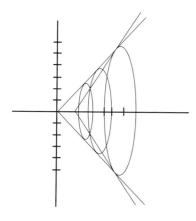

Fig. 17.13

Once the sketch is obtained (see fig. 17.13), it is clear that this volume can be computed using the formula for the volume of a cone; the volume of the solid is the volume of the outer cone minus the volume of the inner cone. The student does several of the same steps that are involved in a calculus volume-of-a-solid-of-revolution problem and must obtain an accurate visual representation of the problem before it is possible to obtain the volume using noncalculus techniques.

In calculus, the correct solution of volume-of-solids-of-revolution problems using "disks or washers" depends on the student's ability to visualize the representative cross section of the solid and to compute the volume of this representative cross section. Sketching the cross sections and labeling their dimensions in a solid-of-revolution problem is a noncalculus activity that can pose considerable problems for students. The final phases of the solution process also can be anticipated by appropriate exercises at the secondary level.

In figure 17.14 are several exercises that could be used at the secondary school geometry level to prepare students for the volume-of-solids-of-revolution problems they will encounter in calculus.

Solids of Revolution

For each of the following regions and for each of the indicated axes, (*a*) sketch the solid of revolution, (*b*) find the volume of the solid of revolution, and (*c*) sketch a cross section of the solid of revolution and label its dimensions for a given *x* or *y*.

1.

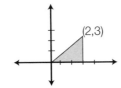

(2,3)

a. Revolve about X-axis
b. Revolve about Y-axis

2.

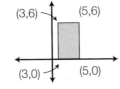

(3,6)   (5,6)
(3,0)   (5,0)

a. Revolve about X-axis
b. Revolve about Y-axis

3.

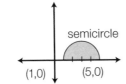

semicircle
(1,0)   (5,0)

a. Revolve about X-axis
b. Revolve about Y = –6
(Requires volume of a torus formula)

4.

4 = 2x   x = 3
4 = 2x = 2

a. Revolve about X-axis
b. Revolve about Y-axis

5.

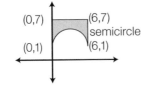

(0,7)   (6,7)
semicircle
(0,1)   (6,1)

a. Revolve about X-axis
b. Revolve about Y-axis

6.

(0,6)
(0,4)   (3,4)
1 2 3

a. Revolve about X-axis
b. Revolve about Y-axis

7.

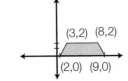

(3,2) (8,2)
(2,0) (9,0)

a. Revolve about X-axis
b. Revolve about X = 1

8.

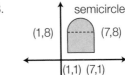

semicircle
(1,8)   (7,8)
(1,1) (7,1)

a. Revolve about Y-axis
b. Revolve about X = 1

Fig. 17.14

A Geometric Look at Calculus Problem Solving

Many physical applications of calculus involve definite integration. Two of the many applications discussed in any standard calculus course are determining the work performed or the force exerted by liquids. As with optimization and related-rates problems, students' difficulties with solving these types of applied problems do not usually stem from the difficulty of the calculus involved. Indeed, the actual integrals encountered are often very simple to evaluate. Rather, it is finding the appropriate integrand and limits of integration that poses the greatest difficulty for the student; this in turn depends on the accuracy and detail of the student's representation of the problem situation. The real "problem" is often one of geometry, not calculus. Consider an example:

> A sheet of metal in the shape of an equilateral triangle of side 3 feet is completely submerged vertically in oil having a density of 50 pounds per cubic foot. The triangle is positioned so that one side is parallel to and 2 feet below the surface of the oil and the opposite vertex points down (see fig. 17.15). What is the force exerted by the oil on one face of the triangle?

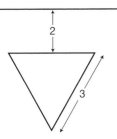

Fig. 17.15

In solving this problem, we must describe the width of the triangle as a function of depth and the range of depths the triangle occupies. In fact, the force exerted on one face of any plane region similarly submerged in liquid of density D pounds per cubic feet is given by the integral

$$D \int_a^b x \cdot w(x)\,dx$$

where x is the depth in feet, $w(x)$ is the width of the region at depth x, and a and b are the shallowest and deepest levels (in feet) respectively of the region.

Finding the function $w(x)$ is an exercise that requires the use of both the Pythagorean theorem and similar triangles. Stated specifically as a geometry problem, the problem above is written thus:

> In figure 17.16, $\triangle ABC$ is an equilateral triangle with sides of length 3. Segment DE is parallel to segment AB. P and Q are the midpoints of AB and DE respectively. If PQ has length h, find the length w of DE in terms of h.

Since PC has length $3\sqrt{3/2}$ (by use of the Pythagorean theorem on right triangle BPC) and since $\triangle DEC$ and $\triangle ABC$ are similar, the following relationship holds for w and h:

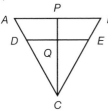

Fig. 17.16

$$\frac{3\sqrt{3}/2 - h}{w} = \frac{\sqrt{3}}{2}$$

Thus $w = 3 - (2\sqrt{3}/3)h$. Applying this result to the previous calculus problem, and noting that $h = (x - 2)$, yields
$$w(x) = 3 + (4\sqrt{3} - 2\sqrt{3}x)/3,\ 2 \le x \le 2 + 3\sqrt{3}/2$$

295

This problem points out a very different way of viewing geometric figures. Instead of being simply a static set of points satisfying one or more properties, a geometric figure can be thought of as defining a dynamic functional relationship between its dimensions. In a very real sense, the function $w(x)$ above "defines" the submerged triangle. We might even graph the function $w(x)$ (see fig. 17.17).

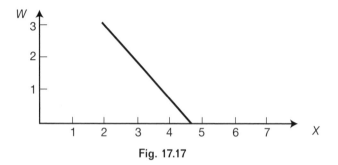

Fig. 17.17

It is interesting to note the different width functions that define different plane regions submerged in a similar manner. For example, if the region is a rectangle with one side parallel to the surface, then $w(x)$ is a constant function over an appropriate interval. If, however, the region is a circle, then the width function involves the square root of a quadratic expression (see the illustrations in fig. 17.18).

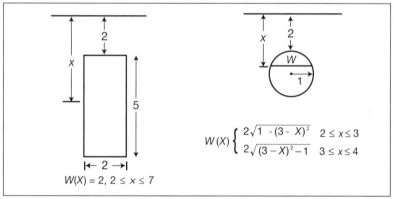

$$W(X) = 2, \; 2 \leq x \leq 7$$

$$W(X) \begin{cases} 2\sqrt{1 - (3 - X)^2} & 2 \leq x \leq 3 \\ 2\sqrt{(3 - X)^2 - 1} & 3 \leq x \leq 4 \end{cases}$$

Fig. 17.18

In figure 17.19 are some exercises that enable high school students to focus on the geometric aspects inherent in many applied calculus problems involving definite integrals.

Geometric Aspects of Definite Integration

1. Find the volume of a right circular cone inscribed in (or circumscribed about) a sphere of radius a as a function of its radius; its height.

2. Find the volume of a right circular cylinder inscribed in a sphere of radius a as a function of its radius; its height.

Find the "width function" that defines—

3. an isosceles triangle of height a and base b;

4. a scalene triangle with sides a, b, c.

Find the "cross-sectional area function" that defines—

5. a right circular cone of radius r and height h;

6. a hemisphere of radius r.

Fig. 17.19

Conclusion

Geometry has a central role to play in the high school curriculum, particularly the formal geometry course with its emphasis on proof. In this article, it is suggested that there are other roles for geometry—it is especially suited for providing calculus readiness. The ideas presented are only a beginning, and any standard calculus textbook is a source for additional problems. These suggestions may be integrated into an existing tenth-grade geometry course or used as part of a half-year, senior-level course in mathematical problem solving.

The approach suggested in this article may be more useful for most students in high school than an introduction to polynomial calculus.

BIBLIOGRAPHY

Anton, Howard. *Calculus with Analytic Geometry.* New York: John Wiley & Sons, 1984.

Lean, G., and Ken Clements. "Spatial Ability, Visual Imagery, and Mathematical Performance." *Educational Studies in Mathematics* 12 (1981): 267–99.

Meserve, Bruce. "Geometry as a Gateway to Mathematics." In *Developments in Mathematical Education: Proceedings of the Second International Congress on Mathematical Education*, edited by A. G. Howson. Cambridge: Cambridge University Press, 1973.

Mundy, Joan F. "Spatial Ability, Mathematics Achievement, and Spatial Training in Male and Female Calculus Students." (Doctoral dissertation, University of New Hampshire, 1980). *Dissertation Abstracts International* 41 (1981): 4633A (University Microfilms No. 8108871).

Purcell, Edwin J. *Calculus with Analytic Geometry.* Englewood Cliffs, N.J.: Prentice-Hall, 1978.

Swokowski, Earl W. *Calculus with Analytic Geometry.* Boston: Prindle, Weber & Schmidt, 1984.

13

From *The Ideas of Algebra, K–12,* NCTM's Fiftieth Yearbook (1988)

Gail Burrill, NCTM president from 1996 to 1998, recommended Zalman Usiskin's chapter, **"Conceptions of School Algebra and Uses of Variables,"** published in 1988. She was not alone in her choice. Past presidents John Dossey (1986–88) and Shirley Frye (1988–90) also recommended this selection.

This work appeared as chapter 2 in *The Ideas of Algebra, K–12,* the Council's Fiftieth Yearbook, which focused specifically on the subject of algebra across the grades. Usiskin's chapter presents different conceptions of algebra, which relate to the different levels of import given to the use of variables. These include algebra as generalized arithmetic, as a study of procedures for solving certain kinds of problems, as the study of relationships among quantities, and as the study of structures. With the attention paid to algebra and its importance, however defined—as a course or courses or as essential standards within a K–12 mathematics curriculum—the continued popularity and importance of this chapter is not a surprise.

Conceptions of School Algebra and Uses of Variables

Zalman Usiskin

What Is School Algebra?

ALGEBRA is not easily defined. The algebra taught in school has quite a different cast from the algebra taught to mathematics majors. Two mathematicians whose writings have greatly influenced algebra instruction at the college level, Saunders Mac Lane and Garrett Birkhoff (1967), begin their *Algebra* with an attempt to bridge school and university algebras:

> Algebra starts as the art of manipulating sums, products, and powers of numbers. The rules for these manipulations hold for all numbers, so the manipulations may be carried out with letters standing for the numbers. It then appears that the same rules hold for various different sorts of numbers . . . and that the rules even apply to things . . . which are not numbers at all. An algebraic system, as we will study it, is thus a set of elements of any sort on which functions such as addition and multiplication operate, provided only that these operations satisfy certain basic rules. (P. 1)

If the first sentence in the quote above is thought of as arithmetic, then the second sentence is school algebra. For the purposes of this article, then, school algebra has to do with the understanding of "letters" (today we usually call them *variables)* and their operations, and we consider students to be studying algebra when they first encounter variables.

However, since the concept of variable itself is multifaceted, reducing algebra to the study of variables does not answer the question "What is school algebra?" Consider these equations, all of which have the same form—the product of two numbers equals a third:

1. $A = LW$
2. $40 = 5x$
3. $\sin x = \cos x \cdot \tan x$
4. $1 = n \cdot (1/n)$
5. $y = kx$

Each of these has a different feel. We usually call (1) a formula, (2) an equation (or open sentence) to solve, (3) an identity, (4) a property, and (5) an equation of a function of direct variation (not to be solved). These different names reflect different uses to which the idea of variable is put. In (1), A, L, and W stand for the quantities area, length, and width and have the feel of knowns. In (2), we tend to think of x as unknown. In (3), x is an argument of a function. Equation (4), unlike the others, generalizes an arithmetic pattern, and n identifies an instance of the pattern. In (5), x is again an argument of a function, y the value, and k a constant (or parameter, depending on how it is used). Only with (5) is there the feel of "variability," from which the term *variable* arose. Even so, no such feel is present if we think of that equation as representing the line with slope k containing the origin.

Conceptions of variable change over time. In a text of the 1950s (Hart 1951a), the word *variable* is not mentioned until the discussion of systems (p. 168), and then it is described as "a changing number." The introduction of what we today call variables comes much earlier (p. 11), through formulas, with these cryptic statements: "In each formula, the letters represent numbers. *Use of letters to represent numbers is a principal characteristic of algebra*" (Hart's italics). In the second book in that series (Hart 1951b), there is a more formal definition of variable (p. 91): "A variable is a literal number that may have two or more values during a particular discussion."

Modern texts in the late part of that decade had a different conception, represented by this quote from May and Van Engen (1959) as part of a careful analysis of this term:

> Roughly speaking, a variable is a symbol for which one substitutes names for some objects, usually a number in algebra. A variable is always associated with a set of objects whose names can be substituted for it. These objects are called values of the variable. (P. 70)

Today the tendency is to avoid the "name-object" distinction and to think of a variable simply as a symbol for which things (most accurately, things from a particular replacement set) can be substituted.

The "symbol for an element of a replacement set" conception of variable seems so natural today that it is seldom questioned. However, it is not the only view possible for variables. In the early part of this century, the formalist school of mathematics considered variables and all other mathematics symbols merely as marks on paper related to each other by assumed or derived properties that are also marks on paper (Kramer 1981).

Although we might consider such a view tenable to philosophers but impractical to users of mathematics, present-day computer algebras such as MACSYMA and muMath (see Pavelle, Rothstein, and Fitch [1981]) deal with letters without any need to refer to numerical values. That is, today's computers can operate as both experienced and inexperienced users of algebra do operate, blindly manipulating variables without any concern for, or knowledge of, what they represent.

Many students think all variables are letters that stand for numbers. Yet the values a variable takes are not always numbers, even in high school mathematics. In geometry, variables often represent points, as seen by the use of the variables A, B, and C when we write "if $AB = BC$, then $\triangle ABC$ is isosceles." In logic, the variables p and q often stand for propositions; in analysis, the variable f often stands for a function; in linear algebra the variable A may stand for a matrix, or the variable $\mathbf{v}$ for a vector, and in higher algebra the variable $*$ may represent an operation. The last of these demonstrates that variables need not be represented by letters.

Students also tend to believe that a variable is always a letter. This view is supported by many educators, for

$$3 + x = 7 \text{ and } 3 + \Delta = 7$$

are usually considered algebra, whereas

$$3 + \_\_\_ = 7 \text{ and } 3 + ? = 7$$

are not, even though the blank and the question mark are, in this context of desiring a solution to an equation, logically equivalent to the x and the Δ.

In summary, variables have many possible definitions, referents, and symbols. Trying to fit the idea of variable into a single conception oversimplifies the idea and in turn distorts the purposes of algebra.

Two Fundamental Issues in Algebra Instruction

Perhaps the major issue surrounding the teaching of algebra in schools today regards the extent to which students should be required to be able to do various manipulative skills by hand. (Everyone seems to acknowledge the importance of students having *some* way of doing the skills.) A 1977 NCTM-MAA report detailing what students need to learn in high school mathematics emphasizes the importance of learning and practicing these skills. Yet more recent reports convey a different tone:

> The basic thrust in Algebra I and II has been to give students moderate technical facility. . . . In the future, students (and adults) may not have to do much algebraic manipulation. . . . Some blocks of traditional drill can surely be curtailed. (CBMS 1983, p. 4)

A second issue relating to the algebra curriculum is the question of the role of functions and the timing of their introduction. Currently, functions are treated in most first-year algebra books as a relatively insignificant topic and first become a major topic in advanced or second-year algebra. Yet in some elementary school curricula (e.g., CSMP 1975) function ideas have been introduced as early as first grade, and others have argued that functions should be used as the major vehicle through which variables and algebra are introduced.

It is clear that these two issues relate to the very purposes for teaching and learning algebra, to the goals of algebra instruction, to the conceptions we have of this body of subject matter. What is not as obvious is that they relate to the ways in which variables are used. In this paper I try to present a framework for considering these and other issues relating to the teaching of algebra. My thesis is that the purposes we have for teaching algebra, the conceptions we have of the subject, and the uses of variables are inextricably related. ***Purposes for algebra*** *are determined by, or are related to,* ***different conceptions of algebra***, *which correlate with the different relative importance given to various* ***uses of variables***.

303

Conception 1: Algebra as generalized arithmetic

In this conception, it is natural to think of variables as pattern generalizers. For instance, $3 + 5.7 = 5.7 + 3$ is generalized as $a + b = b + a$. The pattern

$$3 \cdot 5 = 15$$
$$2 \cdot 5 = 10$$
$$1 \cdot 5 = 5$$
$$0 \cdot 5 = 0$$

is extended to give multiplication by negatives (which, in this conception, is often considered algebra, not arithmetic):

$$-1 \cdot 5 = -5$$
$$-2 \cdot 5 = -10$$

This idea is generalized to give properties such as

$$-x \cdot y = -xy.$$

At a more advanced level, the notion of variable as pattern generalizer is fundamental in mathematical modeling. We often find relations between numbers that we wish to describe mathematically, and variables are exceedingly useful tools in that description. For instance, the world record T (in seconds) for the mile run in the year Y since 1900 is rather closely described by the equation

$$T = -0.4Y + 1020.$$

This equation merely generalizes the arithmetic values found in many almanacs. In 1974, when the record was 3 minutes 51.1 seconds and had not changed in seven years, I used this equation to predict that in 1985 the record would be 3 minutes 46 seconds (for graphs, see Usiskin [1976] or Bushaw et al. [1980]). The actual record at the end of 1985 was 3 minutes 46.31 seconds.

The key instructions for the student in this conception of algebra are *translate* and *generalize*. These are important skills not only for algebra but also for arithmetic. In a compendium of applications of arithmetic (Usiskin and Bell 1984), Max Bell and I concluded that it is impossible to adequately study arithmetic without implicitly or explicitly dealing with variables. Which is easier, "The product of any number and zero is zero" or "For all n, $n \cdot 0 = 0$"? The superiority of algebraic over English language descriptions of number relationships is due to the similarity of the two syntaxes. The algebraic description looks like the numerical description; the English description does not. A reader in doubt of the value of variables should try to describe the rule for multiplying fractions first in English, then in algebra.

Historically, the invention of algebraic notation in 1564 by François Viète (1969) had immediate effects. Within fifty years, analytic geometry had been invented and brought to an advanced form. Within a hundred years, there was calculus. Such is the power of algebra as generalized arithmetic.

304

Conception 2: Algebra as a study of procedures for solving certain kinds of problems

Consider the following problem:

When 3 is added to 5 times a certain number, the sum is 40. Find the number.

The problem is easily translated into the language of algebra:

$$5x + 3 = 40$$

Under the conception of algebra as a generalizer of patterns, we do not have unknowns. We generalize known relationships among numbers, and so we do not have even the feeling of unknowns. Under that conception, this problem is finished; we have found the general pattern. However, under the conception of algebra as a study of procedures, we have only begun.

We solve with a procedure. Perhaps add –3 to each side:

$$5x + 3 + -3 = 40 + -3$$

Then simplify (the number of steps required depends on the level of student and preference of the teacher):

$$5x = 37$$

Now solve this equation in some way, arriving at $x = 7.4$. The "certain number" in the problem is 7.4, and the result is easily checked.

In solving these kinds of problems, many students have difficulty moving from arithmetic to algebra. Whereas the arithmetic solution ("in your head") involves subtracting 3 and dividing by 5, the algebraic form $5x + 3$ involves multiplication by 5 and addition of 3, the inverse operations. That is, to set up the equation, you must think precisely the opposite of the way you would solve it using arithmetic.

In this conception of algebra, variables are either *unknowns* or *constants*. Whereas the key instructions in the use of a variable as a pattern generalizer are translate and generalize, the key instructions in this use are *simplify* and *solve*. In fact, "simplify" and" solve" are sometimes two different names for the same idea: For example, we ask students to solve $|x - 2| = 5$ to get the answer $x = 7$ or $x = -3$. But we could ask students, "Rewrite $|x - 2| = 5$ without using absolute value." We might then get the answer $(x - 2)^2 = 25$, which is another equivalent sentence.

Pólya (1957) wrote, "If you cannot solve the proposed problem try to solve first some related problem" (p. 31). We follow that dictum literally in solving most sentences, finding equivalent sentences with the same solution. We also simplify expressions so that they can more easily be understood and used. To repeat: simplifying and solving are more similar than they are usually made out to be.

Conception 3: Algebra as the study of relationships among quantities

When we write $A = LW$, the area formula for a rectangle, we are describing a relationship among three quantities. There is not the feel of an unknown, because we are not solving for anything. The feel of formulas such as $A = LW$ is different from the feel of generalizations such as $1 = n \cdot (1/n)$, even though we can think of a formula as a special type of generalization.

Whereas the conception of algebra as the study of relationships may begin with formulas, the crucial distinction between this and the previous conceptions is that, here, variables *vary*. That there is a fundamental difference between the conceptions is evidenced by the usual response of students to the following question:

What happens to the value of $1/x$ as x gets larger and larger?

The question seems simple, but it is enough to baffle most students. We have not asked for a value of x, so x is not an unknown. We have not asked the student to translate. There is a pattern to generalize, but it is not a pattern that looks like arithmetic. (It is not appropriate to ask what happens to the value of $1/2$ as 2 gets larger and larger!) It is fundamentally an algebraic pattern. Perhaps because of its intrinsic algebraic nature, some mathematics educators believe that algebra should first be introduced through this use of variable. For instance, Fey and Good (1985) see the following as the key questions on which to base the study of algebra:

For a given function $f(x)$, find—
1. $f(x)$ for $x = a$;
2. x so that $f(x) = a$;
3. x so that maximum or minimum values of $f(x)$ occur;
4. the rate of change in f near $x = a$;
5. the average value of f over the interval (a,b). (P. 48)

Under this conception, a variable is an *argument* (i.e., stands for a domain value of a function) or a *parameter* (i.e., stands for a number on which other numbers depend). Only in this conception do the notions of independent variable and dependent variable exist. Functions arise rather immediately, for we need to have a name for values that depend on the argument or parameter x. Function notation (as in $f(x) = 3x + 5$) is a new idea when students first see it: $f(x) = 3x + 5$ looks and feels different from $y = 3x + 5$. (In this regard, one reason $y = f(x)$ may confuse students is because the function f, rather than the argument x, has become the parameter. Indeed, the use of $f(x)$ to name a function, as Fey and Good do in the quote above, is seen by some educators as contributing to that confusion.)

That variables as arguments differ from variables as unknowns is further evidenced by the following question:

Find an equation for the line through (6,2) with slope 11.

The usual solution combines all the uses of variables discussed so far, perhaps explaining why some students have difficulty with it. Let us analyze the usual solution. We begin by noting that points on a line are related by an equation of the form

$$y = mx + b.$$

This is both a pattern among variables and a formula. In our minds it is a function with domain variable x and range variable y, but to students it is not clear which of m, x, or b is the argument. As a pattern it is easy to understand, but in the context of this problem, some things are unknown. All the letters look like unknowns (particularly the x and y, letters traditionally used for that purpose).

Now to the solution. Since we know m, we substitute for it:

$$y = 11x + b$$

Thus m is here a constant, not a parameter. Now we need to find b. Thus b has changed from parameter to unknown. But how to find b? We use one pair of the many pairs in the relationship between x and y. That is, we select a value for the argument x for which we know y. Having to substitute a pair of values for x and y can be done because $y = mx + b$ describes a general pattern among numbers. With substitution,

$$2 = 11 \cdot 6 + b,$$

and so $b = -64$. But we haven't found x and y even though we have values for them, because they were not unknowns. We have only found the unknown b, and we substitute in the appropriate equation to get the answer

$$y = 11x - 64.$$

Another way to make the distinction between the different uses of the variables in this problem is to engage quantifiers. We think: For all x and y, there exist m and b with $y = mx + b$. We are given the value that exists for m, so we find the value that exists for b by using one of the "for all x and y" pairs, and so on. Or we use the equivalent set language: We know the line is $\{(x,y): y = mx + b\}$ and we know m and try to find b. In the language of sets or quantifiers, x and y are known as dummy variables because any symbols could be used in their stead. It is rather hard to convince students and even some teachers that $\{x: 3x = 6\} = \{y: 3y = 6\}$, even though each set is $\{2\}$. Many people think that the function with $f(x) = x + 1$ is not the same as the function g with the same domain as f and with $g(y) = y + 1$. Only when variables are used as arguments may they be considered as dummy variables; this special use tends to be not well understood by students.

Conception 4: Algebra as the study of structures

The study of algebra at the college level involves structures such as groups, rings, integral domains, fields, and vector spaces. It seems to bear little resemblance to the study of algebra at the high school level, although the fields of real numbers and complex numbers and the various rings of polynomials underlie the theory of algebra, and properties of integral domains and groups explain why certain equations can be solved and others not. Yet we recognize algebra as the study of structures by the properties we ascribe to operations on real numbers and polynomials. Consider the following problem:

Factor $3x^2 + 4ax - 132a^2$.

The conception of variable represented here is not the same as any previously discussed. There is no function or relation; the variable is not an argument. There is no equation to be solved, so the variable is not acting as an unknown. There is no arithmetic pattern to generalize.

The answer to the factoring question is $(3x + 22a)(x - 6a)$. The answer could be checked by substituting values for x and a in the given polynomial and in the factored answer, but this is almost never done. If factoring were checked that way, there would be a wisp of an argument that here we are generalizing arithmetic. But in fact, the student is usually asked to check by multiplying the binomials, exactly the same procedure that the student has employed to get the answer in the first place. It is silly to check by repeating the process used to get the answer in the first place, but in this kind of problem students tend to treat the variables as marks on paper, without numbers as a referent. In the conception of algebra as the study of structures, the variable is little more than an arbitrary symbol. There is a subtle quandary here. We want students to have the referents (usually real numbers) for variables in mind as they use the variables. But we also want students to be able to operate on the variables without always having to go to the level of the referent. For instance, when we ask students to derive a trigonometric identity such as $2\sin^2 x - 1 = \sin^4 x - \cos^4 x$, we do not want the student to think of the sine or cosine of a specific number or even to think of the sine or cosine functions, and we are not interested in ratios in triangles. We merely want to manipulate $\sin x$ and $\cos x$ into a different form using properties that are just as abstract as the identity we wish to derive.

In these kinds of problems, faith is placed in properties of the variables, in relationships between x's and y's and n's, be they addends, factors, bases, or exponents. The variable has become an arbitrary object in a structure related by certain properties. It is the view of variable found in abstract algebra.

Much criticism has been leveled against the practice by which "symbol pushing" dominates early experiences with algebra. We call it "blind" manipulation when we criticize; "automatic" skills when we praise. Ultimately everyone desires that students have enough facility with algebraic symbols to deal with the appropriate skills abstractly. The key question is, What constitutes "enough facility"?

It is ironic that the two manifestations of this use of variable—theory and manipulation—are often viewed as opposite camps in the setting of policy toward the algebra curriculum, those who favor manipulation on one side, those who favor theory on the other. They come from the same view of variable.

Variables in Computer Science

Algebra has a slightly different cast in computer science from what it has in mathematics. There is often a different syntax. Whereas in ordinary algebra, $x = x + 2$ suggests an equation with no solution, in BASIC the same sentence conveys the replacement of a particular storage location in a computer by a number two greater. This use of variable has been identified by Davis, Jockusch, and McKnight (1978, p. 33):

> Computers give us another view of the basic mathematical concept of variable. From a computer point of view, the name of a variable can be thought of as the address of some specific memory register, and the value of the variable can be thought of as the contents of this memory register.

In computer science, variables are often identified strings of letters and numbers. This conveys a different feel and is the natural result of a different setting for variable. Computer applications tend to involve large numbers of variables that may stand for many different kinds of objects. Also, computers are programmed to manipulate the variables, so we do not have to abbreviate them for the purpose of easing the task of blind manipulation.

In computer science the uses of variables cover all the uses we have described above for variables. There is still the generalizing of arithmetic. The study of algorithms is a study of procedures. In fact, there are typical algebra questions that lend themselves to algorithmic thinking:

> Begin with a number. Add 3 to it. Multiply it by 2. Subtract 11 from the result. . . .

In programming, one learns to consider the variable as an argument far sooner than is customary in algebra. In order to set up arrays, for example, some sort of function notation is needed. And finally, because computers have been programmed to perform manipulations with symbols without any referents for them, computer science has become a vehicle through which many students learn about variables (Papert 1980). Ultimately, because of this influence, it is likely that students will learn the many uses of variables far earlier than they do today.

Summary

The different conceptions of algebra are related to different uses of variables. Here is an oversimplified summary of those relationships:

Conception of algebra	Use of variables
Generalized arithmetic	Pattern generalizers (translate, generalize)
Means to solve certain problems	Unknowns, constants (solve, simplify)
Study of relationships	Arguments, parameters (relate, graph)
Structure	Arbitrary marks on paper (manipulate, justify)

309

Earlier in this article, two issues concerning instruction in algebra were mentioned. Given the discussion above, it is now possible to interpret these issues as a question of the relative importance to be given at various levels of study to the various conceptions.

For example, consider the question of paper-and-pencil manipulative skills. In the past, one had to have such skills in order to solve problems and in order to study functions and other relations. Today, with computers able to simplify expressions, solve sentences, and graph functions, what to do with manipulative skills becomes a question of the importance of algebra as a structure, as the study of arbitrary marks on paper, as the study of arbitrary relationships among symbols. The prevailing view today seems to be that this should not be the major criterion (and certainly not the only criterion) by which algebra content is determined.

Consider the question of the role of function ideas in the study of algebra. It is again a question of the relative importance of the view of algebra as the study of relationships among quantities, in which the predominant manifestation of variable is as argument, compared to the other roles of algebra: as generalized arithmetic or as providing a means to solve problems.

Thus some of the important issues in the teaching and learning of algebra can be crystallized by casting them in the framework of conceptions of algebra and uses of variables, conceptions that have been changed by the explosion in the uses of mathematics and by the omnipresence of computers. No longer is it worthwhile to categorize algebra solely as generalized arithmetic, for it is much more than that. Algebra remains a vehicle for solving certain problems but it is more than that as well. It provides the means by which to describe and analyze relationships. And it is the key to the characterization and understanding of mathematical structures. Given these assets and the increased mathematization of society, it is no surprise that algebra is today the key area of study in secondary school mathematics and that this preeminence is likely to be with us for a long time.

REFERENCES

Bushaw, Donald, Max Bell, Henry Pollak, Maynard Thompson, and Zalman Usiskin. *A Sourcebook of Applications of School Mathematics*. Reston, Va.: National Council of Teachers of Mathematics, 1980.

Comprehensive School Mathematics Program. *CSMP Overview*. St. Louis: CEMREL, 1975.

Conference Board of the Mathematical Sciences. *The Mathematical Sciences Curriculum K–12: What Is Still Fundamental and What Is Not*. Report to the NSB Commission on Precollege Education in Mathematics, Science, and Technology. Washington, D.C.: CBMS, 1983.

Davis, Robert B., Elizabeth Jockusch, and Curtis McKnight. "Cognitive Processes in Learning Algebra." *Journal of Children's Mathematical Behavior* 2 (Spring 1978): 1–320.

Fey, James T., and Richard A. Good. "Rethinking the Sequence and Priorities of High School Mathematics Curricula." In *The Secondary School Mathematics Curriculum*, 1985 Yearbook of the National Council of Teachers of Mathematics, pp. 43–52. Reston, Va.: NCTM, 1985.

Hart, Walter W. *A First Course in Algebra*. 2d ed. Boston: D. C. Heath & Co., 1951a.

———. *A Second Course in Algebra*. 2d ed., enlarged. Boston: D. C. Heath & Co., 1951b.

Kramer, Edna E. *The Nature and Growth of Modern Mathematics*. Princeton, N.J.: Princeton University Press, 1981.

Mac Lane, Saunders, and Garrett Birkhoff. *Algebra*. New York: Macmillan Co., 1967.

May, Kenneth O., and Henry Van Engen. "Relations and Functions." In *The Growth of Mathematical Ideas, Grades K–12*, Twenty-fourth Yearbook of the National Council of Teachers of Mathematics, pp. 65–110. Washington, D.C.: NCTM, 1959.

National Council of Teachers of Mathematics and the Mathematical Association of America. *Recommendations for the Preparation of High School Students for College Mathematics Courses*. Reston, Va.: NCTM; Washington, D.C.: MAA, 1977.

Papert, Seymour. *Mindstorms: Children, Computers, and Powerful Ideas*. New York: Basic Books, 1980.

Pavelle, Richard, Michael Rothstein, and John Fitch. "Computer Algebra." *Scientific American*, December 1981, pp. 136–52.

Pólya, George, *How to Solve It*. 2d ed. Princeton, N.J.: Princeton University Press, 1957.

Usiskin, Zalman. *Algebra through Applications*. Chicago: Department of Education, University of Chicago, 1976.

Usiskin, Zalman, and Max Bell. *Applying Arithmetic*. Preliminary ed. Chicago: Department of Education, University of Chicago, 1984.

Viète, François. "The New Algebra." In *A Source Book on Mathematics, 1200–1800*, edited by D. J. Struik, pp. 74–81. Cambridge, Mass.: Harvard University Press, 1969.

14

From *Multicultural and Gender Equity in the Mathematics Classroom: The Gift of Diversity,* NCTM's Fifty-ninth Yearbook (1997)

NCTM's Fifty-ninth Yearbook, *Multicultural and Gender Equity in the Mathematics Classroom: The Gift of Diversity,* was published in 1997, and it was the Council's first yearbook with a clear focus on equity in the classroom. From that yearbook, Lee Stiff, NCTM's president from 2000 to 2002, selected **chapter 7, "Students' Voices: African Americans and Mathematics,"** by Erica N. Walker and Leah P. McCoy.

In this chapter, Walker and McCoy summarize an ethnographic study using structured interviews with African American high school students. These interviews revealed that students' perceptions of mathematics were related to their experiences and surroundings at home, in school, and in their communities. Of the many findings that this study reported, the one that may convince readers of this chapter's continued importance is the following: "Without seriously restructuring how we view African American students, we will continue to channel them through the mathematics pipeline in ways that limit their opportunities."

Students' Voices: African Americans and Mathematics

Erica N. Walker
Leah P. McCoy

T HERE is perhaps no greater problem in American education than the lagging academic achievement of minority students, particularly African Americans. Despite years of "compensatory" education programs spawned by Lyndon B. Johnson's War on Poverty and the 1954 decision in *Brown v. Board of Education* that affirmed that segregated schooling was detrimental to African Americans' achievement, the academic gap between African American and white students in important subjects still exists. Research documents that "the American educational system is differentially effective for students depending on their social class, race, ethnicity, . . . and other demographic characteristics" (Secada 1992, p. 623). Numerous studies have been conducted to attempt to explain why the public education system seems to be failing minority students in general and African American students in particular.

Although African American children "bring to the formal classroom setting [in kindergarten] the same basic intellectual competencies in mathematical thought and cognitive processes as their white counterparts" (Stiff and Harvey 1988, p. 191), from age nine African Americans and Hispanics do not perform as well as whites on national surveys of mathematics achievement (Secada 1992; Stiff and Harvey 1988). This gap continues to widen through high school (Oakes 1988; Secada 1992).

We cannot continue to allow a significant number of our population to be innumerate; we must acknowledge that what occurs in school has a detrimental effect on African American students. Reasons for this failure must be explored, and ideas for combating whatever obstacles exist for African American students in mathematics must be implemented.

African American students' perceptions of their mathematics performance and the influence of their teachers, families, and peers on the development of that perception may be important factors that have been overlooked in our quest to determine the reasons for disappointing performances in mathematics.

Background

Few studies have investigated African American students' beliefs and ideas about mathematics. To discern those perceptions, the authors conducted an ethnographic study using structured interviews with African American high school students.

The sample was selected from a high school with an enrollment of 1250 students in a small city. Approximately 30 percent of the students are African American, 65 percent are white, and the remaining 5 percent belong to other ethnic groups. The racial makeup of most of the classes followed these proportions. Exceptions were honors classes in algebra 2 and advanced classes in algebra 3 that consisted primarily of white students. The researcher chose the students with the intent of selecting a group of African American students with differing levels of achievement. Seventeen students—nine females and eight males—in grades 9–12 agreed to participate. Four of those students were enrolled in algebra 1, eleven in geometry, one in algebra 3, and one in honors algebra 2.

Interviews were conducted and recorded either during school (during lunch or study halls) or after school. They lasted approximately twenty-five minutes each. All students were asked a series of open-ended questions about their attitudes toward mathematics, for example, "What influences your mathematics performance?" These general questions were based on the related research and were meant to elicit information about students' attitudes toward mathematics, including their future plans and the causes of these attitudes. The students were assigned pseudonyms to ensure confidentiality. The data were analyzed by identifying common responses and comparing different responses.

Voices of Students

The students' perceptions of mathematics were intricately related to their surroundings at home, in school, and in their communities. The results reveal that most students' beliefs about mathematics are not well defined nor do they remain constant. Indeed, their attitudes toward mathematics are easily changed, depending on what aspect of mathematics is being explored.

"Not Saying Nothing"

Some African American students exhibit a disturbing tendency to be silent in the mathematics classroom. Whether the reason is the embarrassment of "not knowing" or knowing too well that one is the only African American person in the classroom, this silence certainly affects students' interest in mathematics and their productivity. Kesha mentioned that she is often silent in mathematics class, even if she has a question, because she doesn't want to call attention to herself as the only African American in her algebra 3 class.

Rick doesn't speak up in mathematics class because he doesn't really "catch up to" the work. In other words, he doesn't speak because he perceives that everyone else understands what's going on better than he. Although Rick is not confident in his mathematics ability, he reveals that when he was younger, he was "real good" in mathematics. What happened to Rick during his mathematics career remains a mystery but is important for researchers to discover.

Others in the sample who were good mathematics students reflected that in many of their mathematics classes they felt intimidated because they were the "only black student." Consequently, they rarely spoke up in mathematics class unless they were directly asked a question by their teacher. Linda and Theresa mentioned that white students had made fun of their use of "black English" and looked at them as if to say, "Where [did] she get that from?" Kesha talked about being the only black person in her algebra 3 class and wanting never to be wrong because she was "representing her race." These concerns, expressed by several others, are prevalent perceptions among many African American students (Gilbert and Gay 1985; Stiff 1990).

At the time of the interviews, Linda and Theresa were geometry students in Ms. Taylor's class (Ms. Taylor is a pseudonym for an African American teacher). Unlike in some of their white teachers' classrooms, they were an integral part of Ms. Taylor's classroom. With pride, they revealed that Ms. Taylor "sits [them] up front" so that she can keep an eye on them. In short, they felt as though they belonged in her classroom.

The teacher's response can be an important weapon against African American silence in mathematics class. If the teacher is content to allow African American students like Rick, Linda, and Theresa—whether they are mediocre or excellent—to sit in the back of the classroom and "not say nothing" (as Linda puts it), then we can expect African American students to feel as though they do not matter. If, however, the teacher takes measures against voicelessness in his or her classroom, then students will thrive and respond, as Linda and Theresa have done in Ms. Taylor's classroom.

Can You Relate?

Whether their mathematics teachers care or not seems very important to African American students. This caring characteristic appears to transcend race, although African American students often indirectly mentioned the desire for teachers to be able to "relate" to them as African Americans.

Students' views of mathematics teachers are most often directly linked to how the teacher interacts with them on a personal level. Of the students interviewed, those who felt a personal relationship with their teacher felt confident in the classroom and strongly desired to perform well in that teacher's class. Angie's contention that African American students work harder for African American teachers is exemplified by Jeff, who performs academically for Ms. Taylor but not for his other teachers. Other examples make up the steady stream of African American students who are not in Ms. Taylor's classes but stay after school with her, preferring her help to their own teacher's.

It is probable that since African American teachers can relate to the culture of African American students, many of these students feel more comfortable in an African American teacher's classroom. There seems to be more at stake in an African American teacher's class, since most students perceived that these teachers care more about, and have positive influences on, African American students. Although white teachers were not as a group denigrated by these students, research supports Michelle's belief that white teachers are willing to turn a blind eye to African American students as long as they aren't disruptive (Gilbert and Gay 1985; Secada 1992). Some students believed that white teachers care less about their academic performance than African American teachers. It appears to these students that white teachers' interest in African American students is directly related to behavior: good behavior merits nothing whereas bad behavior merits attention (albeit negative) from the teacher.

Gordon, an African American male who comes from a two-parent, upper-middle-class home, believes that white teachers have difficulty relating to African American students because the teachers are from "good" homes. This is quite disturbing but not surprising, given the media's propensity to perpetuate stereotypes about both African Americans and white Americans. It is highly doubtful that all white teachers come from "good" homes and equally doubtful that all African American teachers come from "bad" homes. However, this belief underscores the contention that cultural dissonance between teachers and students plays an important role in the dynamics of the mathematics classroom (Ford and Harris 1992; Gilbert and Gay 1985).

Gordon's view regarding white teachers' inability to relate to students from poor homes stems from his inherent belief that most poor families are African American. His suggestion that social workers help "bad" families (who appear to be all African American) reveals his prejudice toward families of lower socioeconomic status. That Gordon has bought into the myth that "white" is "good" and "black" is "bad" indicates that we have not come very far from the historical stereotypes of African Americans.

Run, Dribble, or Solve Equations?

Although a few students who cited extracurricular activities as obstacles to their mathematics success were working students, many more cited sports as a diversion. Angie, Todd, and Wayne are student-athletes. Sports play a major part in their lives, admittedly more than mathematics. For example, Wayne estimated that 75 percent of his time was spent on sports and did not consider his sports participation as a negative influence on his course grade, which was a C. Todd thought that he was preparing for his future by spending most of his time practicing basketball, even though he realized that he was neglecting his studies. Angie, realizing that her schoolwork was suffering, decided to give up track, although she was a stellar runner. When asked if her coach (a white mathematics teacher) supported her decision, Angie said, "No. He pressured me not to quit . . . and after I explained to him that my grades were dropping, he still didn't want me to quit. But I quit anyway. . . ." It appears that this teacher's instincts as a coach outweigh his instincts as a teacher. Is African American students' sole value their sports ability? Are these students being encouraged to excel academically?

Given Angie's, Todd's, and Wayne's attitudes toward schooling—in particular, mathematics—the answers to those questions in their minds are yes and no respectively. Angie and Todd see mathematics as somewhat important to their future careers, but their attention is not focused on it and they do not fully understand its potential impact. Wayne, however, is totally confused about how his school performance relates to what he does later. Todd's and Wayne's dreams are to go to college on sports scholarships rather than doing mathematics, when they both have the full ability to do mathematics. It is their prerogative to want to do something else—Wayne wants to major in political science—but the fact that they are so accepting of society's role for them is disheartening. As Stiff and Harvey (1988) warned, they seem to think that they are valued more for their athletic ability than for their academic prowess, and we apparently have not given them a different message.

"The Easy Way Out"

When asked why more African American students weren't in upper-level mathematics classes, most students said that their peers wanted to take the "easy way out." This statement implies alternatively that students don't want their GPA to drop, they don't want a challenge, or they are scared. Shawn told us after we had stopped recording that he thought a lot of African American students were scared of the standardized tests and didn't perform well because of the fear. This assessment appears to be consistent with the attributes of mathematics anxiety and learned helplessness.

However, the fact that so many African American students were so willing to label their African American peers "lazy" may indicate stereotypical racism. Often their use of the word *lazy* was erroneous; they did not say that the students didn't want to do the work but that their interest was in other things. Apathy toward learning mathematics is perhaps a more accurate description.

Many African American students said that their peers' apathy or laziness was due to not realizing that mathematics was important to achieving their future goals. Not surprisingly, the students in the sample who wanted to be lawyers, accountants, engineers, architects, and computer programmers realized the importance of mathematics and consequently took more mathematics courses—behavior that was also discovered by Matthews (1984) and Reyes and Stanic (1988). Fortunately, these students constitute a majority of the sample. Those students who were somewhat undecided or ambivalent about their careers were the ones who did not see the importance of mathematics. Although they constitute a minority of the sample, it is possible that they may represent the majority of African American students in mathematics. Most students, regardless of their career goals, wanted to take mathematics to get into college and recognized that they needed it for that reason. For the most part, their interest in mathematics did not extend beyond this goal.

"Have You Done Your Math Homework?"

Parents were the most important influence on students' mathematics performance, only slightly ahead of the students' own motivation. Although counselors had been found to be important influences in 1982 (Johnson), no students mentioned counselors as being influential in their taking more mathematics courses; parental encouragement and support appear to have surpassed the influence of counselors. Mothers were most often mentioned as being helpers and instigators, asking, "Have you done your math homework?" or suggesting, "Call Homework Hotline!" Rick, who mentioned that his mother did not

319

play a great role in his life, let alone his mathematics performance, berated her for not making him do his homework and "not being more strict on him" when he was younger. Jeff, whose family problems are the primary reason for his up-and-down performance, did not mention any family member as influential. Todd did not mention any help from home, either. All these African American males were failing at the time of the interview. Lawrence, who is a successful African American male student, mentioned his mother's encouragement and support repeatedly throughout his interview. Other students who were performing well in mathematics class also cited their families as positive influences. These responses support Johnson's (1992) and Ford and Harris's (1992) finding that positive family influences are essential to academic success.

It was clear that successful students' parents would not tolerate less than their children's best. This "do your best" message was repeated over and over again by students who were successful mathematics students.

"Try Harder and Do Better"

Although Ford and Harris (1992) and Ogbu (1988) contend that African Americans may not be performing well in mathematics because of fear of peer reprisal, the students in this study, despite sometimes being teased, try to do well despite their "friends'" attitudes. Parental intervention and influence probably counteract this peer pressure.

Moreover, as discussed by Johnson (1982), students who were successful at mathematics were recruited to help their friends who needed assistance. Their peers were proud of them for being successful. Good students encouraged their friends to "try harder" and "do better."

It appears that positive encouragement from peers is effective and that negative comments from peers might be overrated, at least according to this sample. A combination of factors—teachers, parents, and self-motivation—seems to combat any negativity from small-minded peers.

"You Don't Get Any Experiences, So You Don't Learn"

Although there are many examples to the contrary, students perceived that familial resources such as money and parental education (cultural capital) influenced mathematics achievement. Often mentioned was the belief that "white people . . . have more than African American people" and variations on that theme, including Wayne's astute perception that cultural exposure, "seeing and doing things," is integral to school success. Comments like these were made by students regardless of their grade in mathematics or socioeconomic status. Although many students in the sample whose parents would be considered to have little "cultural capital" are successful mathematics students, those students thought that the lack of advantage might be too much for their less-motivated peers to overcome, especially if other positive influences such as parental support and teacher support are not there. Significantly, Kesha was the only student who exemplified Useem's (1990) theory of parental use of cultural capital, mentioning that her mother came to

320

school and insisted on her course placement in advanced mathematics classes. Kesha was in algebra 3—the highest-level course taken by students in the sample.

Recognizing that there is a tremendous societal emphasis on what one has and hasn't, it is not surprising that the students felt that somehow their family's financial offerings were inadequate. Although this feeling of inadequacy is somewhat pervasive because African Americans are disproportionately poor, these students have shown rather definitively that more than money influences their mathematics achievement.

"I Don't Think They Really Want to Take a Higher Math"

Many factors appear to influence students' motivation to take mathematics courses and maintain an interest in mathematics. When asked why their peers did not want to take more mathematics courses, students said that their peers were intimidated by higher-level mathematics. Successful students said that positive influences—teachers, parents, and their own motivation—attracted them to take more mathematics courses. It seems that if we can increase positive parental and peer influences and decrease negative influences, such as marketing high school students as just sports stars, then the students' own motivation could surface and compete with other distractions, a suggestion made by Stiff and Harvey (1988). Students who were self-motivated and had parents and peers to support them tended to do better in mathematics and realized its importance to their future goals.

Ford and Harris's paradox of underachievement (1992) appeared to be all too prevalent among these students, even those who were "good" students. These students were dissatisfied with their mathematics grades and recognized that they needed to work harder but did not put forth the effort to do so. This paradoxical behavior sideswipes many students with low or misplaced motivation because like Wayne and Jeff, they cannot reconcile their actions with their beliefs. Consequently, they espouse the ideal of working hard to achieve and yet continue to act in a manner completely contradictory to that ideal.

"You Need Math Like You Need Air to Breathe"

The fact that a few students have discerned the importance of mathematics and intuitively realize that mathematics leads to many opportunities otherwise unavailable to them is a positive development. However, far too many do not realize the importance of mathematics in preventing their being squeezed out of the mathematics pipeline into a quagmire of limited opportunities.

Students such as Kesha, who perceived that mathematics is as necessary to her future goals as "air to breathe," planned to take more mathematics courses. Despite their feeling that mathematics was integral to their school success, however, few of the students could see themselves as mathematicians. It was clear that some respondents had little idea of what a mathematician is or does and that we have disseminated a very narrow view of mathematicians to high school students. Even students who had been very successful in mathematics resisted the notion that they could like mathematics enough to want to become mathematicians, which supports the findings of Matthews (1984) and Stiff and

321

Harvey (1988). When we pointed out to some students that their planned careers were very mathematical, it was apparent that they had never thought of their careers in quite that light.

This misapprehension may very well be due to students' having very rarely seen mathematicians, except for their mathematics teachers, in action. It is safe to say that if students saw more practical applications of mathematics, and more people using mathematics, they wouldn't have such a mental block against the possibility that they can be mathematicians. There appears to be a greater emphasis on sports and entertainment role models among African American youth (Stiff and Harvey 1988) because the media portray these role models as African American heroes and heroines. Noteworthy examples of contemporary African American mathematicians, finance wizards, scientists, and professors can certainly be cited. We are doing a grave disservice to African American mathematics students when we limit their horizons to jumping and running. We cannot continue to idolize the athletes and not the actuaries.

Implications

Without seriously restructuring how we view African American students, we will continue to channel them through the mathematics pipeline in ways that limit their opportunities. The fact that some African American students have been so negatively affected by the cultural and social dynamics in their mathematics classrooms indicates that a filtering process, as numerically evidenced by gross disparities in achievement, is occurring. Although many students have been positively influenced by their parents and communities, schools must take an active role in broadening students' perceptions of themselves as mathematics students.

This study suggests that teachers can have a profound effect on the self-image of African American mathematics students. If a teacher actively encourages her students, as Ms. Taylor does by seating them in the front of the classroom and frequently calling on them to solve problems, African American students will thrive. When they are neglected by their teacher or disparaged by other students in the class, they will respond by removing themselves mentally from the classroom. Most of the successful students in this study felt welcomed in their mathematics class because they had a strong relationship with their teacher.

The caring environment exemplified in Ms. Taylor's classroom extended to the homes of successful students. Successful students who expressed discomfort in classes because they were the only African Americans had a safety net of parental encouragement and involvement at home. Parents were responsible for the placement in higher mathematics courses in several instances, directing students to tutors if the parents themselves could not provide assistance and motivating students to "do their best." Students who lacked such positive parental intervention suffered low grades in mathematics courses, were in lower-level classes, and were not at all confident about their mathematics ability.

It is imperative that schools and teachers recognize that what occurs in classrooms can negatively affect students' achievement. The mathematics teacher must realize that his or her classroom environment may be damaging to the confidence of African American students. The classroom should be a positive place where students can and are expected to excel. Schools and parents should work together to ensure that nurturing occurs both in the classroom and in the home.

REFERENCES

Ford, Donna Y, and J. John Harris. "The American Achievement Ideology and Achievement Differentials among Preadolescent Gifted and Nongifted African American Males and Females." *Journal of Negro Education* 61 (Winter 1992): 45–64.

Gilbert, Shirl E., and Geneva Gay. "Improving the Success in School of Poor Black Children." *Phi Delta Kappan* 67 (October 1985): 133–37.

Johnson, Robert C. *Psychosocial Factors Affecting the Mathematical Orientation of Black Americans.* Final Report. St. Louis, Mo.: Institute of Black Studies, 1982. (ERIC Document Reproduction Service no. ED 251 566)

Johnson, Sylvia T. "Extra-School Factors in Achievement, Attainment, and Aspiration among Junior and Senior High School-Age African American Youth." *Journal of Negro Education* 61 (Winter 1992): 99–119.

Matthews, Westina. "Influences on the Learning and Participation of Minorities in Mathematics." *Journal for Research in Mathematics Education* 15 (March 1984): 84–95.

Oakes, Jeannie. "Tracking in Mathematics and Science Education: A Structural Contribution to Unequal Schooling." In *Class, Race, and Gender in American Education,* edited by Lois Weis. Albany, N.Y.: State University of New York Press, 1988.

Ogbu, John. "Class Stratification: Racial Stratification and Schooling." In *Class, Race, and Gender in American Education,* edited by Lois Weis. Albany, N.Y.: State University of New York Press, 1988.

Reyes, Laurie Hart, and George M. A. Stanic. "Race, Sex, Socioeconomic Status, and Mathematics." *Journal for Research in Mathematics Education* 19 (January 1988): 26–43.

Secada, Walter G. "Race, Ethnicity, Social Class, Language, and Achievement in Mathematics." In *Handbook of Research on Mathematics Teaching and Learning: A Project of the National Council of Teachers of Mathematics,* edited by Douglas A. Grouws. New York: Macmillan Publishing Co., 1992.

Steen, Lynn Arthur. "Mathematics for All Americans." In *Teaching and Learning Mathematics in the 1990s,* 1990 Yearbook of the National Council of Teachers of Mathematics, edited by Thomas J. Cooney, pp. 130–43. Reston, Va.: National Council of Teachers of Mathematics, 1990.

Stiff, Lee V. "African-American Students and the Promise of the *Curriculum and Evaluation Standards.*" In *Teaching and Learning Mathematics in the 1990s,* 1990 Yearbook of the National Council of Teachers of Mathematics, edited by Thomas J. Cooney, pp. 152–58. Reston, Va.: National Council of Teachers of Mathematics, 1990.

Stiff, Lee V., and William B. Harvey. "On the Education of Black Children in Mathematics." *Journal of Black Studies* 19 (December 1988): 190–203.

Useem, Elizabeth L. "Social Class and Ability Group Placement in Mathematics in the Transition to Seventh Grade: The Role of Parental Involvement." Paper presented at the annual meeting of the American Educational Research Association, Boston, 1990.

15

From *The Teaching and Learning of Algorithms in School Mathematics*, NCTM's Sixtieth Yearbook (1998)

Questions about algorithms and their importance and use are frequent topics of discussion in mathematics education. Perhaps this is why NCTM past president Cathy Seeley (who served from 2004 to 2006) recommended the inclusion of two chapters from *The Teaching and Learning of Algorithms in School Mathematics,* the Council's Sixtieth Yearbook.

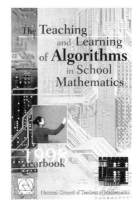

Chapter 3, "What Is an Algorithm? What Is an Answer?" by Stephen B. Maurer, defines an algorithm as a precise, systematic method for solving a class of problems. This chapter provides examples of how algorithms are used, with or without the use of a technological tool, when considering the multiplication of whole numbers, the quadratic formula, and sequences of heads and tails. Maurer also includes an interesting case study on how we use algorithms, and he explores the role of technological tools in expressing solutions and considering what is a "good" answer.

Chapter 5, "Teaching Mental Algorithms Constructively," by Alistair McIntosh, emphasizes the importance of teaching mental computation. In this view of mental computation strategies and algorithms, students need to discover many ways to use mental math while recognizing that there is rarely a single best mental strategy.

Both these chapters, in very different ways, address algorithms and their role across elementary, middle, and high school mathematics.

What Is an Algorithm? What Is an Answer?

Stephen B. Maurer

TODAY, with calculators and computers, the solution to a mathematical problem is often algorithmic. The process of solution, and sometimes even what counts as an answer, differ from before. This paper illustrates the differences with a simple case study—a problem that came up in my own work that I solved with an algorithm. I show how I started, how I reflected on what I did, and how my idea of a solution changed. First, however, I explain what is meant by an algorithm and by modern algorithmic thinking, or algorithmics (Maurer 1992).

What Is an Algorithm?

An algorithm is a precise, systematic method for solving a class of problems. An algorithm takes *input,* follows a *determinate* set of rules, and in a *finite* number of steps gives *output* that provides a *conclusive* answer. *Determinate* means that for each allowed input, the first action is precisely specified and, more generally, that after each action in the sequence the next action is precisely specified. *Conclusive* usually means that the output correctly solves the problem, but it can mean that the algorithm either solves the problem correctly or announces that it cannot solve it.

Example 1. Multiplication

The traditional paper-and-pencil algorithm for multiplying two numbers expressed in Indo-Arabic numerals is brilliant. It is too bad we take it for granted. The algorithm is brilliant because it reduces a general problem to a small subcase—how to multiply two single-digit integers—and does so in a small amount of space. Here is the result of applying the algorithm to 432×378:

$$
\begin{array}{r}
432 \\
\underline{378} \\
3456 \\
3024 \\
\underline{1296} \\
163296
\end{array}
$$

Each row of intermediate calculation is obtained by multiplying the top factor (432) by one digit of the bottom factor—a many-digits-times-one-digit subprocedure. To see the basic one-digit-by-one-digit steps, expand the first intermediate row to show more detail:

$$432$$
$$\underline{\quad 8}$$
$$16$$
$$24$$
$$\underline{32\quad}$$
$$3456$$

Of course, the calculation is not usually written this way. To save space, the "carries" are all either done mentally or marked with small digits.

This method is indeed an algorithm: it is determinate, finite, and conclusive. To demonstrate these characteristics formally would take some work. To begin with, we would have to write down the steps of the algorithm carefully, say, using algorithmic language, as in the next example. We never do so with children because a formal description is quite complicated.

Notice that this algorithm involves *iteration* (loops); that is, some subprocess is applied repeatedly. In this instance the subprocess is to multiply two single-digit numbers and carry.

Example 2. The Quadratic Formula

The traditional formula for solving $ax^2 + bx + c = 0$ seems simple enough; where is the algorithm and why bother with it? In example 1 we needed an algorithm because the "formula" for multiplication, $a \times b$, is too simple; it doesn't tell us how to do the multiplication. But for quadratics, if we already know the four basic arithmetic operations, the quadratic formula seems to tell us everything we have to do.

Well, not quite. There are several cases—two distinct real roots, one repeated real root, and no real roots. When there is just one root, we probably don't want to output it twice. When there are complex roots and we are using a calculator that balks at negative square roots, we must compensate. So we can use the following algorithm:

Input: a, b, c	(coefficients of $ax^2 + bx + c$, with $a \neq 0$)
Algorithm	
Let $D = b^2 - 4ac$	
If	(three cases follow)
$D > 0$ then	(two real roots)
let $s = \sqrt{D}$	
let $x_1 = (-b + s)/2a$	
let $x^2 = (-b - s)/2a$	
else if $D = 0$ then let $x = -b/2a$	(one repeated real root)
else if $D < 0$ then	(two complex roots)
let $s = \sqrt{-D}$	
let $x_1 = (-b + is)/2a$	
let $x^2 = (-b - is)/2a$	
endif	
Output: the roots x_1 and x_2 or the root x	

This procedure is an algorithm: it is determinate, finite, and conclusive. Although it is longer than the quadratic formula, we would argue that it yields a better answer.

Notice the algorithmic notation. To express an algorithm precisely, one needs a special language. Computer scientists call such notation *pseudocode* because it does not run on any actual computer. I prefer to accent the positive: it is a good notation for human communication about algorithms. Hence I call it *algorithmic language*.

Notice that our algorithm has several if-statements (and no loops). If we want to be even more comprehensive and allow input with $a = 0$, then we have to include several more cases and thus several more if-statements. Procedures in the everyday world—calculating taxes, for instance—often involve many cases. Thus algorithms for everyday problems often include many if-statements.

Example 3. Sequences of Heads and Tails

An important role of mathematics is to guide us in making decisions under uncertainty. We can do so using probability theory, but often the simplest approach is simulation. Suppose we flip a fair coin until we get two heads in a row. How many flips should this take? If we actually carry out this experiment many times, we find out what to expect. The following is an algorithm to carry out the experiment one time. Running it on a computer is faster than flipping coins, though with a whole class we could do the latter (parallel processing!). Rand(0,1) is a command for flipping a coin; the output 1 means heads, 0 means tails. The algorithm could be run a thousand times inside a loop of a bigger algorithm, which could then analyze the output data in various ways (e.g., determine the average and the variance, draw graphs, etc.).

```
Input: (none)
Algorithm
      Let count = headcount = 0
Repeat
count = count + 1
      let flip = Rand(0,1)                    (0 or 1, at random)
      if flip = 1 then let headcount
      = headcount + 1
      else let headcount = 0
endrepeat when headcount = 2
Output: count                                 (total flips to get two heads in a row)
```

This algorithm violates our definition in two ways. First, when the command generates a random number, it cannot be said to be precisely specified. Second, it is theoretically possible that this procedure will not terminate; we might get 0s forever. Nonetheless, we certainly want to study such "algorithms," and so it is traditional to relax the three defining conditions. The hard part, actually, is to get computers to perform such procedures, since computers really are determinate machines. In other words, how can computers be made to produce what appear to be random numbers? Fortunately, there are good answers, which we will not pursue here.

Algorithmics

The phrase *algorithmic mathematics* has two meanings, traditional and contemporary (Maurer 1984). The traditional meaning emphasizes carrying out algorithms; the contemporary meaning emphasizes developing them, understanding them, and choosing intelligently among different algorithms for the same task.

This contemporary meaning has its own name: *algorithmics*. At the advanced level algorithmics is quite sophisticated. Developing algorithms expands to whole courses in algorithm design (bottom up, top down, backtrack, recursive, etc.). Understanding becomes algorithm validation (mathematical induction is the basic tool). Choosing intelligently becomes complexity theory (precise ways to count or bound the number of steps in an algorithm or needed by any algorithm for a specified problem). In schools algorithmics can be more informal.

In any event, algorithmics does *not* mean performing algorithms over and over by hand. Algorithms will be carried out more and more by machines or by person-machine combinations. Algorithmics is thinking *about* algorithms, not thinking *like* algorithms. For more on the difference between modern algorithmic thinking and traditional problem solving, see, for instance, Hart (1991), Maurer (1984, 1992), Maurer and Ralston (1991), and NCTM (1989, 1990).

The Case Study

While preparing a contest book (Berzsenyi and Maurer 1997), I needed to create figures to illustrate plane geometry constructions. We were producing the book with the mathematics typesetting system TEX. For figures we were using a macro package called PICTEX; this package has considerable production advantages for use with TEX, but some of the PICTEX commands for geometric constructions are quite primitive.

Consider figure 3.1. In this triangle, D is the midpoint of BC and E is the trisection point of AC closer to A. With PICTEX it is easy to get all the line segments shown. Just name the coordinates of the endpoints (in some units, say, 0.1 cm). For instance, I declared $A = (0, 0)$, $B = (100, 0)$, and C = (60, 90), in which case $D = (80, 45)$ and $E = (20, 30)$. However, suppose I also needed to mark the intersection point F. Of course, PICTEX draws the intersection as it draws the line segments, but it doesn't automatically put down a large dot and a letter. To do so, PICTEX needs to know the coordinates of the intersection point, and it does not compute them itself. So a problem I faced over and over was to find the coordinates of the intersection of two lines when each line was given by two points (see note 1 in appendix 2).

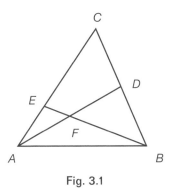

Fig. 3.1

Figure 3.2 is a more complicated example, directly from the contest book. Circle *O* is inscribed in equilateral triangle *ABC*, and *DE* is tangent to the circle at *T* and perpendicular to side *AC* at *E*. I needed to find the coordinates of every labeled point in order to instruct PICTEX to draw the lines and insert the labels. To begin with, I knew the coordinates of *A*, *B*, and *C* because I had created the triangle by choosing them. To illustrate how I found the other points, consider *E*. It is the intersection of the lines *AC* and *DE*. I knew two points on *AC*, namely, *A* and *C*. As for line *DE*, I knew how to find the coordinates of *T* (see appendix 1). From this information it was easy to name another point on *DE*; use the fact that the slope of *DE* is the negative reciprocal of the known slope of *AC*, and go out from *T* along this slope by, say, one unit. Sure enough, I had reduced finding *E* to the problem of finding the intersection point of two lines when each line was defined by two points.

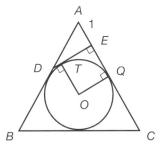

Fig. 3.2

After solving this type of problem by hand once or twice (tedious!), I decided I really needed a little program to do it. I could run this program and edit the book on the same personal computer and paste the program's results into PICTEX.

I planned to write a three-line program in BASIC. The first line would ask for the eight coordinates of the two pairs of defining points as input and assign them to variables. The next two lines would print the value of the formulas for the *x*- and *y*-coordinates of the intersection, formulas expressed in terms of the input variables. I would merely have to derive these formulas once, algebraically by hand, and be done with it.

How would I derive these formulas? Finding the intersection of two lines is a standard problem of algebra 1 if the equations of the lines are given. Using "reduce to the previous case" (a basic strategy of algorithmics), I planned as follows:

1. Use the two defining points for each straight line to find an equation in the form $y = mx + b$.

2. Solve these two equations for x by setting $m_1 x + b_1 = m_2 x + b_2$.

3. Plug the x solution into $y = m_1 x + b_1$ to get the y solution.

All I had to do was substitute the results of each step into the next step to get explicit formulas in terms of the initial inputs. I got out a piece of paper and started.

Well, it was easy in principle but not in fact. The expressions for m_1 and m_2 were modest, but b_1 and b_2 were messier. The x formula was even messier. I didn't want to do step 3.

I could have done it. But I would probably have had to check it three times to get it right. And then I would still probably have made a mistake when I tried to type it into the computer in a form acceptable to the software. I don't have as much patience for such calculations as I did in my youth, and most youths never have much patience for them. Was there a better way with a modern tool?

For a brief moment I considered doing the whole thing in Mathematica. It could have solved all the algebraic equations. But first I would have had to enter the problem with the correct, complicated syntax. Then I would also have had to use the algebraic solution to create a function in Mathematica to compute the numerical values in each particular case because the algebraic output of Mathematica could not go directly into BASIC. And Mathematica is a big program, slow to start up. Using the program did not seem worth it.

So I thought a little harder, and then—I had to laugh at myself. I had not been following the algorithmicist's advice I had given to others so many times: *Formulas are not so important anymore.* Why bother substituting to get one final algebraic formula when I was after numerical answers anyway? I had already solved the problem; I could just leave the results unsubstituted. The BASIC program would have more lines, but it would be much easier to write.

To see what I mean, look at the final True BASIC program I wrote, Algorithm 1. Variable m_1 is the slope of the first straight line; m_2 is the slope of the second. The equations of the two lines are $y = m_1 x + k_1$ and $y = m_2 x + k_2$. (I used k because I had already used b for the vertical coordinates of the first two points.) I computed k_1 by substituting

$$m_1 = \frac{y - b_1}{x - a_1}$$

into the line equation and rewriting on paper. Next, I set $m_1 x + k_1 = m_2 x + k_2$ and solved for x on paper; the result is the "let x" command. Then y is obtained directly from the formula for the first straight line. The last little touch concerns the output. I had True BASIC produce exactly the expression I had to plug into PICTEX, not just the proper coordinates. This way, as I edited the book, whenever I needed this program, I just switched to BASIC (without closing the editing program), ran this algorithm, and cut and pasted the output back into my editing file (see note 2 in appendix 2).

```
                        Algorithm 1
  do
          Input prompt "first pair of points ": a1,b1, a2,b2
          Input prompt "second pair ": c1,d1,c2,d2
          let m1 = (b2–b1)/(a2–a1)
          let m2 = (d2–d1)/(c2–c1)
          let k1 = b1–m1*a1
          let k2 = d1–m2*c1
          let x = (k2–k1)/(m1–m2)
          let y = m1*x + k1
          print " \put {$\bullet$} at ";
          print using "###.## ": x,y
  loop
  end
```

For more-challenging problems that I faced in computing figures, see appendix 1.

Conclusions

Mine was a little problem, with a simple solution, but it illustrates several important points:

1. This was the most standard of problems—find the intersection of two lines, given two points on each line. The context for solving it was different—an algorithm in a computer language instead of a paper-and-pencil algorithm. When the emphasis is on how to effectively compute the answer with the tools available, creativity will always be needed in even the most elementary mathematics. Therefore, a primary goal of mathematics education should be to prepare students to be flexible problem solvers. They will have to create new algorithms or at least modify old ones.

2. What is easy, and what is hard, depends on the tool being used. PICTEX makes it easy to draw intersections but harder to mark them. Suppose I had been using Geometer's Sketchpad instead. Then marking an intersection would have been easy. Something else would have been hard. (In this instance, the figures would have been external to the TEX code and our electronic files would not have been "device independent.") Or compare PICTEX as a tool with straightedge and compass. With those traditional tools, we can use traditional

333

geometric constructions to determine, say, where E has to be. With the different tools of $\text{P\!\i CT}_{\!E}\text{X}$ and BASIC, constructing the figure was an entirely new problem (and lots of fun).

3. People must be able to express solutions in many forms: symbols on paper, symbols in computer languages, button sequences for calculators, and so forth. Notice, for instance, that mathematical language looks different on computers than on paper; there are no subscripts or fraction bars in computer input. In fact, different computer languages have rather different syntaxes, and we must be able to translate back and forth among them as well as to them. Perhaps someday we will be able to write standard paper-and-pencil mathematical syntax on a pressure pad and all computer programs will understand it. But for the foreseeable future, the number of different forms for expressing mathematics will proliferate.

4. Single formulas are not so important anymore. The appropriate solution was a procedure, not a formula, yet out of habit I was hung up on getting a formula, since I knew one existed (see note 3 in appendix 2).

What Is an Answer?

Several of my conclusions involve the same idea: the concept of "answer" is changing. A good answer is easy to understand and easy to use, as well as somehow elegant. But what is easy to use depends on the tools of the time, and what is easy to understand depends on what tools and symbols people are used to. If algorithms can be evaluated quickly and people are used to reading them, then a longer algorithm may be a better answer than a shorter formula. In the case study, the algorithm was just an unpacking of the formula, but an algorithm may not depend on a formula at all and still be a better answer. This is the gist of the following words from a speech by John Kemeny (1983, pp. 204–5):

> I think mathematicians have tremendous pre-occupation with formulas, because they love finding answers "in closed form". Consider the simplest kind of integration problem, say,
>
> $$\int_0^{13} e^x dx.$$
>
> It's the kind of problem you put on a freshman calculus exam so that every student should get at least one problem right. Everybody knows how to do it. The answer is $e^{13} - 1$. Suppose someone came along and said, "Why didn't you use numerical integration?" You'll say that is absolutely mad. Here is a closed form solution, and you can get the exact answer in two minutes; in 10 seconds if you are good. Why in heaven's name would you want to use numerical integration to get an approximate answer? So my question is, if that's the exact answer, please tell me what it is to one significant figure. . . .

Next you'll argue that it is still important that $e^{13} - 1$ is exact. You know, for years I accepted that, until one day I woke up, and it hit me that the original integral

$$\int_0^{13} e^x dx$$

is also exact! I do not mean that as a joke—I mean that as a deep remark about mathematics, one that I overlooked for a large number of years. Now the question is, if you have two forms that are exact, why is one preferable to the other one? And if you think about that particular example, I think it's very easy to reconstruct the reason. We did not have computers but had tables of e^x. Even today there may be some advantage; you may be able to use a pocket calculator instead of a computer on $e^{13} - 1$. But the numerical integration takes less than a second on a computer. . . . [I]f you put in there $x^2 \sin x \, e^x$ I've got absolutely no doubt whatsoever [what is the better answer]. And it's still less than one second on the computer, and it's the same algorithm that works. . . . The lesson in this particular case is that mathematicians have got to change their way of thinking from exclusive emphasis on formulas to algorithms.

Wilf (1982) made the same point with respect to counting problems. A good listing algorithm, working directly from the problem statement, may be faster for counting the items in question than evaluating a formula with multiple sums, products, combinations, and so on. If so, the problem statement is a better answer than the formula!

Barwise and Etchemendy (1991) have proposed an even more striking change in what constitutes an answer. Given humankind's very high absorption rate for visual information and given the power of current computers to produce graphics and provide graphical user interfaces, then a picture or a movie may be a better answer than anything written in symbols. Quite a proposal when not long ago many mathematicians eschewed any answer that depended on a picture!

What do you think is a good answer?

REFERENCES

Barwise, Jon, and John Etchemendy. "Visual Information and Valid Reasoning." In *Visualization in Mathematics*, edited by Walter Zimmermann and Stephen Cunningham, pp. 9–24. MAA Notes no. 19. Washington, D.C.: Mathematical Association of America, 1991. Reprinted in *Philosophy and the Computer*, edited by Leslie Burkholder, pp. 160–82. Boulder, Colo.: Westview Press, 1992.

Berzsenyi, George, and Stephen B. Maurer. *The Contest Problem Book V.* Washington, D.C.: Mathematical Association of America, 1997.

Hart, Eric W. "Discrete Mathematics: An Exciting and Necessary Addition to the Secondary School Curriculum." In *Discrete Mathematics across the Curriculum, K–12*, 1991 Yearbook of the National Council of Teachers of Mathematics, edited by Margaret J. Kenney, pp. 67–77. Reston, Va.: National Council of Teachers of Mathematics, 1991.

Kemeny, John. "Finite Mathematics—Then and Now." In *The Future of College Mathematics: Proceedings of a Conference/Workshop on the First Two Years of College Mathematics*, edited by Anthony Ralston and Gail S. Young, pp. 201–8. New York: Springer-Verlag, 1983.

Maurer, Stephen B. "Two Meanings of Algorithmic Mathematics." *Mathematics Teacher* 77 (September 1984): 430–35.

———. "What Are Algorithms? What Is Algorithmics?" In *The Influence of Computers and Informatics on Mathematics and Its Teaching,* edited by Bernard Cornu and Anthony Ralston. Science and Technology Series no. 44. Paris: United Nations Education, Scientific, and Cultural Organization, October 1992.

Maurer, Stephen B., and Anthony Ralston. *Discrete Algorithmic Mathematics.* Reading, Mass.: Addison-Wesley Publishing Co., 1991.

National Council of Teachers of Mathematics. *Curriculum and Evaluation Standards for School Mathematics.* Reston, Va.: National Council of Teachers of Mathematics, 1989.

———. *Discrete Mathematics and the Secondary Mathematics Curriculum.* Reston, Va.: National Council of Teachers of Mathematics, 1990.

Wilf, Herbert S. "What Is an Answer?" *American Mathematical Monthly* 89 (May 1982): 289–92.

APPENDIX 1: EXTENSIONS

Algorithm 1 is not the only program I wrote for the contest book. Here are two more-challenging problems. You might like to give all three problems to your students.

Problem 2. Find the coordinates of the intersection of a line and a circle when the line is given by two points and the circle by the center and radius.

I needed an algorithmic solution to problem 2 to find T in figure 3.2. The center of the circle is the known centroid of the equilateral triangle, and thus the radius also was easily computed by hand. The center is a known point on OT. Since line OT is parallel to side AC, a second point on OT could easily be found.

Problem 3. Determine the coordinates of the corners of a little right-angle sign, given the point where the perpendicular lines meet, the direction of one of the perpendicular lines, and the desired side length of the sign.

Completing figure 3.2 required the solution of problem 3 as well.

APPENDIX 2: NOTES

1. In figure 3.1, it is possible to mentally compute the coordinates of F if the right theorem is known. There is a theorem that states the ratio AF/AD in terms of BD/BC and AE/AC. It turns out that $AF/AD = 1/2$ in this case, so $F = (40, 22.5)$. And in other cases I had to deal with, other special methods could surely be devised. But the point in algorithmic thinking is to avoid repeated special-case solutions, no matter how ingenious.

2. The program displayed in this paper is not quite the program I used in editing the problems book. For instance, in going from m_1 and m_2 to x, I didn't define k but rather a slightly different intermediate quantity less in keeping with

traditional notation. But the idea—never actually complete any algebraic substitutions—was the same.

3. The change in emphasis from formulas to procedures involves an interesting historical about-face. Although the word *algebra* derives from the title of a ninth-century Arabic text (*Hisab al-jabr wal-muqabala*), the algebra done at that time is very different from what we do now. The problems were similar to those of today, for instance, solving linear and quadratic equations, and the method was similar—do the same thing to both sides of an equation until the unknown is isolated. But there were no formulas. There weren't even any symbols. Everything was done in words. A solution was a procedure, exemplified by several examples, not a one-line formula. So, in moving from algebra to algorithms, in some sense we are moving back to the original algebra. The about-face is not complete. Our algorithms do contain symbolic expressions. But the solution need not be a single formula.

Teaching Mental Algorithms Constructively

Alistair McIntosh

N everyday life we use mental computation far more than formal written arithmetic. Even in 1957, before electronic calculators were universally available, adults used formal written computation for only 25 percent of their calculations (Wandt and Brown 1957); the rest they computed mentally. And this is true not only of adults. For example, Carraher and Schliemann (1985) have shown that children both tend to prefer informal mental strategies to formal written algorithms and are more proficient in their use.

I regularly ask groups of elementary school teachers how much of the time in their mathematics classes they devote to teaching formal written algorithms. They say that depending on the class they teach, they spend between 50 percent and 95 percent of their mathematics time teaching formal written algorithms; and yet they agree that these are seldom used, particularly in an age of electronic calculators and computers, and that outside the classroom people are much more likely to need and use mental-computation skills.

For this reason alone we should shift our emphasis to the teaching of mental computation. How should we do this? Traditional pedagogical methods would suggest that we should look for the best mental-computation algorithms and teach them. I suggest, however, that doing so may be counterproductive.

The differences between standard written algorithms and mental algorithms have been brilliantly analyzed by Plunkett (1979). He points out that although formal written algorithms have the advantage of providing a standard routine that will work for any numbers—large or small, whole or decimal—they also have disadvantages in that they do not correspond to the ways in which people tend to think about numbers and they discourage students from thinking about the numbers involved or from exercising any active choice or initiative while carrying out the computation. Mental-computation strategies are quite different: they are flexible and can be adapted to suit the numbers concerned; they involve a definite, if not conscious, choice of strategy based on considering the numbers involved; and, almost always, they require understanding. I have found one exception to the last rule, and it is highly significant, as I shall discuss later. But the most important distinction between normal written and mental-computation algorithms at present is exemplified by the following: Suppose we say to a class, "What is 36 + 79? Do it with pencil and paper." Then, by and large, we will expect every child in the classroom who gets the problem right to do the same thing: "6 and 9 are 15; put down the 5 and carry the 1; 3 and 7 are 10, and 1 more is 11. The answer is 115." But ask the same class to do the same calculation mentally and the situation alters radically. In my experience and that of colleagues who have been asking children similar questions over the past ten years in a variety of classrooms, we would expect to hear any or all of the following:

"3 and 7 are 10; 6 and 9 are 15; that's 115."
"30 and 70 are 100; 6 and 9 are 15; that's 115."
"36 and 80 are 116; less 1 is 115."
"36 and 70 are 106, and 9 is 115."
"79 and 6 are 85, and 30 is 115."
"79 and 21 are 100; 36 less 21 is 15; 100 and 15 are 115."

A number of students, particularly those from classrooms in which formal written algorithms have been heavily emphasized, will screw up their eyes, raise their hand as though writing in the air in front of them, and say, "6 and 9 are 15; put down the 5 and carry the 1; 3 and 7 are 10, and 1 more is 11. The answer is 115."

All these methods work, all of them are more or less efficient, they rarely appear to have been taught, and almost all take account of the particular numbers involved, as opposed to being automatically applied. We can tell this because when we ask children to do another similar calculation, for example, 25 + 83, they often use a quite different mental strategy.

Almost all the children we have interviewed have, one way or another, acquired a range of strategies for performing mental computations, but the less confident or competent are particularly handicapped by the lack of specific classroom support in acquiring more-efficient and more-reliable strategies. Moreover, in asking seventy-two children in grades 2 to 7 in three different schools to explain their strategies for a total of more than three thousand mental computations, we found that the same range of mental-calculation strategies was eventually available to almost all the students: but the more able children acquired them up to two years earlier.

The students started by using the most primitive strategies based on counting forward and back by ones. They then progressed to a range of more-sophisticated strategies—counting forward and back in larger numbers, using number properties such as commutativity, bridging 10 or multiples of 10 (with 38 + 7, for example, 38 and 2 is 40 and 5 more is 45), building on known facts, compensating (e.g., 28 + 27 is the same as 30 + 25), using place value to split up numbers, and so on. We are not talking here about highly intelligent children, whom we deliberately excluded from our research, but the general run of ordinary children.

These children have powers of mental computation that they develop themselves without support. The heart of all these self-devised algorithms is that the child tries to turn a difficult calculation into an easy one: 28 + 27 is hard, 30 + 25 is easy; 9 × 17 is hard, 10 × 17 − 17 is easier. The more able children develop their ability to devise algorithms and so make life easy. Unfortunately, the less able children are less capable of helping themselves and so are left with much more difficult tasks. Try starting with 28 and adding on 27 in ones without error, and you will appreciate their problem.

340

We must find ways of helping children in classrooms improve their mental-computation abilities, but it appears that teaching them particular methods may be unhelpful. We are led to that belief because when we interviewed the students, we found only one strategy that had been deliberately taught, either by parents or teachers, to a significant number of children. It was the only strategy the children tried to use without understanding. The strategy is commonly referred to by the students as "taking off the zeros." "What is 70 + 20?" "90." "How did you do it?" "I took off the noughts, 7 + 2 = 9; put back the noughts, 90." "Why does it work?" "Don't know." "My mother [or father or teacher] showed me." "How many noughts did you take off?" "Two." "How many did you put back?" "One." "Why?" "Don't know. It just works."

This lack of understanding leads to few errors in addition or subtraction, but it causes problems in multiplication and division. When asked to calculate mentally 150 ÷ 30, most children who have been taught this rule take off the noughts, then "put back" one or both of them, which produces an answer of 50 or 500. Because they have learned to do this algorithm without understanding, they rarely query the answer, even when it is put in a real-life context, for example, ordering thirty-seat buses to take 150 children on an outing.

I recall one child who had no problems with the mental calculation because she approached it in an entirely different way. She was a capable eight-year-old child who had not been taught about removing zeros, and she answered, almost straight away, "5." "Why?" "Well, 30 and 30 is 60, and 30 is 90, and 30 is 120, and 30 is 150. Five 30s." This child understood the problem, thought about the numbers, hadn't been taught how to do the calculation, and used a method she was in control of. In trying to help the other children by teaching them what to do, teachers and parents had quite unconsciously prevented them from using their own number sense.

So what do we do?

We must stop restricting our mental-arithmetic sessions to bursts of short, unrelated calculations in which we emphasize accuracy and speed. If children are given time, they try—often with success—to invent an algorithm. If we emphasize speed, we remove this possibility. So first, don't pressure children by emphasizing speed.

Second, when a calculation has been done, do not concentrate only on whether the students have the right or the wrong answer. Ask at least three or four children to explain how they did the mental calculation, and show interest and enthusiasm for the variety of algorithms presented. Students need to realize that there are many ways of doing a calculation mentally and that there is rarely one "best" method. Of the mental algorithms that children hear discussed, they will use the one that they trust and understand best.

Often you will notice a heightening of interest in a method described by a child, which indicates that the algorithm has been understood and appreciated by several children. At such times "teaching an algorithm" is valid. You might say, "Let's see if we can all use Amanda's method to do these calculations." Making such a suggestion is quite different from imposing a single algorithm from outside. The situation is more like what happens when someone shows off a new toy. A child might ask, "Can I have a go with that?" We all want to try it out a few times to see if we would like it for ourselves. So third, let the students' interest be the springboard for practicing mental algorithms.

Finally, make experiences with mental arithmetic pleasurable for all the children by shifting the emphasis from testing their performance to supporting and encouraging their attempts to think for themselves. Here are three suggestions to get you going. They are based on the work of hundreds of Australian primary school teachers whom we have observed adapting and using these ideas in their classrooms (McIntosh, De Nardi, and Swan 1994):

1. Instead of asking students the answer to a calculation, do the reverse! Say, "I did a calculation, and the answer was 12. What might the calculation have been?" Stick to a simple number—between 10 and 30—so that all the children can engage at their own level. Record all the answers and have the children discuss which are alike. Later you can challenge the students to write down as many ways as they can think of to make 12.

2. Give a calculation, give children plenty of time to work out their answer, establish the answer, then ask, "Who can tell me how he or she did it?" "Do you understand Dan's way?" "Who did it the same way as Dan?" "Who did it a different way?"

3. Students very often deal with abstract calculations in class that they never relate to real life. To help the students make this vital link, give them a calculation, for example, 20 − 7 or 15 × 4, let them give the answer, then invite them to suggest situations like the following that incorporate that calculation: "I had $20, but I spent $7. I have only $13 left." "I have 15 toy cars. All together they have 60 wheels." Often children restrict themselves to stories about money or lollipops, so widen their horizons by asking them to compose a fantasy story or a situation in the kitchen or the zoo.

Activities such as these help your students view the development of their mental-calculation algorithms as an enjoyable, relevant, and approachable challenge. What I am proposing is not new. In fact it is simply elaborating on the advice Warren Colburn gave more than one hundred fifty years ago (Colburn 1970):

> The learner should never be told directly how to perform any operation in arithmetic. . . . Nothing gives scholars so much confidence in their own powers and stimulates them so much to use their own efforts as to allow them to pursue their own methods and to encourage them in them.

REFERENCES

Carraher, Terezinha Nunes, and Analúcia Dias Schliemann. "Computation Routines Prescribed by Schools: Help or Hindrance?" *Journal for Research in Mathematics Education* 16 (January 1985): 37–44.

Colburn, Warren. "Teaching of Arithmetic." In *Readings in the History of Mathematics Education,* edited by James K. Bidwell and Robert G. Clason, pp. 24–37. Washington, D.C.: National Council of Teachers of Mathematics, 1970. Reprinted from *Elementary School Teacher* 12 (June 1912): 463–80. (Text of an address delivered by Warren Colburn before the American Institute of Instruction in Boston, August 1830).

McIntosh, Alistair J., Ellita De Nardi, and Paul Swan. *Think Mathematically.* Melbourne, Victoria: Longman, 1994.

Plunkett, Stuart. "Decomposition and All That Rot." *Mathematics in School* 8 (May 1979): 2–5.

Wandt, Edwin, and Gerald W. Brown. "Non-Occupational Uses of Mathematics: Mental and Written—Approximate and Exact." *Arithmetic Teacher* 4 (October 1957): 151–54.

16

From *Developing Mathematical Reasoning in Grades K–12,* NCTM's Sixty-first Yearbook (1999)

"Twenty Questions about Mathematical Reasoning," by Lynn Arthur Steen, appeared as chapter 23 in NCTM's Sixty-first Yearbook in 1999. It was selected for this volume by Shirley Frye, who served as NCTM's president from 1988 to 1990.

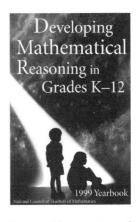

What a timely chapter! For those implementing the Common Core State Standards (National Governors Association Center for Best Practices and Council of Chief State School Officers 2010), the focus of the chapter on questions concerning mathematical reasoning relates to two of the Common Core's Standards for Mathematical Practice: "Reason abstractly and quantitatively" and "Construct viable arguments and critique the reasoning of others." The chapter also considers points of emphasis made in NCTM's *Focus in High School Mathematics: Reasoning and Sense Making* (NCTM 2009).

Just the headings alone in Steen's chapter raise questions that might engage any teacher population in discussing the importance of reasoning in teaching and learning mathematics. Some examples include "Is Mathematical Reasoning Useful?"; "Can Teachers Teach Mathematical Reasoning?"; "Do Skills Lead to Understanding?"; and "Is Proof Essential to Mathematics?" Such questions could serve as professional development discussion starters for grade-level or course-level teams.

Twenty Questions about Mathematical Reasoning

Lynn Arthur Steen

W E begin with two warm-up questions. First, *Why is mathematics an integral part of the K–12 curriculum?* The answers are self-evident and commonplace: to teach basic skills, to help children learn to think logically, to prepare students for productive life and work, and to develop quantitatively literate citizens.

Second, and more problematic, *How does mathematical reasoning advance these goals?* This is not at all self-evident since it depends greatly on the interpretation of "mathematical reasoning." Sometimes this phrase denotes the distinctively mathematical methodology of axiomatic reasoning, logical deduction, and formal inference. Other times, it signals a much broader quantitative and geometric craft that blends analysis and intuition with reasoning and inference, both rigorous and suggestive. This ambiguity confounds any analysis and leaves room for many questions.

1. Is Mathematical Reasoning Mathematical?

Epistemologically, reasoning is the foundation of mathematics. As science verifies through observation, mathematics relies on logic. The description of mathematics as the "science of drawing necessary conclusions," given over a century ago by the philosopher C. S. Peirce, still resonates among mathematicians of today. For example, a contemporary report by mathematicians on school mathematics asserts that "the essence of mathematics lies in proofs" (Ross 1997).

Yet mathematics today encompasses a vast landscape of methods, procedures, and practices in which reasoning is only one among many tools (e.g., Mandelbrot 1994; Thurston 1994; Denning 1997). Computation and computer graphics have opened new frontiers of both theory and application that could not have been explored by previous generations of mathematicians. This frontier has revealed surprising mathematical insights; for example, that deterministic phenomena can exhibit random behavior, that repetition can be the source of chaos as well as accuracy, and that uncertainty is not entirely haphazard since regularity always emerges (Steen 1990).

It took innovative mathematical methods to achieve these insights—methods that were not tied exclusively to formal inference. Does this mean that mathematical reasoning now includes the kind of instinct exhibited by a good engineer who finds solutions that work without worrying about formal proof? Does it include the kinds of inferences from "noisy" data that define the modern practice of statistics? Must mathematical reasoning be symbolic or deductive? Must it employ numbers and algebra? What about visual, inductive, and heuristic inferences? What about the new arenas of experimental mathematics and computer-assisted problem solving? What, indeed, is distinctively *mathematical* about mathematical reasoning?

2. Is Mathematical Reasoning Useful?

For most problems found in mathematics textbooks, mathematical reasoning is quite useful. But how often do people find textbook problems in real life? At work or in daily life, factors other than strict reasoning are often more important. Sometimes intuition and instinct provide better guides; sometimes computer simulations are more convenient or more reliable; sometimes rules of thumb or back-of-the-envelope estimates are all that is needed.

In ordinary circumstances, people employ mathematics in two rather different ways: by applying known formulas or procedures to solve standard problems or by confronting perplexing problems through typically mathematical strategies (e.g., translating to another setting, looking for patterns, reasoning by analogy, generalizing and simplifying, exploring specific cases, or abstracting to remove irrelevant detail). Rarely do they engage in the rigorous deduction characteristic of formal mathematics. At work and in the home, sophisticated multistep calculations based on concrete measurement-based mathematics are far more common than are chains of logical reasoning leading to mathematical proof (Forman and Steen 1995). It is not the methodology of formal deduction that makes mathematics useful for ordinary work so much as the mathematical habits of problem solving and the mathematical skills of calculation (Packer 1997).

Can people do mathematics without reasoning? Many certainly do—using routine methods ingrained as habit. Can people reason without using mathematics? Obviously so, even about situations (e.g., gambling, investing) that mathematicians would see as intrinsically mathematical. Those few people who employ advanced mathematics necessarily engage in some forms of mathematical reasoning, although even for them the role played by reasoning may be unconscious or subordinate to other means of investigation and analysis. But how much mathematical reasoning is really needed for the kinds of mathematics that people do in their lives and work? Does ordinary mathematical practice really require much mathematical reasoning?

3. Is Mathematical Reasoning an Appropriate Goal of School Mathematics?

Mathematics teachers often claim that all types of critical thinking and problem solving are really examples of mathematical reasoning. Nevertheless, employers have a different view, rooted in a paradox: Graduates with degrees in mathematics or computer science are often less successful than other graduates in solving the kinds of problems that arise in real work settings. Often students trained in mathematics tend to seek precise or rigorous solutions regardless of whether the context warrants such an approach. For employers, this distinctively mathematical approach is frequently *not* the preferred means of solving most problems arising in authentic contexts. Critical thinking and problem solving about the kinds of problems arising in real work situations is often better learned in other subjects or in integrative contexts (Brown 1995).

The goals of school mathematics seem to shift every decade, from "conceptual understanding" in the new math '60s to "basic skills" in the back-to-basics '70s, from "problem solving" in the pragmatic '80s to "mathematical power" in the standards-inspired '90s. Will "mathematical reasoning" be next? Not likely. In its strict (deductive) meaning, mathematical reasoning is hardly sufficient to support the public purposes of school mathematics. Everyone needs the practice of mathematics. But who really needs to understand mathematics? Who really needs mathematical reasoning? Can one make the case that every high school graduate needs to be able to think mathematically rather than just perform mathematically?

4. Can Teachers Teach Mathematical Reasoning?

The Third International Mathematics and Science Study (TIMSS) documented that U.S. mathematics teachers focus on teaching students how to do mathematics and not on understanding what they do (NCES 1996). There are many reasons for this, including teachers' image of mathematics as a set of skills, parental demands that children master the basics before advancing to higher-order tasks, and the constraining environment of state-mandated tests that emphasize routine calculations.

Many believe that curricular reform based on mathematical reasoning will never succeed since there are far too few teachers prepared to do justice to such a goal. Even if enough willing and able teachers can be found (or educated), will the public allow them to teach mathematical reasoning in school? Might the fear of "fuzzy mathematics" (Cheney 1997) constrain even those teachers who might want to stress understanding?

5. Can Mathematical Reasoning Be Taught?

Just as we do not really know what mathematical reasoning is, so we also do not know very much about how it develops. Research does support a few general conclusions. First, successful learners are mathematically active (Anderson, Reder, and Simon 1997). Passive strategies (memorization, drill, templates) are much less likely than active tasks (discussion, projects, teamwork) to produce either lasting skills or deep understanding. Second, successful mathematics learners are more likely to engage in reflective (or "metacognitive") activity (Resnick 1987). Students who think about what they are doing and why they are doing it are more successful than those who just follow rules they have been taught.

We also know that students differ: No single strategy works for all students, nor even for the same student in all circumstances. Howard Gardner's theory of multiple intelligences (Gardner 1983, 1995) supports the practice of experienced teachers who create

349

multiple means for students to approach different topics. Diverse experiences provide implicit contexts in which mathematical reasoning may emerge. But can we be sure that it will eventually emerge? Might some students, or some types of reasoning, require explicit instruction? Are there some types of mathematical reasoning that can only develop through students' construction and reflection? If some types of mathematical reasoning cannot be taught explicitly, is it appropriate to require it of all high school graduates?

6. Do Skills Lead to Understanding?

Although mathematical performance generally involves a blend of skills, knowledge, procedures, understanding, reasoning, and application, the public mantra for improving mathematics education focuses on skills, knowledge, and performance—*what* students "know and are able to do." To this public agenda, mathematics educators consistently add reasoning and understanding—why and *how* mathematics works as it does.

Experienced teachers know that knowledge and performance are not reliable indicators of either reasoning or understanding. Deep understanding must be well-connected. In contrast, superficial understanding is inert, useful primarily in carefully prescribed contexts such as those found in typical mathematics classrooms (Glaser 1992). Persons with well-connected understanding attach importance to different patterns and are better able to engage in mathematical reasoning. Moreover, students with different levels of skills may be equally able to address tasks requiring more sophisticated mathematical reasoning (Cai 1995).

Nonetheless, the public values (and hence demands) mathematics education not so much for its power to enhance reasoning as for the quantitative skills that are so necessary in today's world. It is not that adults devalue understanding but that they expect basic skills first (Wadsworth 1997). They believe in a natural order of learning—first skills, then higher order reasoning. But, do skills naturally lead to understanding? Or is it the reverse—that understanding helps secure skills? Does proficiency with mathematical facts and procedures necessarily enhance mathematical reasoning? Conversely, can mathematical reasoning develop in some students even if they lack a firm grasp of facts and basic skills? Might the relation of skills to reasoning be like that of spelling to writing—where proficiency in one is unrelated to proficiency in the other?

7. Can Drill Help Develop Mathematical Reasoning?

Critics of current educational practice indict "drill and kill" methods for two crimes against mathematics: disinterest and anxiety. Both cause many students to avoid the subject as soon as they are given a choice. Yet despite the earnest efforts to focus mathematics on reasoning, one out of every two students thinks that learning mathematics is mostly memorization (Kenney and Silver 1997).

350

And they may have a point. Research shows rather convincingly that real competence comes only with extensive practice (Bjork and Druckman 1994). Nevertheless, practice is certainly not sufficient to ensure understanding. Both the evidence of research and the wisdom of experience suggest that students who can draw on both recalled and deduced mathematical facts make more progress than those who rely on one without the other (Askew and William 1995).

However, children who can recite multiplication facts may still not understand why the answers are as they are or recognize when multiplication is an appropriate operation, much less understand how ratios relate to multiplication. High school students who memorize proofs in a traditional geometry course may show good recall of key theorems but be totally unable to see how the ideas of these proofs can be used in other contexts. Is there, indeed, any real evidence that practiced recall leads to reasoning and understanding?

8. Is Proof Essential to Mathematics?

Despite the dominance of proof as the methodology of advanced mathematics courses, contemporary advances in applied, computer-aided, and so-called "experimental" mathematics have restored to mathematical practice much of the free-wheeling spirit of earlier eras. Indeed, these recent innovations have led some to proclaim the "death" of proof— that although proof is still useful in some contexts, it may no longer be the *sine qua non* of mathematical truth (Horgan 1993). Although this claim is hotly disputed by many leading mathematicians, it resonates with diverse pedagogical concerns about the appropriateness (or effectiveness) of proof as a tool for learning mathematics. Uncertainty about the role of proof in school mathematics caused NCTM in its *Standards* (NCTM 1989) to resort to euphemisms—"justify," "validate," "test conjectures," "follow logical arguments." Rarely do the *Standards* use the crystalline term "proof."

In fact, most people understand *proof* in a pragmatic rather than a philosophical way: provide just enough evidence to be convincing. For many people, proof is tantamount to the civil legal test of "preponderance of evidence"; others require the stricter standard of "beyond reasonable doubt." In routine uses of mathematics, what works takes precedence over what is provable. So how much understanding of formal proof is necessary for the routine practice of mathematics? Probably not very much. But how much is needed for advanced study of mathematics? Undoubtedly a great deal.

9. Does Learning Proofs Enhance Mathematical Reasoning?

Nothing divides research mathematicians and mathematics educators from each other as do debates about the role of proof in school mathematics. Proof is central to mathematical reasoning, yet there is precious little agreement on how, when, why, or to whom to teach it. Its suitability for school mathematics has always been open to question, both on the grounds of pedagogy and relevance.

351

The vocabulary of mathematical truth, rigor, and certainty is not a natural habitat for most students. Their world is more empirical, relying on modeling, interpretation, and applications. Only a very few students in high school comprehend proof as mathematicians do—as a logically rigorous deduction of conclusions from hypotheses (Dreyfus 1990). Students generally have very little comprehension of what proof means in mathematics, nor much appreciation of its importance (Schoenfeld 1994). Might early introduction of proof actually do more to hinder than enhance the development of mathematical reasoning?

Although mathematicians often advocate including proof in school curricula so students can learn the logical nature of mathematics (Ross 1997), the most significant potential contribution of proof in mathematics education may be its role in communicating mathematical understanding (Hanna and Jahnke 1996). The important question about proof may not be whether it is crucial to understanding the nature of mathematics as a deductive, logical science, but whether it helps students and teachers communicate mathematically. Is, perhaps, proof in the school classroom more appropriate as a means than as an end?

10. Does "Math Anxiety" Prevent Mathematical Reasoning?

Mathematics is perhaps unique among school subjects in being a major cause of anxiety. Many students believe deeply that they cannot do mathematics and so learn to avoid it; a few are so paralyzed by the prospect that they exhibit physiological evidence of acute anxiety (Buxton 1991; Tobias 1993). It may seem obvious that anyone suffering even mildly from "math anxiety" would not engage in much mathematical reasoning. But this is not at all the case. Many students (and adults) who fear mathematics are, in fact, quite capable of thinking mathematically, and do so quite often—particularly in their attempts to avoid mathematics! What they really fear is not mathematics itself, but school mathematics (Cockcroft 1982).

Both research and common sense say that anxiety is reduced when individuals can control uncertainties (Bjork and Druckman 1994). When percentages and ratios appear as impossible riddles, panic ensues. But, when self-constructed reasoning—under the control of the individual—takes over, much valid mathematical reasoning may emerge, often in a form not taught in school. How can schools respect each student's unique approach to mathematical reasoning while still teaching what society expects (and examines)? Would reduced panic result in improved reasoning? Is this a case where less may be more—where reduced instruction might yield deeper understanding?

11. Do Cooperative Activities Enhance Individual Understanding?

Arguments for cooperative learning and teamwork come from two rather different sources: first, from those (primarily in the education world) who view these activities as effective strategies for learning mathematical reasoning and, second, from those (primarily in the business world) who view cooperative activities as essential for productive employees (SCANS 1991). Advocates envision mathematics classes as communities where students engage in collaborative mathematical practice with each other and with their teachers (Silver, Kilpatrick, and Schlesinger 1990). In such classes students would regularly engage in authentic forms of mathematical practice by inventing strategies, arguing about approaches, and justifying their work.

Parents often object to educators' rationale for teamwork since they view mathematics as an ideal subject in which individual accomplishment can be objectively measured and rewarded. They worry both that children who are above average will be held back by slower students and that those who are behind will be instructed, not by teachers, but by other children. Ironically, despite their distrust of teamwork in subjects like mathematics, most parents and students admire teamwork in sports and musical organizations. (Of course, in sports and music—as in the workplace—success accrues not to individuals but to the team as a whole.) Despite these objections, there is considerable evidence that cooperative learning is effective, especially for children (Bjork and Druckman 1994). For high school students and adults, however, the evidence is more mixed. Older students bring to cooperative groups stronger individual motivations, complex experiences in social interactions, and often some defensiveness or embarrassment about learning.

Employers value teamwork because it produces results that no individual can accomplish alone. But can teamwork in the classroom also produce reasoning at a higher level than could be accomplished by any single member of a team? Will individual members of a team learn more mathematics as a result? Just how do group activities promote mathematical reasoning in individuals? Even more difficult and important: How can mathematics educators gain public support for cooperative activities?

12. Can Calculators and Computers Increase Mathematical Reasoning?

At home and at work, calculators and computers are "power tools" that remove human impediments to mathematical performance. For example, spreadsheets and statistical packages are used by professionals both to extend the power of mind as well as to substitute for it—by performing countless calculations without error or effort. Students certainly need to learn these empowering uses of technology.

But, in addition, calculators and computers are responsible for a "rebirth of experimental mathematics" (Mandelbrot 1994). They provide educators with wonderful tools for generating and validating patterns that can help students learn to reason mathematically. Computer games can help children master basic skills; intelligent tutors can help older students master algebraic procedures. Many educators have argued that since programming enforces logical rigor, computer languages such as Logo and ISETL can help students learn to reason.

353

Calculators and computers hold tremendous potential for mathematics. Depending on how they are used, they can either enhance mathematical reasoning or substitute for it, either develop mathematical reasoning or limit it. However, judging from public evidence, the actual effect of calculators in school is as often negative as positive: For every student who learns to use spreadsheets, there seem to be several who reach for a calculator to add single digit numbers or to divide by 10. Why are the consequences of calculators in school mathematics so mixed? Why is there such a big gap between aspirations and accomplishment?

13. Why Do So Many Students Feel That Mathematics Is a Foreign Culture?

A substantial number of children find school mathematics opaque. Part of children's difficulty in learning school mathematics lies in their failure to reconcile the rules of school-math with their own independently developed mathematical intuition (Freudenthal 1983; Resnick 1987). Too often, entrenched assumptions—like regular grammar applied in contexts where irregularity rules—impede learning.

To what extent does the mathematical environment in a child's home affect how the child responds to mathematics in school? Many people believe that certain peoples or cultures are better suited to mathematics than others. The thriving—and controversial—specialty of ethnomathematics documents beyond reasonable doubt that all societies have developed some form of mathematics and that these forms reflect the cultures in which they emerge. Historically alert mathematicians can recognize similarities and differences in the mathematics of different cultures and can trace the influence of cultures on one another in the evolution of mathematics (Joseph 1992). Thus, there are undeniable cultural differences in mathematics.

But are there cultural differences in the development of mathematical reasoning? Here the evidence is less definitive. World-class mathematicians have emerged from societies all around the globe, yet certain cultures put greater emphasis on the kinds of rigor and reasoning that give mathematics its special power. Students growing up in these cultures are more likely to recognize a zone of comfort in school mathematics, whereas students growing up in cultures that view the world through other lenses may feel as if school mathematics is a foreign culture. Why do some students see mathematics as the only welcoming subject in school, whereas others see it as the most foreign of cultures? Why, indeed, do some children find mathematics so unreasonably hard?

14. Is Context Essential for Mathematical Reasoning?

For at least a decade, both educational researchers and reformers have been preaching the message of *situated cognition* or *contextualized learning*. For much longer, scientists and engineers have fussed at mathematicians for persisting with context-free instruction

(Rutherford 1997). Recently vocational educators have joined the chorus, citing persistent lack of context in mathematics courses as one of the chief impediments to student learning (Bailey 1997; Hoachlander 1997). Yet, according to a National Research Council report, there is no consistent evidence that performance is enhanced when learning takes place in the setting in which skills will be performed (Bjork and Druckman 1994).

Context can affect learning in two opposing ways: Generally, it enhances motivation and long-term learning, but it can also can limit the utility of what is learned. Knowledge often becomes context-bound when it is taught in just one context. Anyone who has ever taught mathematics has heard complaints from teachers in other subjects that students don't appear to know any of the mathematics they were supposed to have learned in mathematics class. The pervasive problem of compartmentalized knowledge has led many educators to assume that transfer of knowledge from one subject to another is atypical. In fact, transfer does occur, but not nearly as systematically or as predictably as we would like.

Just how situated is mathematical cognition? Does instruction in context facilitate learning mathematics? Does it limit or enhance the likelihood of transfer to other domains? When, if ever, does mathematical reasoning transfer to other domains?

15. Must Students Really Construct Their Own Knowledge?

One of the most widely accepted goals of the mathematics community is that students should understand the mathematics they perform. For centuries, educators have known that understanding grows only with active learning. This has led, in the argot of mathematics educators, to a widespread belief that students *construct* their own understanding (Davis, Maher, and Noddings 1990; Hiebert and Carpenter 1992). In this view, understanding cannot be delivered by instructors, no matter how skillful, but must be created by learners in their own minds.

The constructivist posits that children learn as they attempt to solve meaningful problems. In this view, understanding emerges from reflection catalyzed by questions (Campbell and Johnson 1995). The teacher's primary role is not to instruct but to pose problems and ask questions that provoke students to reflect on their work and justify their reasoning. In this way, activities such as explaining, justifying, and exemplifying not only demonstrate understanding but also help create it.

According to supporters, constructivism focuses education on the learner (what happens in students' minds), on inquiry (seeking the right questions, not just the right answers), on relevance (questions of natural interest to children), and on activity (learning with both hand and mind) (Brooks and Brooks 1993). Yet critics (e.g., Anderson et al. 1997; Wu 1996) contend that constructivist methods too easily slight the importance both of didactics (systematic instruction) and drill (systematic practice). What is the appropriate balance between teacher-directed and student-inspired learning? Do students need to construct everything for themselves? What should be memorized and what constructed?

355

16. How Many Mathematics Are There?

Mathematics lives in many environments—home math, school math, street math, business math, work math—and many students who succeed in one mathematical world fail in another. Even though these are all mathematics, these environments offer fundamentally different contexts in which students learn and use mathematics. One might well imagine that, like multiple intelligences (Gardner 1983), there may be multiple mathematics (Grubb 1996).

Evidence of multiple mathematics abounds. Research documents what parents and teachers know from rueful experience—that many children see school mathematics as disconnected from sense-making and the world of everyday experience (Silver, Kilpatrick, and Schlesinger 1990; Schoenfeld 1991). The widespread separation of symbols from meaning and of calculation from reasoning is an inheritance of an educational system whose historic purpose was to separate the practical from the abstract and workers from scholars (Resnick 1987). Only for the elite was abstract or higher-order reasoning a goal (much less an accomplishment) of education. School has helped foster the public's view of different mathematics for different purposes.

This history encourages a pervasive myth about mathematics learning—that mathematical reasoning is appropriate only for the ten percent of students who are destined for mathematically rich careers in science and engineering. Yet in today's workplace, mathematical thinking is needed by more students than ever before. Nonetheless, some students learn mathematics better in mathematics classes, some in science or shop courses, and some on the job or at home. Do these settings offer different mathematics? In what circumstances is abstract mathematics appropriate? When is concrete mathematics better? Can we trust students to know which type of mathematics is best for them in particular contexts? Do teachers know enough to decide? Does anyone?

17. How Does Our Brain Do Mathematics?

Recent research in neuroscience has begun to open a window into what has heretofore been largely beyond the reach of science: the neural mechanism of cognition. Intriguingly, this research suggests a Darwinian mechanism of diversity and selection that operates within the brain just as it does among species in an ecosystem (Edelman 1992; Abbott 1994; Changeux and Connes 1995). Such a mechanism may help explain the stages of mathematical creativity noted in the classic work of Jacques Hadamard (1945) of preparation (trial and error), incubation (often subconscious), illumination (frequently sudden), and verification (requiring reasoning). According to this theory, mathematical reasoning depends on the same two forces as the evolution of species: a mechanism for generating diversity (alternatives) and a strategy for selection that stabilizes optimal choices among this diversity.

What, indeed, is the neural mechanism of mathematical thought? This is now a researchable question, and the implications of such research are profound. For the first time, we may be able to connect mathematical thinking to the biology of the brain. We now know, for instance, that memory involves several anatomically different structures.

As improved understanding of physiology has moved athletes' performances to the edge of human potential, might we soon be able to scientifically improve individuals' mathematical performance? Can we identify the biochemistry of mathematical reasoning? Might neuroscience help educators understand the vexing problem of transfer—or of the relation of skills to reasoning?

18. Is Our Brain Like a Computer?

We tend naively to think of the brain as a computer—especially when it is engaged in mathematical activity. Store basic facts in memory; provide key algorithms for calculation; then push a button. Much of the drill-oriented pedagogy of traditional mathematics education is rooted in this metaphor. In fact, as contemporary neuroscience reveals, the brain is less like a computer to be programmed or a disk to be filled than like an ecosystem to be nourished (Abbott 1996; ECS 1996, 1997).

Although the evidence against the brain-as-computer metaphor is overwhelming (e.g., recovery patterns of stroke victims), the paradigm persists in large measure for lack of a compelling alternative. But that may be about to change. Research in the intersection of evolutionary genetics and neuroscience suggests potentially important neurological differences between those cognitive capacities that are evolutionarily primitive (e.g., counting) and those such as arithmetic (not to mention algebra!) that are more recent social constructs (Geary 1995). Capacity for reasoning is created by a continually changing process of natural selection of neuronal groups responding to an individual's goals (called "values" by Edelman [1992]). Thus both the processes of cognition and the elements on which these processes act—if you will, procedures and facts—are subject to the evolutionary pressures of diversity and selection within the living brain.

19. Is The Capacity for Mathematics Innate?

For years, linguists and neuroscientists have studied the way babies learn language in an effort to understand the relation of human language to the genetic endowment of our species. As children naturally develop their own rules of grammar—regularizing irregular verbs, for example—so they also invent rules to explain patterns they see around them. To the extent that making patterns is a mathematical activity (Steen 1988; Devlin 1994), young children learning language are doing mathematics!

There is abundant evidence that young children, on their own, develop simple mathematical rules that they use to solve problems in their environment (Resnick 1987). Yet these patterns often lead to mathematical misconceptions—e.g., that multiplication makes things bigger—that persist despite subsequent contrary evidence and instruction (Askew and William 1995). Does this mean that young children have the same innate capacity to learn mathematics as they have to learn language? How might mathematical reasoning be enhanced if babies were bathed in an environment as rich in mathematical patterns as it is in natural language?

20. Is School Too Late?

Although certain aspects of the brain are determined by genetics and by the environment in the womb, both neurons and synapses grow and change rapidly during the early years of life. How they grow is determined by the environment of the infant. What they become—after five or six years—determines to a considerable degree the cognitive capacity of the child and adult. Although much of the brain is formed at birth, much remains plastic, amenable to being shaped by experience. The capacity for abstract thinking is particularly plastic. Synapse growth occurs at a phenomenal rate until age two or three and then gradually diminishes for the rest of life (ECS 1997). "Use it or lose it" is a fitting description of the early brain.

Everyone knows the importance of aural stimulation for the learning of language in the first years of life. Recent research has provided rather firm evidence that musical stimulation in these early years enhances capacity for spatial and mathematical abstraction later in life (Rauscher and Shaw 1997). (Whether early musical stimulation enhances musicality is less clear.) Apparently the acoustical bath of aural structure provided by classical music does for the abstract centers of the brain what hearing phonemes does for language learning.

This research leads to many questions that are hardly touched on in mathematics education. Are there "windows" for learning arithmetic or algebra, or for mathematical reasoning, as there surely are for learning language? What, besides music, can enhance the young brain's capacity for mathematical thinking? How sensitive is mathematical ability to the sensory environment of a baby? Just how does learning change the brain's physiology? Might we someday be able to sculpt children's capacity for mathematical reasoning?

REFERENCES

Abbott, John. "The Search for Next-Century Learning." *AAHE Bulletin* 48, no. 7 (March 1996): 3–6.

———. *Learning Makes Sense: Recreating Education for a Changing Future.* Letchworth, U.K.: Education 2000, 1994.

Anderson, John R., Lynne M. Reder, and Herbert A. Simon. "Applications and Misapplications of Cognitive Psychology to Mathematics Education." Carnegie Mellon University Web Site, <sands.psy.cmu.edu/ACT/papers/misapplied-absja.html> (1997).

Askew, Michael, and Dylan William. *Recent Research in Mathematics Education 5–16.* London: Her Majesty's Stationery Office, 1995.

Bailey, Thomas. "Integrating Vocational and Academic Education." In *High School Mathematics at Work.* Washington, D.C.: National Academy Press, 1997.

Bjork, Robert A., and Daniel Druckman, eds. *Learning, Remembering, Believing: Enhancing Human Performance.* Washington, D.C.: National Research Council, 1994.

Brooks, Jacqueline G., and Martin G. Brooks. *In Search of Understanding: The Case for Constructivist Classrooms.* Alexandria, Va.: American Society for Curriculum Development, 1993.

Brown, Patricia, ed. *Promoting Dialogue on School-to-Work Transition.* Washington, D.C.: National Governors' Association, 1995.

Buxton, Laurie. *Math Panic.* Portsmouth, N.H.: Heinemann Educational Books, 1991.

Cai, Jinfa. "A Cognitive Analysis of U.S. and Chinese Students' Mathematical Performance on Tasks Involving Computation, Simple Problem Solving, and Complex Problem Solving." *Journal for Research in Mathematics Education,* Monograph No. 7. Reston, Va.: National Council of Teachers of Mathematics, 1995.

Campbell, Patricia F., and Martin L. Johnson. "How Primary Students Think and Learn." In *Prospects for School Mathematics,* edited by Iris M. Carl, pp. 21–42. Reston, Va.: National Council of Teachers of Mathematics, 1995.

Changeux, Jean-Pierre, and Alain Connes. *Conversations on Mind, Matter, and Mathematics.* Princeton, N.J.: Princeton University Press, 1995.

Cheney, Lynne. "Creative Math or Just 'Fuzzy Math'? Once Again, Basic Skills Fall Prey to a Fad." *The New York Times* (11 August 1997): A13.

Cockcroft, Sir Wilfred. *Mathematics Counts.* London: Her Majesty's Stationery Office, 1982.

Davis, Robert B., Carolyn A. Maher, and Nel Noddings, eds. "Constructivist Views on the Teaching and Learning of Mathematics." *Journal for Research in Mathematics Education,* Monograph No. 4. Reston, Va.: National Council of Teachers of Mathematics, 1990.

Denning, Peter J. "Quantitative Practices." In *Why Numbers Count: Quantitative Literacy for Tomorrow's America,* edited by Lynn Arthur Steen, pp. 106–17. New York: The College Board, 1997.

Devlin, Keith. *Mathematics: The Science of Patterns.* New York: W. H. Freeman, 1994.

Dreyfus, Tommy. "Advanced Mathematical Thinking." In *Mathematics and Cognition,* edited by Pearla Nesher and Jeremy Kilpatrick, pp. 113–34. Cambridge U.K.: Cambridge University Press, 1990.

Edelman, Gerald. *Bright Air, Brilliant Fire.* New York: Basic Books, 1992.

Education Commission of the States. *Bridging the Gap Between Neuroscience and Education.* Denver, Colo.: Education Commission of the States, 1996.

———. *Brain Research Implications for Education* 15, no. 1 (Winter 1997).

Forman, Susan L., and Lynn Arthur Steen. "Mathematics for Life and Work." In *Prospects for School Mathematics,* edited by Iris M. Carl, pp. 219–41. Reston, Va.: National Council of Teachers of Mathematics, 1995.

Freudenthal, Hans. "Major Problems of Mathematics Education." In *Proceedings of the Fourth International Congress on Mathematical Education,* edited by Marilyn Sweng, et al., pp. 1–7. Boston, Mass.: Birkhäuser, 1983.

Gardner, Howard. *Frames of Mind: The Theory of Multiple Intelligences.* New York: Basic Books, 1983.

———. "Reflections on Multiple Intelligences: Myths and Messages." *Phi Delta Kappan* (November 1995): 200–209.

Geary, David C. "Reflections on Evolution and Culture in Children's Cognition: Implications for Mathematical Development and Instruction." *American Psychologist* 50, no. 1 (1995).

Glaser, Robert. "Expert Knowledge and the Processes of Thinking." In *Enhancing Thinking Skills in the Sciences and Mathematics,* edited by Diane. F. Halpern, pp. 63–75. Hillsdale, N.J.: Lawrence Erlbaum Associates, 1992.

Grubb, Norton. "Exploring Multiple Mathematics." Project EXTEND Web Site <www.stolaf .edu/other/extend/Expectations/grubb.html> (1996).

Hadamard, Jacques. *The Psychology of Invention in the Mathematical Field.* Princeton, N.J.: Princeton University Press, 1945.

Hanna, Gila, and H. Niels Jahnke. "Proof and Proving." In *International Handbook of Mathematics Education,* edited by Alan J. Bishop, et al., pp. 877–908. Dordrecht: Kluwer Academic Publishers, 1996.

Hiebert, James, and Thomas P. Carpenter. "Learning and Teaching with Understanding." In *Handbook of Research on Mathematics Teaching and Learning,* edited by Douglas A. Grouws, pp. 65–97. New York: Macmillan, 1992.

Hoachlander, Gary. "Organizing Mathematics Education Around Work." In *Why Numbers Count: Quantitative Literacy for Tomorrow's America,* edited by Lynn Arthur Steen, pp. 122–36. New York: The College Board, 1997.

Horgan, John. "The Death of Proof." *Scientific American* 269 (April 1993): 93–103.

Joseph, George Gheverghese. *The Crest of the Peacock: Non-European Roots of Mathematics.* London: Penguin Books, 1992.

Kenney, Patricia Ann, and Edward A. Silver, eds. *Results from the Sixth Mathematics Assessment of the National Assessment of Educational Progress.* Reston, Va.: National Council of Teachers of Mathematics, 1997.

Mandelbrot, Benoit. "Fractals, the Computer, and Mathematics Education." In *Proceedings of the 7th International Congress on Mathematical Education,* edited by Claude Gaulin, et al. pp. 77–98. Sainte-Foy, Que.: Les Presses de l'Université Laval, 1994.

National Center for Educational Statistics. *Pursuing Excellence.* Washington, D.C.: U.S. Department of Education, 1996.

National Council of Teachers of Mathematics (NCTM). *Curriculum and Evaluation Standards for School Mathematics.* Reston, Va.: NCTM, 1989.

———. *Focus in High School Mathematics: Reasoning and Sense Making.* Reston, Va.: NCTM, 2009.

National Governors Association Center for Best Practices and Council of Chief State School Officers (NGA Center and CCSSO). *Common Core State Standards for Mathematics. Common Core State Standards (College- and Career-Readiness Standards and K–12 Standards in English Language Arts and Math).* Washington, D.C.: NGA Center and CCSSO, 2010. http://www. corestandards.org.

Packer, Arnold. "Mathematical Competencies that Employers Expect." In *Why Numbers Count: Quantitative Literacy for Tomorrow's America,* edited by Lynn Arthur Steen, pp. 137–54. New York: The College Board, 1997.

Rauscher, Frances, and Gordon Shaw. "Music Training Causes Long-Term Enhancement of Preschool Children's Spatial-Temporal Reasoning." *Neurological Research* 19 (February 1997): 2–8.

Resnick, Lauren B. *Education and Learning to Think*. Washington, D.C.: National Research Council, 1987.

Ross, Kenneth A. "Second Report from the MAA Task Force on the NCTM Standards." Mathematical Association of America Web Site <www.maa.org/past/nctmupdate.html> (1997).

Rutherford, F. James. "Thinking Quantitatively about Science." In *Why Numbers Count: Quantitative Literacy for Tomorrow's America*, edited by Lynn Arthur Steen, pp. 60–74. New York: The College Board, 1997.

Schoenfeld, Alan H. "On Mathematics as Sense-Making." In *Informal Reasoning and Education*, edited by J. F. Voss, D. N. Perkins, and J. W. Segal, pp. 311–43. Hillsdale, N.J.: Lawrence Erlbaum Associates, 1991.

———. "What Do We Know About Curricula?" *Journal of Mathematical Behavior* 13 (1994): 55–80.

Secretary's Commission on Achieving Necessary Skills (SCANS). *What Work Requires of Schools: A SCANS Report for America 2000*. Washington, D.C.: U.S. Department of Labor, 1991.

Silver, Edward A., Jeremy Kilpatrick, and Beth Schlesinger. *Thinking Through Mathematics*. New York: College Entrance Examination Board, 1990.

Steen, Lynn Arthur. "The Science of Patterns." *Science* 240 (29 April 1988): 611–16.

———. ed. *On the Shoulders of Giants: New Approaches to Numeracy*. Washington, D.C.: National Academy Press, 1990.

Thurston, William P. "On Proof and Progress in Mathematics." *Bulletin of the American Mathematical Society* 30 (1994): 161–77.

Tobias, Sheila. *Overcoming Mathematics Anxiety* (rev. ed.). New York: W. W. Norton, 1993.

Wadsworth, Deborah. "Civic Numeracy: Does the Public Care?" In *Why Numbers Count: Quantitative Literacy for Tomorrow's America*, edited by Lynn Arthur Steen, pp. 11–22. New York: The College Board, 1997.

Wu, Hung-Hsi. "The Mathematician and the Mathematics Education Reform." *Notices of the American Mathematical Society* 43, no. 12 (December 1996): 1531–37.

17

From *Making Sense of Fractions, Ratios, and Proportions,* NCTM's Sixty-fourth Yearbook (2002)

Fractions of all types, including decimals and percent, have been receiving a lot of attention lately. Francis (Skip) Fennell served as NCTM president from 2006 to 2008, and during the same period he served as a member and task group chair of the National Mathematics Advisory Panel. The work of that panel, along with Skip's prior experiences and efforts related to the importance of number sense, heightened his interest in the teaching and learning of fractions.

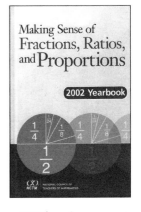

Skip found that many of the chapters in the Sixty-fourth Yearbook, *Making Sense of Fractions, Ratios, and Proportions,* are particularly relevant for the implementation of the Common Core State Standards (National Governors Association Center for Best Practices and Council of Chief State School Officers 2010) and as support for courses in mathematics and pedagogy related to fractions of all types. His selection of **chapter 5, "Using Manipulative Models to Build Number Sense for Addition of Fractions,"** by Kathleen Cramer and Apryl Henry, and **chapter 8, "Examining Dimensions of Fraction Operation Sense,"** by DeAnn Huinker, was based on the importance of the development of rational number sense, which both of these chapters emphasize and support.

Using Manipulative Models to Build Number Sense for Addition of Fractions

Kathleen Cramer
Apryl Henry

T HE Rational Number Project (RNP) has reported on several long-term teaching experiments concerning the teaching and learning of fractions among fourth- and fifth-grade students (Bezuk and Cramer 1989; Post et al. 1985). A curriculum created for these teaching experiments and then revised on the basis of what was learned reflects the following four beliefs: (1) children's learning about fractions can be optimized through active involvement with multiple concrete models, (2) most children need to use concrete models over extended periods of time to develop mental images needed to think conceptually about fractions, (3) children benefit from opportunities to talk to one another and with their teacher about fraction ideas as they construct their own understandings of fraction as a number, and (4) teaching materials for fractions should focus on the development of conceptual knowledge prior to formal work with symbols and algorithms (Cramer et al. 1997).

A decade of research on the teaching and learning of fractions among fourth and fifth graders has shown us that of the four pedagogical beliefs listed above, the second is the most important. In order to develop fraction sense, most children need extended periods of time with physical models such as fraction circles, Cuisenaire rods, paper folding, and chips. These models allow students to develop mental images for fractions, and these mental images enable students to understand about fraction size. Students can use their understanding of fraction size to operate on fractions in a meaningful way. The multiple models mentioned above were used in RNP teaching experiments. The fraction circle model used in combination with the RNP activities (see fig. 5.1) was the most powerful of the models. During interviews, students consistently referred to fraction circles as the model that helped them order fractions and estimate the reasonableness of fraction operations.

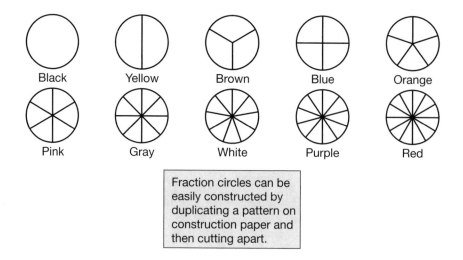

Black Yellow Brown Blue Orange

Pink Gray White Purple Red

Fraction circles can be easily constructed by duplicating a pattern on construction paper and then cutting apart.

Fig. 5.1. Fraction circles

We know a traditional approach to fraction instruction does not foster students' number sense. Consider the following two fourth-grade students. Both were tracked into the middle mathematics group in a suburban school district south of Minneapolis. They were interviewed the same day as a part of the district's project to field-test the RNP curriculum. The interview was the final one given to both students at the end of five weeks of instruction. Jeremy was in a class using the RNP curriculum, whereas Annie used a traditional textbook program. Jeremy had not learned any procedures for ordering fractions but had interacted daily with fraction manipulatives, exploring fraction concepts and ideas involving ordering at the concrete level. Annie had few opportunities with manipulative materials, since the traditional program moved quickly into procedural skills dealing with fractions. Both children were asked to select the smaller of this pair of fractions: 4/35 and 4/29. Jeremy responded by reasoning, "4/29 is greater. Pieces are bigger, and there's 4 of the smaller pieces and 4 of the bigger piece." Annie reasoned, "4/29 is less. Well, nothing times 29 equals 35, and it's lower numbers."

Annie demonstrated what the RNP documented many times: When students do not have mental images for fractions, they resort to whole-number strategies to try and make sense of the problem (Post et al. 1985). In this instance, Annie's learned procedure did not help her with this problem, since one denominator was not a factor of the other denominator; she therefore used a whole-number ordering strategy. Jeremy, by contrast, was able to reason about the fraction pair. Even though he did not have a concrete model to show 35ths or 29ths, he was able to transfer his thinking about ordering fractions with same numerators—which he did experience concretely in the RNP lessons—to a new situation. We can see that his language is tied to a manipulative model when he states that the "pieces are bigger." During the lessons he used circular pieces of different colors as the main model. Jeremy's thinking is typical of students who have used the RNP curriculum. Table 5.1 shows how he used mental images of fractions in three other ordering tasks from the same interview. Annie's responses to the same questions are also shown. Notice Jeremy's reliance on his mental pictures of the fraction circles.

Table 5.1

Other Examples of Students' Ordering Strategies

Problem Task	Jeremy [RNP]	Annie [TEXT]
Order smallest to largest: 1/5, 1/3, and 1/4	1/5, 1/4, 1/3. Pieces for 1/5 are the smallest; for 1/4 there's four to cover the circle; 1/3 there's three.	1/3, 1/4, 1/5. I was thinking of the lowest numbers since all these [numerators] are the same.
5/12 and 3/4: Are they equal or is one less?	3/4 is greater than 5/12. 5/12's are red and not as big. It takes 3 reds to cover a fourth and there's only five of them so it wouldn't cover two of them.	5/12 is less. If you took 4 * 3 to get 12, common denominators and 3 * 3 = 9, it would be 9/12 and it would be more.
4/5 and 9/10: Are they equal or is one less?	9/10 is greater. The pieces are smaller because it takes 10, but there's a smaller gap.	4/5 is less. 5 * 2 = 10 and 4 * 2 = 8 which gives 10/8 and 8 is less than 9.

One purpose of this article is to share with teachers other examples of students' thinking to give them a picture of what it means for students to exhibit number sense with fractions. Interviews with fourth graders who used the RNP curriculum and with fourth-grade students who used a traditional curriculum were conducted by RNP staff as well as classroom teachers. By looking at examples of conceptual and procedural thinking, teachers can construct for themselves a clearer picture of what they will want to observe in their own students. Focusing on children's thinking and on the effect the use of manipulative models has on students' thinking is a way teachers can learn to offer fraction lessons that involve their students with the use of manipulative models over an extended period of time.

Addition of Fractions

A child having fraction sense should be able to estimate a reasonable answer to fraction addition problems. Karen is an example of a fourth grader who was taught the RNP curriculum. She was asked to consider this problem on the day before she worked on fraction addition in her mathematics class (Cramer et al. 1997):

Jon calculated a problem as follows: 2/3 + 1/4 = 3/7. Do you agree?

Karen: I don't agree. He did it weird. You don't add the top and bottom numbers.

Teacher: What would be an estimate?

Karen: It would be . . . greater than 1/2 because 2/3 is greater than 1/2.

Teacher: Would it be greater or less than 1?

Karen: Less than 1. You'd need 1/3, and 1/4 is less than 1/3.

Teacher: What about 3/7?

Karen: 3/7 is less than 1/2.

Teacher: How do you know?

Karen: Because 3/7 isn't 1/2. I just know.

Notice how Karen, in thinking through this problem, used her knowledge of the relative size of fractions to bring meaning to the problem. She knew 2/3 was greater than 1/2. She did not do any symbolic procedure to compare fractions. She may have known from her many experiences with models for fractions that 2/3 is greater than 1/2. She also ordered 1/3 and 1/4. She realized that 2/3 + 1/3 would equal 1, and since 1/4 is less than 1/3, the total would have to be less than 1. She was less clear about why 3/7 was less than 1/2, but she was sure that was true.

The RNP has documented how children who have had extended periods of time using manipulative models for fractions—and in particular, a circular model—construct for themselves informal ordering strategies to compare fractions (Post et al. 1985). We are using the term *informal strategies* to contrast with traditional ordering procedures of finding decimal equivalents or changing both fractions to equivalent fractions with like denominators. The traditional ordering algorithm is not as helpful in estimation problems as students' constructed strategies.

Karen used two different ordering strategies that the RNP has documented and which reflect students' use of mental images to judge the relative size of fractions (Bezuk and Cramer 1989). She used the same numerator and the transitive strategies. In the earlier example, Jeremy demonstrated the same numerator strategy, which involves coordinating an inverse relationship between the size of the denominator and the size of the fraction. In concrete terms, students realize that the more equal parts a unit is partitioned into, the smaller each part is. Jeremy noted that since the numerator was 4 in both fractions, then four of the bigger parts would be larger than four of the smaller parts. Jeremy's thinking noted the role of both parts of the fraction, numerator and denominator. This is important, since often children who learn this relationship as an abstract rule inappropriately apply this thinking when the numerators are different. Karen also used this strategy to conclude that the sum of 2/3 and 1/4 was less than 1 because 1/4 is less than 1/3.

The transitive strategy involves using a reference point such as 1/2. For example, Karen in another interview compared 5/12 and 3/4. She tried to apply an algorithm she had learned along the way to the problem, but she had difficulty. When asked to just think about the fractions in her mind, she responded, "5/12 is less than 1/2; it's one less; 3/4 is more toward the whole than 1/2." What she meant by "one less" was that it was one red piece of a fraction circle away from 6/12, which is 1/2 (twelve reds equaled one circle). Karen used the transitive strategy when estimating fractions when she said that the sum had to be greater than 1/2, since one of the addends, 2/3, was already greater than 1/2.

The use of the transitive and the same-numerator strategies was common among RNP students. Table 5.2 compares RNP and textbook students' estimation strategies (or lack of strategies) for a fraction addition problem. Students were randomly selected to participate in interviews given by project staff during the five instructional weeks. The problem in table 5.2 was on the final interview.

Table 5.2

Examples of Students' Fraction Estimation Strategies for the Number Line Problem

Problem
Tell me about where 2/3 + 1/6 would be on this number line: 0   1   2

Responses

RNP Students	Textbook Students
Two browns are more than a half . . . 1/6 is not much more . . . less one.	2/3 equals 4/6 . . . 4/6 is less than 1/2, so 2/3 + 1/6 is less than 1/2.
Between 1/2 and 1, closer to 1 . . . 2/3 is almost one; 1/6 is not equal to 1/3, it is less than 1/3.	I have no idea.
Between 1/2 and 1 . . . 2/3 is a little more than 1/2 . . . 2/6 equals 1/3.	Student wouldn't estimate. He tried to find the exact answer. He was incorrect and used that as his estimate.
2/3 is greater than 1/2, 1/6 is a little more . . . you need 1/6 more to make a real whole.	Between 1/2 and 1 . . . just a guess.
Between 1/2 and 1 . . . 2/3 is more than 1/2; then add 1/6. It's not greater than 1 because 1/6 is less than 1/3.	Maybe one . . . I don't know.
Not a whole . . . there's 2/3, and 2/6 is not enough to make 1/3.	Between 1/2 and 1, closer to 1; just a guess.
2/6 fits over 1/3, and there are 2 of them, and you add 1/6 more . . . 5/6.	Student wouldn't estimate. He found the exact answer correctly and used that as his estimate.

Despite the same amount of instructional time and an emphasis on teaching the operations, textbook students did not have an adequate understanding of fraction size to find a reasonable estimate to a simple problem. RNP students' thinking depended on mental images for fractions and was directly related to their use of fraction circles. RNP students were better able to verbalize their thoughts. But then, the RNP lessons emphasized students' discussions (belief 3). In the lessons the manipulative model became the focal point of the discussion; students talked about their actions with the fraction circles. The extended use of manipulatives fostered students' verbal skills.

When estimating the sum of 2/3 and 1/6, RNP students reflected on how far 2/3 is from 1. Often RNP students considered the residual, or "leftover part," when they added fractions or when they ordered particular fraction pairs. The power the fraction circle model has on students' images for fractions is particularly evident in this type of reasoning. Consider Jeremy's use of residuals when he ordered 4/5 and 9/10 (see table 5.1). He imagined fraction circles and saw that both fractions had one part missing to make a whole. Since 1/10 is smaller than 1/5, less is missing for 9/10; thus it is the bigger fraction. From this reasoning he knew that 9/10 was greater than 4/5. This type of reasoning required Jeremy to work through several steps mentally. His strong mental images were developed through extensive experiences with fraction circles and supported this multistep thought process.

Since the first RNP teaching experiment, we have documented this type of thinking among fourth graders using fraction circles (Cramer, Post, and delMas 2002). As part of the field testing of the RNP curriculum, each classroom teacher interviewed two or three students at the end of the five weeks of instruction. One of the questions was to order 4/5 and 11/12. An example of a residual strategy for this problem is shown by an RNP student. She stated that 4/5 was less than 11/12. Her reasoning was as follows: "4/5 has bigger pieces, but one piece is left; 11/12 has smaller pieces, but only one piece is left. Fifth piece that is missing is bigger." The RNP curriculum did not directly teach students to order fractions by looking at the residual part, nor did it teach students to find common denominators. Results showed that 63 percent of the 53 RNP students interviewed ordered the pair correctly; 21 out of the 53 RNP students constructed for themselves an ordering strategy that relied on residuals. Only 3 of the 53 RNP students used a standard algorithm for ordering these fractions. A few correct solutions relied on drawings. Of the 57 textbook students interviewed, 67 percent ordered the fraction pair correctly. Only 9 out of 57 textbook students used a residual strategy to order these fractions; 18 out of 57 used a cross-product or common denominator method. A few used pictures and found the correct answer but could not explain their reasoning.

Conclusion

Often you hear teachers argue that there is not enough time to use manipulative materials. Even when manipulatives are used, teachers often make the transition to symbols too soon. The RNP students discussed in this article used manipulative models virtually every day during five weeks of instruction. The predominant model used was fraction circles. The samples of students' thinking presented here show the benefits of using manipulative models for five instructional weeks. RNP students developed number sense for fractions. In general, they had an understanding of fraction size evidenced by the type of ordering strategies they comfortably used. They were able to estimate reasonable answers to fraction addition problems. They were also able to verbalize their thinking. Students using a traditional program did not develop number sense.

Developing an understanding of fraction size and estimating a reasonable answer to fraction operation problems are appropriate goals for elementary school–aged children. Much of the symbolic manipulation of fraction symbols done in fourth and fifth grades can be adequately addressed in the middle grades. Students will be more successful if teachers in elementary school invest their time building meaning for fractions using concrete models and emphasizing concepts, informal ordering strategies, and estimation.

REFERENCES

Bezuk, Nadine, and Kathleen Cramer. "Teaching about Fractions: What, When, and How?" In *New Directions for Elementary School Mathematics,* 1989 Yearbook of the National Council of Teachers of Mathematics (NCTM), edited by Paul R. Trafton, pp. 156–67. Reston, Va.: NCTM, 1989.

Cramer, Kathleen, Merlyn J. Behr, Richard Lesh, and Thomas Post. *The Rational Number Project Fraction Lessons: Level 1 and Level 2.* Dubuque, Iowa: Kendall/Hunt Publishing Co., 1997.

Cramer, Kathleen A., Thomas R. Post, and Robert C. delMas. "Initial Fraction Learning by Fourth- and Fifth-Graders: A Comparison of the Effects of Using Commercial Curricula with the Effects of Using the Rational Number Project Curriculum." *Journal for Research in Mathematics Education* 33 (March 2002): 111–44.

National Governors Association Center for Best Practices and Council of Chief State School Officers (NGA Center and CCSSO). *Common Core State Standards for Mathematics. Common Core State Standards (College- and Career-Readiness Standards and K–12 Standards in English Language Arts and Math).* Washington, D.C.: NGA Center and CCSSO, 2010. http://www.corestandards.org.

Post, Thomas R., Ipke Wachsmuth, Richard Lesh, and Merlyn J. Behr. "Order and Equivalence of Rational Numbers: A Cognitive Analysis." *Journal for Research in Mathematics Education* 16 (January 1985): 18–36.

Examining Dimensions of Fraction Operation Sense

DeAnn Huinker

NUMBER sense and operation sense are similar to problem solving in that they "represent a certain way of thinking rather than a body of knowledge that can be transmitted to others" (Sowder 1992, p. 3). The development of these constructs evolve over time while strengthening intuition about and flexibility with numbers, operations, and their relationships. The need to develop number sense is well documented and its characteristics have received extensive analysis, but the need to develop operation sense has not received the same attention or scrutiny. This discussion presents seven dimensions of operation sense that are applicable to fractions, as well as to whole numbers. These dimensions incorporate a synthesis of the attempts to define and characterize number sense and operation sense (Howden 1989; Sowder 1992). Examples of students' work and reasoning are drawn from work with a class of fifth-grade students from a large, urban school system.

Operation Sense

Fundamental to operation sense is an understanding of the meanings and models of operations. Addition is usually introduced as putting together quantities, and subtraction is presented as separating an amount into subgroups. Multiplication is introduced as combining equal groups, and division is discussed as separating an amount into equal groups. Students should become familiar with a variety of models for fraction operations, such as area models using circular or rectangular regions, set models using counters or cubes, and linear models with number lines.

> A class of fifth-grade students was presented with this word problem: Three students are helping the teacher with a special task. He plans to give each student 5/8 of a candy bar to thank them for their help. How much candy will the teacher need so he can give each student 5/8 of a candy bar?

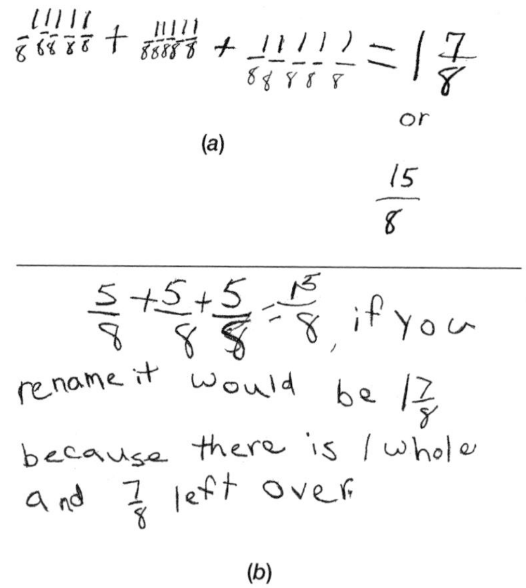

The two students' responses in figure 8.1 show how these students used symbols to model and solve a multiplication situation using repeated addition. Other students used paper-strip models to represent each group demonstrating an understanding of multiplication as combining equal groups.

A second dimension of operation sense is the ability to recognize and describe real-world situations for specific operations. Operation sense involves a familiarity with a variety of situations for each operation. For example, addition and subtraction include combining, separating, and comparison situations. Multiplication and division include situations with equal groups, arrays, rates, comparisons, and Cartesian products. Students need to explore many different situations with varied problem structures. They should also regularly describe or pose their own situations for specific operations. The fifth-grade students were presented with a variety of problem structures through daily solving of word problems and were regularly asked to describe real-world situations and pose their own word problems. The word problem in figure 8.2 was posed by a student to illustrate a division situation using fractions.

A third dimension of operation sense involves having meaning for symbols and formal mathematical language. Meaning for symbols and formal mathematical language develops when connections are made to students' conceptual understandings and informal language. Symbols become tools for thinking when students use them as records of actions and things they already know. Without this understanding, students manipulate symbols without meaning rather than thinking of symbols as quantities and actions to be performed or records of actions already performed.

Kieren (1988) noted that premature symbolism leads to symbolic knowledge that students cannot connect to the real world, resulting in virtual elimination of any possibility for students to develop number and operation sense. Symbolic knowledge that is not based on understanding is "highly dependent on memory and subject to deterioration" (p. 178). Thus, operation sense involves having meaning for symbols that contributes to the robustness of knowledge.

(a)

(b)

Fig. 8.1. Two students' approaches for solving 3 × 5/8

Operation sense is strengthened through an emphasis on connecting real-world, oral language, concrete, pictorial, and symbolic representations of fractions. Thus a fourth dimension of operation sense is the ability to translate easily among these representations. Given any one of these situations, students should be able to produce the others. Students who are able to make these connections have demonstrated lasting ability to use their mathematical knowledge flexibly to solve problems. Students who can easily translate among these representations are more likely to reason with fraction symbols as quantities and not as two whole numbers when solving word problems (Towsley 1989).

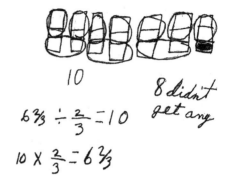

Fig. 8.2. Student-posed word problem for division with fractions

Students who cannot easily connect these representations lack the power to make sense of fraction concepts and operations and to see the usefulness of fractions in the world around them. The instruction with the fifth-grade students continually emphasized translations from real-world situations to concrete, pictorial, and symbolic representations, thus drawing on varied connections to make sense of and solve word problems. After students examined a situation such as that shown in figure 8.2, explored it using models, and shared solution strategies, the class discussed how to translate the real-world representation into appropriate symbolic representations or equations.

A fifth dimension of operation sense is an understanding of relationships among operations. Addition and subtraction are inverse operations, as are multiplication and division. Multiplication can be viewed as repeated addition, and division can be viewed as repeated subtraction or repeated addition. The use of these relationships contribute to the development of thinking strategies for basic facts and student-generated algorithms for computation. In figure 8.2, note that the student wrote two number sentences. This demonstrates some understanding of the inverse relationship between multiplication and division of fractions.

Operation sense provides a basis for the development of student-generated strategies for computation. Implicit in these strategies is the ability to take numbers apart and put them back together flexibly or to create an equivalent problem. Also implicit in student-generated strategies is the application of the commutative, associative, and distributive

375

properties (Schifter 1999). Thus, a sixth dimension of operation sense involves the ability to compose and decompose numbers and to use properties of operations. For example, Shontae applied the distributive property to solve 2 1/2 ÷ 4. She used two and one half paper strips to solve (2 ÷ 4) + (1/2 ÷ 4).

Teacher: You have 2 1/2 candy bars, and there are four people. You are going to share the candy bars so that each person gets the same amount. How much of a whole candy bar would each person get?

Shontae: There would be four 1/2 pieces for everyone. Then I thought how many—well, there's still 1/2 left over, and I need to divide it into four pieces. Then I thought that 4/8 is the same as 1/2. So I made it into four pieces, and then if I gave everyone 1/8, they would all have 5/8.

A seventh dimension of operation sense is knowledge of the effects of an operation on a pair of numbers. Operation sense interacts with number sense and enables students to make thoughtful decisions about the reasonableness of results. Understanding an operation includes being able to reason about the effect it will have on the numbers to which it is applied. When you add two numbers, does the answer get larger or smaller? When you subtract, what can you say for sure about the answer? Does the answer always get larger when you multiply two numbers? Does the answer always get smaller when you divide? Can you subtract a larger number from a smaller number? Can you divide a smaller number by a larger number?

Individuals develop expectations about the results of operations from their work with whole numbers. Operating with fractions requires these expectations to be reconsidered. No longer does the answer always get smaller when you divide or larger when you multiply. The students in this class developed an understanding of multiplication and division with fractions through the exploration of problem situations. They reported the results of their explorations in the context of situations that gave meaning to the quantities and relationships among the quantities. Thus they readily reconsidered their expectations regarding the effects of an operation on a pair of numbers and were

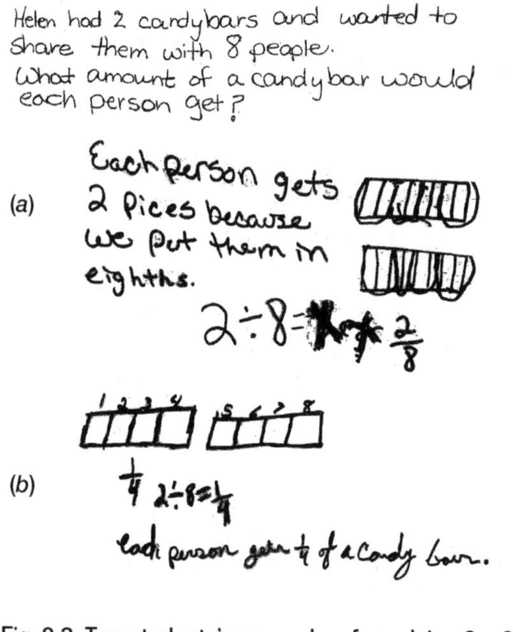

Fig. 8.3. Two students' approaches for solving 2 ÷ 8

able to accept and describe, for example, situations in which the answer from a division situation was larger (see fig. 8.2).

Helen, a fifth-grade student, posed the word problem in figure 8.3 for her classmates to solve. She and the other students meaningfully explored the division of a smaller number by a larger number. The response in part *a* shows how a student partitioned each candy bar into eight sections and then gave two pieces to each person. The response in part *b* shows how a student realized that each candy bar only needed to be partitioned into four sections. Because the computation was embedded in a problem situation, the students readily accepted the division of a smaller number by a larger number. During whole-class discussions, the teacher explicitly found opportunities to challenge students' preconceived notions about the effects of operations on a pair of numbers. For example, in the situation shown in figure 8.3, the students had to justify and prove the accuracy of the number sentence, $2 \div 8$, that they had written to represent the situation.

Concluding Comments

The work with the fifth-grade students on developing meaning for fraction operations deliberately emphasized the dimensions of operation sense. Instruction used a problem-solving approach and encouraged students to explore operations on fractions with models. The students were encouraged to generate their own strategies for solving problems and to communicate their reasoning by sharing their approaches with one another.

Each student was individually interviewed prior to and following a four-week instructional unit on fractions. The students explored addition, subtraction, multiplication, and division situations with the goal of developing operation sense. Due to time constraints, not all students were asked all items. Table 8.1 lists selected items from the interviews. The preassessments revealed that the fifth-grade students overwhelmingly manipulated symbolic representations with little understanding. Some students tried to recall previously learned symbolic procedures but were unable to do so successfully. Students made significant progress in their fraction operation sense by the end of the four weeks. Even though the students were not taught procedures for solving fraction computation problems, the students could draw on their operation sense for fractions. The lower results for $2\ 1/2 \div 4$ reflect both the difficulty of this item and the limited instructional time that focused on situations involving division of a mixed number by a whole number.

Table 8.1

Interview results with fifth-grade students (percent correct)

Topic	Preassessment Item	Postassessment Item	Number of Students	Preassessment Results	Postassessment Results
Add related fractions	3/4 + 7/8	1/2 + 3/4	25	0%	84%
Subtract a fraction from a whole number	3 – 1/4	3 – 5/8	25	4%	92%
Divide a whole number by a fraction	Not assessed	2 ÷ 2/3	19	----	89%
Divide a mixed number by a whole number	Not assessed	2 1/2 ÷ 4	15	----	60%

Fractions are a major area of study in upper elementary school mathematics. The traditional instructional approach for fractions has been heavily symbolic and procedural. The rush to tell students how to perform procedures prevents them from establishing a solid foundation of operation sense for fractions (Kamii 1999). It is time to shift the emphasis and redefine the goal of fraction instruction in elementary school from learning computational rules to developing fraction operation sense. Students with richly connected knowledge of fraction operations are able to develop flexible student-generated strategies for computation and work with problem situations meaningfully.

REFERENCES

Howden, Hilde. "Teaching Number Sense." *Arithmetic Teacher* 36 (February 1989): 6–11.

Kamii, Constance, and Mary Ann Warrington. "Teaching Fractions: Fostering Children's Own Reasoning." In *Developing Mathematical Reasoning in Grades K–12,* 1999 Yearbook of the National Council of Teachers of Mathematics (NCTM), edited by Lee V. Stiff, pp. 82–92. Reston, Va.: NCTM, 1999.

Kieren, Thomas E. "Personal Knowledge of Rational Numbers." In *Number Concepts and Operations in the Middle Grades,* edited by James Hiebert and Merlyn Behr, pp. 162–81. Hillsdale, N.J.: Lawrence Erlbaum Associates; Reston, Va.: National Council of Teachers of Mathematics, 1988.

National Governors Association Center for Best Practices and Council of Chief State School Officers (NGA Center and CCSSO). *Common Core State Standards for Mathematics. Common Core State Standards (College- and Career-Readiness Standards and K–12 Standards in English Language Arts and Math).* Washington, D.C.: NGA Center and CCSSO, 2010. http://www.corestandards.org.

Schifter, Deborah. "Reasoning about Operations: Early Algebraic Thinking in Grades K–6." In Developing Mathematical Reasoning in Grades K–12, 1999 Yearbook of the National Council of Teachers of Mathematics (NCTM), edited by Lee V. Stiff, pp. 62–81. Reston, Va.: NCTM, 1999.

Sowder, Judith T. "Making Sense of Numbers in School Mathematics." In *Analysis of Arithmetic for Mathematics Teaching,* edited by Gaea Leinhardt, Ralph Putnam, and Rosemary A. Hattrup, pp. 1–51. Hillsdale, N.J.: Lawrence Erlbaum Associates, 1992.

Towsley, Ann. "The Use of Conceptual and Procedural Knowledge in the Learning of Concepts and Multiplication of Fractions in Grades 4 and 5." Ph.D. diss., University of Michigan, 1989.

18

From *Thinking and Reasoning with Data and Chance,* NCTM's
Sixty-eighth Yearbook (2006)

From the Council's Sixty-eighth Yearbook, J. Michael
Shaughnessy (who served as NCTM president from 2010
to 2012) has recommended **chapter 22, "Some Important
Comparisons between Statistics and Mathematics, and
Why Teachers Should Care,"** by Allan Rossman, Beth
Chance, and Elsa Medina.

The initial paragraph in this chapter establishes statistics
as a field of study that is a separate discipline, and not just a
branch of mathematics. The authors address the critical differences and similarities between the two disciplines of mathematics and statistics, with particular attention to the role of
context, issues of measurement, the importance of data collec-
tion, the lack of definitive conclusions, and the communication of statistical knowledge.
Important instructional considerations are raised, including the final query posed in the
chapter: Why should teachers care? (about the differences). This selection is particularly
timely, given the 2012 release of the *Mathematical Education of Teachers II* report from the
Conference Board of the Mathematical Sciences.

Some Important Comparisons between Statistics and Mathematics, and Why Teachers Should Care

Allan Rossman
Beth Chance
Elsa Medina

TATISTICS is a mathematical science. That sentence is likely to be the shortest in this entire article, but we want to draw your attention to several things about it:

- We use the singular *is* and not the plural *are* to emphasize that statistics is a field of study, not just a bunch of numbers.
- We use *mathematical* as an adjective because although statistics certainly makes use of much mathematics, it is a separate discipline and not a branch of mathematics.
- We use the noun *science* because statistics is the science of gaining insight from data.

This article highlights some of the differences between statistics and mathematics and suggests some implications of these differences for teachers and students. We realize that these distinctions may not be universal but hope our broad strokes can help highlight some fundamental distinctions in the disciplines. Our aim is not to provide a philosophical discussion of these issues but rather to provoke thought and to present ideas that will guide classroom practice. Toward this end we facilitate and illustrate our points with concrete examples.

The Crucial Role of Context

A primary difference between the two disciplines is that in statistics, *context is crucial.* Mathematics is an abstract field of study; it can exist independently of context. Mathematicians often strive to "strip away" the context that can get in the way of studying the underlying structure. For example, linear functions can be studied for their own mathematical properties, without considering their applications. Indeed, one could argue that worrying about the complications of real data diverts students' attention from the underlying mathematical ideas. But in statistics, one cannot ignore the context when analyzing data. Consider the dot plot in fig. 22.1.

Fig. 22.1. Dot plot without a scale

This plot reveals virtually nothing; it's just a bunch of dots! Granted, we can see that one dot appears far to the left of the others, and there is a cluster of six dots less far to the left that seems to be separated from the majority of the dots. But we can't interpret the plot or draw any conclusions from it. Let's include a scale (see fig. 22.2).

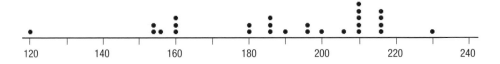

Fig. 22.2. Dot plot including a scale

This is better, and we now know that the outlier is at 120, and the lower cluster is between 150 and 160. But we still cannot gain any insights from this display. Let's include the units (see fig. 22.3). Now we know that this graph reveals weights in pounds, but we still do not know whether we're analyzing weights of statistics students or of horses or furniture or . . . .

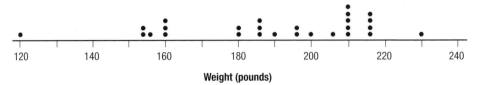

Fig. 22.3. Dot plot including a scale and units

Finally, we'll tell you the context: these are the weights of the rowers on the 2004 U.S. men's Olympic rowing team. Armed with that knowledge, we can summarize what the dot plot reveals and even suggest some explanations for its apparent anomalies. It makes sense that the data include an outlier who weighs much less than the others: he's the cox-swain, the team member who calls out instructions to keep the rowers in synch but does not put an oar in the water himself. He needs to be light so as not to add much weight to the boat. The cluster in the 150s also has an explanation: those six rowers participate in "lightweight" events with strict weight limitations, two in a pairs event and the rest in a fours event. As for the rest, the majority of the rowers are big, strong athletes in the 180-to-230 pound range.

For another example of how context is paramount in statistics, consider the following scatterplot of data from a study about whether there is a relationship between the age (in months) at which a child first speaks and his or her score on a Gesell aptitude test taken later in childhood (Moore and McCabe 1993, p. 132). The least-squares regression line is drawn on the plot (see fig. 22.4).

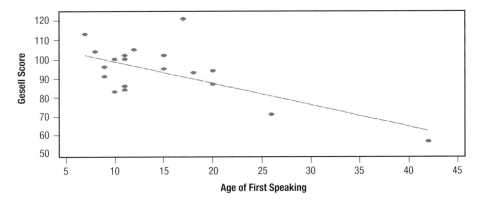

Fig. 22.4. Scatterplot revealing an apparent negative association

This scatterplot and line reveal a negative association between the variables, indicating that a large value of one variable tends to appear with a small value of the other. Moreover, the slope of the line is statistically significant (p-value = .002, r^2 = .410). But on closer inspection, we see that this apparent negative association is driven largely by the two extreme cases in the bottom right of the plot. What do we make of this? Should we conclude that there's a negative association here or not? Should we discount the outliers or not? Of course, to answer this question, we must first consider the context. We see that those two outliers are exceptional children who take a very long time to speak (3.5 years and a bit longer than 2 years) and who also have very low aptitude as measured by Gesell. To get a sense for whether the negative association between speaking age and aptitude score holds for more "typical" children, we can delete the outliers and refit the line (see fig. 22.5). This scatterplot and line reveal essentially no association between the variables. The slope coefficient is no longer statistically significant (p-value = .890, r^2 = .001).

So, our conclusion here contains two parts: Children who take an exceptionally long time to speak tend to have low aptitude, but otherwise there is virtually no relationship between when a child speaks and his or her aptitude score. (We should learn more about how these data were collected before deciding whether these results generalize to a larger group, and we could gather more data to examine whether those two exceptional children are indicative of a larger pattern.) Once again, the context drives our analysis and conclusions. Fitting a line to these data without considering the context would have blinded us to much of what the data reveal about the underlying question of speaking and intelligence.

385

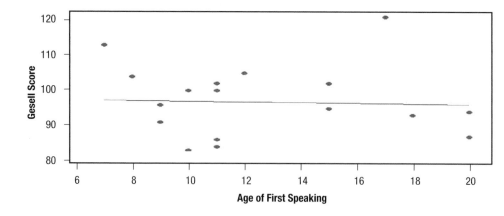

Fig. 22.5. Scatterplot with outliers deleted

Issues of Measurement

Another important issue that distinguishes statistics from mathematics is that measurement issues play a large role in statistics. Measurement is also important in mathematics; in fact, it is one of the standards in the *Principles and Standards for School Mathematics* (National Council of Teachers of Mathematics [NCTM] 2000), but the focus is different. In mathematics, measurement includes getting students to learn about appropriate units to measure attributes of an object such as length, area, and volume and to use formulas to measure those attributes. In statistics, drawing conclusions from data depends crucially on taking valid measurements of the properties being studied. Measuring a rower's weight is quite straightforward, but measuring a child's aptitude is quite challenging, not to mention controversial. Many other properties of interest in statistical studies of human beings are hard to measure accurately; examples include unemployment, intelligence, memory ability, and teaching effectiveness.

Another example of a hard-to-measure property comes from a study involving a city's pace of life. Researchers studied whether a city's "pace of life" is associated with its heart disease rate (Ramsey and Schafer 1997, p. 251). They measured a city's pace of life in three different ways:

- Average walking speed of pedestrians over a distance of 60 feet during business hours on a clear summer day along a main downtown street

- Average time a sample of bank clerks take to make change for two $20 bills or to give $20 bills for change

- Average ratio of total syllables to time of response when asking a sample of postal clerks to explain the difference between regular, certified, and insured mail

386

The scatterplots in figure 22.6 reveal a slight positive association between heart rate and these three measures of a city's pace of life. Although these researchers quantified "pace of life" in their study, it is also important to remember that measurement categorizations (e.g., fast pace of life) or proxies can differ across settings. For example, a "long commute" can mean many different things to people in different cities and cultures.

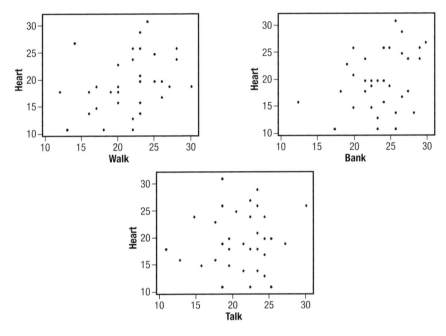

Fig. 22.6. Scatterplots revealing a slight positive correction

The Importance of Data Collection

Mathematics can be studied and carried out without analyzing data, but even when mathematicians examine data, they typically focus on detecting and analyzing patterns in the data. How the data were collected is not relevant to purely mathematical analyses, but this is a crucial consideration in statistics. The design of the data-collection strategy determines the scope of conclusions that can be drawn. Can we generalize a study's results to a larger population? It depends on whether the sample was randomly selected. Can we draw a cause-and-effect conclusion from a study? It depends on whether it was a controlled, randomized experiment.

For example, consider two studies that asked women, "Do you give more emotional support to your husband or boyfriend than you receive in return?" In study A, 96 percent of a sample of 4500 women answered yes, but in study B 44 percent of a sample of 767 women answered yes. How do we reconcile these results? In which study do we place more confidence for representing the beliefs of all American women? The answers depend on how the data were collected. Study A was conducted by sociologist Shere Hite, who distributed more than 100,000 questionnaires through women's groups (Hite 1987). Study B was conducted on a random sample of women, sponsored by ABC News and the *Washington Post* (Moore 1992). Despite its larger sample, the Hite study does not accurately reflect the opinions of all American women. The two primary sources of bias are distributing the survey only through women's groups, which tend to appeal to certain kinds of women, and the voluntary response reflected in the low response rate, which suggests that women with a strong opinion on the issue were more likely to voice their opinion. Thus, even with the smaller sample, study B provides more credible data because it involved a random sample, giving all kinds of women the same chance of being selected. Moreover, the response rate in this study was roughly 80 percent, indicating much less of a voluntary response bias than in the Hite study.

For another example, consider two studies A and B comparing "success" rates between two groups. The data are summarized in table 22.1, including the p-value for comparing the two success rates with each study. (The p-value is the probability of observing sample proportions at least as discrepant as those observed if there were no underlying difference between the groups.) The methods used for calculating these p-values are identical, and the p-values are very similar despite the much smaller difference in success rates in study B. The small p-values indicate strong evidence of a statistically significant difference in success rates between the two groups. But does this mean that we draw identical conclusions from the two studies? One issue is still the size of the difference. It is important to remember that with large sample sizes, as in study B, even small differences in success rates will be classified as "statistically significant." The other primary issue in drawing conclusions from these studies is how the data were collected.

Table 22.1
Two Studies Comparing Success Rates

Study	Group	Successes	Trials	Success Rate	p-value
A	1	42	561	.689	.011
	2	30	62	.484	
B	1	806	908	.888	.015
	2	614	667	.920	

Study A is a social experiment in which three- and four-year-old children from poverty-level families were randomly assigned to either receive preschool instruction or not, with a response of whether they were arrested for a crime by the time they were nineteen years old (Ramsey and Schafer 1997, p. 533). Because the children were randomly assigned to a treatment and because the p-value turned out to be so small, we can legitimately draw a cause-and-effect conclusion between the absence of preschool instruction and the higher arrest rate. However, study B is an observational study in which researchers examined court records of people who had been abused or not as children, comparing their rates of committing a violent crime as an adult (Ramsey and Schafer 1997, p. 533). Because this was not a randomized experiment, no cause-and-effect conclusion between the child abuse and the violent crime rate can be drawn, despite the very small p-value. So although the calculations in two studies can be identical, the conclusions drawn can differ substantially, depending on how the data were collected.

The Lack of Definitive Conclusions

Statistics and mathematics ask different types of questions and therefore reach different kinds of conclusions, with different thought processes to link questions to conclusions. Mathematics involves rigorous deductive reasoning, proving results that follow logically from axioms and definitions. The quality of a solution is determined by its correctness and succinctness, and there is often an irrefutable correct answer.

In contrast, statistics involves inductive reasoning and uncertain conclusions. Statisticians often come to different but reasonable conclusions when analyzing the same data. In fact, within these types of judgments lies the art of data analysis. All statistical inference requires the use of inductive reasoning because informed inferences are made from observed results to defensible, but ultimately uncertain, conclusions. In statistics we summarize conclusions with phrases such as "We have strong evidence that . . ." and "The data strongly suggest that . . ." but steadfastly resist saying things like "The data prove that. . . ." The quality of conclusions lies in the analysts' ability to support and defend their arguments.

For example, we often ask students to collect data for comparing prices between two different grocery stores. This project raises many practical issues of measurement and data collection, such as whether students should record sale prices or regular prices and how students should obtain a random sample of grocery items. Consider some sample data, collected by students, on price differences between two stores on 37 items (see fig. 22.7).

Fig. 22.7. Student-collected data on price differences between two stores

How should we analyze these data? One reasonable approach is to calculate the mean of these differences and test whether it differs significantly from zero; a t test yields a p-value of .308. Thus, although the sample mean price difference is below zero, the not-so-small *p*-value reveals that this distance is consistent with what we expect to see from random sampling. Therefore these data do not provide convincing evidence that the population mean price difference differs from zero. But we have not established that the average price is the same between the two stores; we have concluded only that the sample data do not provide compelling evidence to reject that hypothesis.

Not only is this conclusion uncertain, but we could have selected a different analysis altogether. We could instead perform a sign test of whether the median price differs significantly from zero, which is equivalent to asking whether the proportion of items costing less in one particular store differs significantly from one-half. Twenty-one of the items are cheaper in one store, with only eight items cheaper in the other store (and eight "ties"). The *p*-value for this sign test turns out to be .024, which does suggest a statistically significant difference. But even this conclusion is uncertain, for the *p*-value reveals that sample data this extreme could have arisen by chance even if there was no difference between the median prices in these stores. Although the price differences are not large enough to discount chance as an explanation, when we find that 21 of 29 items are cheaper in one store, without worrying about how much cheaper, we do get a slightly different story.

So, do the stores' prices differ significantly, as the sign test suggests, or not, as the *t* test suggests? Which of these two conflicting conclusions is correct? Which is reasonable? Well, neither and both. It depends on what question we want to ask (about the mean or median), and even then neither conclusion is certain. To complicate matters still further, statisticians might disagree on whether a normal model for the price differences is reasonable enough to justify applying the *t* test, and statisticians might also disagree on how the "ties" should be handled when conducting the sign test. Furthermore, we need to consider whether the students took a random sample of all items at the store or a random sample of items that students really buy, perhaps from recent receipts. Another issue is that even if a statistically significant difference is obtained, it may not be large enough to convince a consumer to drive to the cheaper store that is farther away. An individual consumer might well be interested only in certain items and may want to consider other factors such as the store's location, parking, and checkout speed. This lack of definitive conclusions, and even the lack of a single appropriate method of analysis, is common in statistical investigations.

But this lack of definitiveness does not mean that all analyses are equally reasonable or that statistics can be used to prove any desired conclusion, an unfortunately commonly held belief. Just as some verbal arguments are more persuasive and soundly reasoned than others, so, too, with statistical arguments.

Communicating Statistical Knowledge

Because of the need to make persuasive arguments, terminology is essential in statistics. One complication, as in mathematics, is that many common terms from everyday language have technical meanings in statistics. Examples include words such as *bias, sample, statistic, accuracy, precision, confound, correlation, random, normal, confidence,* and *significant.* Students are very tempted to use these words loosely, without considering their technical meanings. The casual and technical meanings of these terms are similar enough for students to falsely believe that they can get by with only a casual understanding. We suspect that this confusion is less of an issue with words like *root* and *power* in mathematics. Consequently, studying statistics is akin to studying a foreign language, but where students first need to unlearn words that they think they know and have been using informally for many years. Students need lots of practice to become comfortable using these terms correctly, and they often stumble at first before acquiring enough familiarity to use the language well.

Communication is essential in both statistics and mathematics, but in some sense even more so in statistics because of its collaborative nature. Statisticians routinely must interact with clients whose technical skills vary greatly, from eliciting a clear statement of the problem from those clients through communicating to them the results and conclusions of the analysis. Although introductory students are far from professional statisticians, the ability to communicate statistical ideas in layperson's terms is essential and an important component of many courses. Communication is important in mathematics also, but that communication is more often done symbolically in mathematics.

Why Should Teachers Care?

We have argued that statistics is a different discipline from mathematics, that it involves a different type of reasoning and different intellectual skills. Even if you find our case persuasive, the question remains: Why should classroom teachers care? We see two primary reasons:

- A different type of instructional preparation is needed for teaching statistics.
- Students react differently to statistics than to mathematics.

In order to help students see the relevance of context, measurement issues, and data collection strategies in statistics, it's imperative that teachers present real data, in meaningful contexts, and from genuine studies. Fortunately, there are a plethora of resources available now to help teachers with this, from books to CD-ROMs to Web sites (Moore 2000).

Instructors also need to help students learn to relate their comments to the context and to always consider data collection issues when stating their conclusions. The examples that we present above illustrate how crucial these issues are in statistics. Although this is sometimes done in mathematics courses, it's not nearly as prevalent or as essential as it is with statistics. Many students do not initially expect this type of focus, and teachers need to be prepared for students' discomfort. Students also need to be reminded that there can be multiple correct approaches to analyzing data and even different reasonable conclusions and that they will also be evaluated on how well they explain and support their conclusions.

Proficiency with calculations is necessary but not sufficient. This is also true in mathematics, of course, where students need to develop their mathematical reasoning skills and communicate that reasoning process. But many components of statistical reasoning, such as understanding when cause-and-effect conclusions can be drawn, are not mathematical, and again students need time to become comfortable with these ideas and modes of thinking.

The experiences and reactions of students to studying statistics are different from studying mathematics. Educational research shows that students (and others) have tremendous difficulties with reasoning under uncertainty (Garfield 1995; Shaughnessy 1992; Garfield and Ahlgren 1988). Also, many students (and others) are very uncomfortable with uncertainty, with the lack of definitive conclusions, and with the need for detailed interpretations and explanations that are integral to studying statistics. Helping students develop a healthy skepticism about numerical arguments, without allowing them to slip to the extremes of cynicism or naïve acceptance, is a great challenge.

Another difference regarding instructional preparation is that many teachers do not have ample opportunities to develop their own statistical skills and understanding of statistical concepts before teaching them to students. This challenge is especially acute because few programs in mathematics teacher preparation offer much instruction in statistics, and much of the instruction that is provided concentrates on the mathematical aspects of statistics. The recent *The Mathematical Education of Teachers* report makes these points quite forcefully (Conference Board of the Mathematical Sciences 2001). Helping students develop their communication skills and statistical judgment, so crucial in the practice of statistics, is also very challenging and an area in which mathematics teachers are provided with little instruction.

Conclusion

Many of the points that we have made about teaching statistics apply equally well to the teaching of mathematics. For example, real data and applications can help encourage students' interest in mathematics as well as statistics; students can construct their own knowledge of mathematical as well as statistical ideas through engagement with learning activities; and mathematics as well as statistics can help students develop general problem-solving approaches that allow nonunique solutions. Our contention is that these approaches are inherent to the discipline of statistics and arise naturally in considering statistical concepts. The intention of this article is to provoke the reader to consider how to improve mathematical as well as statistical discussions with students. In particular, statistical thinking requires an often distinct frame of thinking, which a teacher can use in designing instructional activities and evaluating students' performance.

Because of the differences between statistics and mathematics, teachers should expect that some mathematically strong students may be frustrated while studying statistics. But on the bright side, many students who may not be initially excited by mathematics will be intrigued and empowered by their experiences with statistics. Despite creative efforts in the teaching of mathematics, many students still develop a staid impression of the discipline. Statistical topics offer the opportunity of "starting fresh" and opening students' eyes to a new perspective on mathematical sciences.

REFERENCES

Conference Board of the Mathematical Sciences. *The Mathematical Education of Teachers.* Providence, R.I., and Washington, D.C.: American Mathematical Society and Mathematical Association of America, 2001.

Garfield, Joan. "How Students Learn Statistics." *International Statistical Review* 63 (1995): 25–34.

Garfield, Joan, and Andrew Ahlgren. "Difficulties in Learning Basic Concepts in Probability and Statistics: Implications for Research." *Journal for Research in Mathematics Education* 19 (January 1988): 44–63.

Hite, Shere. *Women and Love: A Cultural Revolution in Progress.* New York: Alfred A. Knopf, 1987.

Moore, David S., and George P. McCabe. *Introduction to the Practice of Statistics.* 2nd ed. New York: W. H. Freeman & Co., 1993.

Moore, David W. *The Superpollsters: How They Measure and Manipulate Public Opinion in America.* New York: Four Walls Eight Windows, 1992.

Moore, Thomas L., ed. *Teaching Statistics.* Washington, D.C.: Mathematical Association of America, 2000.

National Council of Teachers of Mathematics (NCTM). *Principles and Standards for School Mathematics.* Reston, Va.: NCTM, 2000.

Ramsey, Fred, and Daniel Schafer. *The Statistical Sleuth: A Course in Methods of Data Analysis.* Pacific Grove, Calif.: Duxbury, 1997.

Shaughnessy, J. Michael. "Research in Probability and Statistics: Reflections and Directions." In *Handbook of Research on Mathematics Teaching and Learning,* edited by Douglas A. Grouws, pp. 465–94. New York: Macmillan Publishing Co., 1992.

19

From *The Learning of Mathematics*, NCTM's Sixty-ninth Yearbook (2007)

For the final selection in this final yearbook, Lee Stiff, NCTM president from 2000 to 2002, has recommended a chapter from *The Learning of Mathematics*, the Council's Sixty-ninth Yearbook. **Chapter 6, "Examining School Mathematics through the Lenses of Learning and Equity,"** by Celia Rousseau Anderson, builds on the consideration of an equity lens, as suggested by NCTM's Research Committee (2005), to gain greater understanding of the implications of research for equity.

Using the Equity Principle found in *Principles and Standards for School Mathematics* (National Council of Teachers of Mathematics 2000) as a guiding framework, Anderson examines considerations related to equity and learning with a particular focus on classroom and policy issues. She also explores other factors that shape students' opportunities to learn mathematics with understanding. Readers are encouraged to examine equity as it relates to learning, teaching, curriculum, technology, and assessment. As states, school districts, and schools move toward full implementation of the Common Core State Standards, as well as the assessments from the Partnership for Assessment of Readiness for College and Careers and the Smarter Balanced Assessment Consortium, equity becomes more—so much more—than one consideration among many. As we have learned, a culture of equity maximizes the learning potential of all students (NCTM Equity Position 2008).

Examining School Mathematics through the Lenses of Learning and Equity

Celia Rousseau Anderson

> Imagine a classroom, a school, or a district where all students have access to high quality, engaging mathematics instruction. There are ambitious expectations for all, with accommodation for those who need it. Knowledgeable teachers have adequate resources to support their work and are continually growing as professionals. The curriculum is mathematically rich, offering students opportunities to learn important mathematical concepts and procedures with understanding.
>
> —*Principles and Standards for School Mathematics*

THE authors of *Principles and Standards for School Mathematics* (National Council of Teachers of Mathematics [NCTM] 2000) open the document with a vision—a mental picture of the end goal of mathematics education, as they see it. It is a powerful image in which *"all students* . . . have the opportunity and support necessary to learn significant mathematics with depth and understanding" (p. 5; emphasis added). It is an exciting image in which all students are able to experience success in school mathematics. It is an image that involves the interrelated goals of equity and learning.

Yet, as the authors of *Principles and Standards* acknowledge, that vision is highly ambitious (NCTM 2000). Several indicators demonstrate how far we currently are from achieving those goals, particularly with respect to equity. Despite some gains, gaps in opportunity to learn and achievement persist (Oakes et al. 2000; Tate 2005; Tate and Rousseau 2002). According to the NCTM Research Committee (2005), equity in mathematics education has proved to be an elusive goal. "No simple solutions exist [to the problems of inequity]. The difficulties in improving the situation make it apparent that the issues are complex and resistant to easy solutions" (p. 95). Thus, in some ways, we have a vision of the ends that we seek without a clear picture of how to get there.

As the NCTM Research Committee (2005) has argued, achieving equity in mathematics education will require attention and effort on the part of all involved. They assert that researchers, for example, can play a crucial role in the search for equity through applying a "critical equity lens" to research in mathematics education. Such a lens offers a means to gain more understanding of the implications of research for equity. In this article, I build on the idea of a "critical equity lens" to highlight various considerations that are important for all those who seek to achieve the vision of mathematics learning described by the authors of *Principles and Standards for School Mathematics*.

The View through an Equity Lens

What is an equity lens? As the members of the NCTM Research Committee (2005) describe it, an equity lens, when applied to a research study or other educational situation, would highlight equity-related concerns or issues. The lens would delineate those conditions and bring them to the fore. Rather than fade into the background, where they have a chance (if not a likelihood) of remaining unnoticed, those concerns would become the focal point.

An important consideration, however, is to make clear exactly what aspects of school mathematics that lens would highlight. No single generally accepted definition for equity exists. Thus, I state at the outset the definition from which I am working, because it will determine the conditions on which I train my lens. The different conceptions of equity (Gutierrez 2002b; NCTM Research Committee 2005; Secada 1989; Tate and Rousseau 2002) are beyond the scope of this article to examine. For that reason, I look to the Equity Principle of *Principles and Standards* (NCTM 2000) for the framework to be used for the purposes of this article. Although the Equity Principle does not provide an explicit definition of the meaning of equity, it focuses specific attention on access and opportunities to learn. For example, the authors of *Principles and Standards* (NCTM 2000) assert that "all students, regardless of their personal characteristics, backgrounds, or physical challenges, must have opportunities to study—and support to learn—mathematics. Equity . . . demands that reasonable and appropriate accommodations be made to promote access and attainment for all students" (p. 12). Following this introduction to the concept of equity, the authors cite the importance of access and opportunity to learn several more times in the text of the Principle (NCTM 2000). Thus, in an effort to be consistent with the view of equity reflected in *Principles and Standards*, I ground my discussion of equity in the issues of access and opportunity.

Equity and Learning: The View through Both Lenses

I teach the mathematics methods course for preservice elementary school teachers at my university. Each semester I begin the course with an activity related to the six Principles from *Principles and Standards* (NCTM 2000). I assign the students to groups, with each group responsible for a specific Principle.[1] After they discuss their assigned Principle and summarize what it means to them, they watch a video of a middle school classroom. As they view the video, one of their tasks is to look for evidence related to their Principle. They are to process what is happening in the lesson as a whole, but with a specific focus on the aspects of classroom interaction related to their Principle. Essentially, they watch the events of the classroom through a particular lens. Some are assigned to watch with

---

[1] I first experienced this activity in a professional development session led by Linda McQuillan of the Madison Metropolitan School District.

an "equity" lens and others with a "learning" lens. During the discussion that follows, the groups have the opportunity to share the view through their assigned lens and to note the interactions or connections between those perspectives. For example, both the equity group and the learning group might highlight the same classroom practice or action of the teacher. They are then able to see how the same practice can serve dual purposes. Although the students begin the activity focused only on the view through a single lens, to see the whole picture with clarity, they must overlay a second lens.

The need to combine lenses applies beyond that introductory activity to the examination of all aspects of mathematics education. The application of only one lens is problematic because the potential exists for the picture to become distorted. For example, we can view classrooms through only a learning lens. With that lens, we ask a particular set of questions: Are students learning mathematics with understanding? Are they connecting new knowledge with existing knowledge? Can they apply what they are learning in a variety of situations? If we view the same class through an equity lens, we ask a potentially different set of questions: Who has access to the learning that is occurring? Are all students able to participate in the learning process? Who has access to the resources that support learning? Our understanding of the situation is more complete when we begin to combine those questions to examine learning and equity in conjunction.

We have several examples in which the learning lens has been applied to situations in the absence of the equity lens. However, when the equity lens was superimposed, the picture became clearer. Members of the NCTM Research Committee (2005) cite the example of the research involving Cognitively Guided Instruction (CGI). The CGI studies provided evidence that teachers can use knowledge extracted from studies of learners' thinking to strategically influence students' learning (Carpenter et al. 1988; Carpenter et al. 1989). Thus, the picture, as seen through the learning lens, appeared in a very positive light. However, when the CGI researchers overlaid an equity lens and examined more closely the strategies of boys and girls, they found gender-related differences in the abstractness of the strategies used (Fennema et al. 1998). The application of the equity lens shifted the image to reveal a slightly different picture, at least with respect to the learning of the girls in those classes. Similarly, Secada (1996) applied an equity lens to a classroom that is often viewed as an exemplar in the use of student discourse to promote learning. Viewed through a learning lens, the picture of that classroom appeared one way. When the equity lens was applied, a different image emerged. Secada's application of the equity lens revealed that limited English proficiency was a potentially marginalizing factor in the classroom, excluding students from activities intended to support learning. Thus, situations in school mathematics should not be viewed exclusively through a learning lens. An equity lens must also be superimposed to clarify the picture with respect to the learning of *all* students.

Conversely, the importance of the learning lens is highlighted in an examination of achievement trends. Reviews by Secada (1992) and by Tate (1997) have noted increased

399

achievement by all student groups and a reduction in the achievement gaps on national assessments over the past two decades. Through an equity lens, that trend would appear to be positive. However, an overlay of the learning lens qualifies that progress. As both Secada and Tate note, the growth and reduction of the achievement gaps have occurred largely on basic skills assessment items. Viewed through a learning lens, success on basic skills is an insufficient indicator of understanding; thus the picture is not as rosy as it might initially have appeared. As we imagine those lenses, then, we must keep in mind that they cannot operate in isolation. The consistent application of both lenses is necessary not only for researchers but for all those involved in mathematics education.

Seeing the Whole Picture

Another characteristic of a critical equity lens, as I imagine it, is the necessity to apply that lens at different levels. In other words, the lens must have the capacity not only to zoom in but also to zoom out to consider broader policy-level issues and conditions. That capability is particularly important as we conceptualize equity with respect to learning.

Research related to students' learning in mathematics is often framed within a psychological paradigm and focused on individual students' cognition (Tate and Rousseau 2002). However, according to Secada (1991), a focus on the individual is problematic from an equity standpoint. "As the external and social worlds get transformed into the inner and personal worlds, external and social issues are transformed into internal and personal states" (p. 22). Secada (1993) argued, in fact, that the psychological focus on the individual fails to adequately explain the nature of student learning, in part because external factors that shape student learning fall outside the view of a lens focused strictly on the individual. Secada asserted that a full picture would require attention to larger social and cultural forces. To get a clear picture, we must shift the focus beyond the individual student to the larger spheres in which the student's learning is situated.

Classroom Practices

Although mathematics education has traditionally maintained cognitive focus on the individual, more recent scholarship has begun to attend to the social dimension of learning (e.g., Carpenter et al. 1999; Cobb and Yackel 1996). Work attending to the social dimension of learning has generally broadened the focus from the individual to include the classroom practices that influence students' learning. Attention to social processes is also embedded in the recommendations included in *Principles and Standards* (NCTM 2000). For example, the Learning Principle explicitly names classroom discourse and social interaction as means of promoting students' understanding. Thus, a learning lens, applied at the classroom level, would likely take into consideration the role of discourse in students' learning.

Yet we must also examine interactions at the classroom level through an equity lens. Secada and Berman (1999) suggest, for example, that the classroom norm of having students articulate their thinking is of potential significance from an equity standpoint.

They note that the expectations for communication can put some students at a disadvantage. Similarly, Cobb (2001) argues that "students' home communities can involve differing norms of participation, language, and communication, some of which might actually be in conflict with those that the teacher seeks to establish in the classroom" (p. 471).

Such a scenario prevailed in a study by Murrell (1994) of middle school mathematics classrooms. In his study, he focused on the mathematics learning experiences of African American males. He characterized the classes as implementing the kind of discourse, or "math talk," advocated in the NCTM (1989, 1991) *Standards* documents. Murrell found that those communication expectations were not benefiting African American males in the ways intended. He notes that the students did not interpret math talk in a manner consistent with the goals of the teacher. As teachers engaged in more math talk as a means of exploring and elaborating mathematics principles and concepts, the students did not regard the discussion as an increased focus on mathematical learning. Rather, they tended to regard the emphasis on math talk simply as a new regimen to be mastered to meet their teachers' requirements. Murrell goes on to suggest that this mismatch between the students' views and the teacher's intentions would likely have occurred not only for African American males but for any student who "has not been socialized into the mathematics discourse forms assumed by mainstream teachers" (p. 565). Lubienski (1996) found similar results in her study of the impact of classroom discussion on her students' learning experiences. In particular, Lubienski noted that the higher-SES (socioeconomic status) students in her class shared more of her expectations for discussion, seeming to share her beliefs about the purpose of discussion as an arena to "create, share, analyze, and validate mathematical ideas" (p. 185). The lower-SES students were more apt to focus on right and wrong answers and to limit their participation to those instances in which they could share answers rather than ideas.

Lubienski (1996) suggested that such differences between the teacher's intentions and students' perceptions are a manifestation of "cultural confusion." She asserted that the changes in the curriculum and classroom practices advocated by NCTM are cultural in nature. "They involve central beliefs and norms regarding ways of knowing and communicating" (p. 246). However, that culture is more closely aligned with that of the middle class. Zevenbergen (2000) makes a similar assertion, arguing that success in the mathematics classroom requires cultural knowledge that can serve to privilege the middle class or those who are in the linguistic majority. Thus, an equity lens reveals that classroom practices intended to promote understanding have the potential to maintain, rather than eliminate, learning differences.

An important point to note, however, is that the potentially negative impact of "cultural confusion" can be mitigated when teachers make explicit the cultural knowledge and expectations of the classroom and attend to any potential factors that might restrict student participation. For example, Boaler (2002) examined the classrooms of two different reform programs in which access to mathematics learning appeared to be equitably distributed. She found that teachers in those classrooms focused attention on teaching students how to explain and justify their thinking. Her findings suggest that such explicit attention to the norms of participation can ensure that all students have access to the mathematical learning promoted by those practices. Similarly, Gutierrez (2002a) presented evidence that language differences need not be a constraint in the learning of mathematics. She notes that teachers can discover how language influences the specific participation patterns and learning of individual students. Such attention to the impact of language can allow the teacher to develop strategies that better support students' learning of mathematics.

Secada (1991) has argued that a cognitive focus on student learning, in the absence of considerations of culture or language, can cast students' nonparticipation in the classroom in a negative light, characterizing the student as deficient. In other words, a failure to apply an equity lens can lead to a distorted picture of classroom learning experiences. However, with an equity lens, students' participation can be understood within the broader context of the cultural and linguistic norms that students bring to the classroom. Moreover, working with those larger influences in mind, teachers can create classrooms that promote access for all students (Boaler 2002).

Knowledge Policies

However, student learning is shaped by more than the interactions in the classroom. Thus, the equity lens must be able to take in a wide-angle picture that includes broader impacts on students' opportunity to learn with understanding. For example, even as we examine mathematics learning in classrooms, we must acknowledge that students' opportunities to learn with understanding are not equitably distributed across those classrooms. Some of those inequities relate to what Tate and Johnson (1999) refer to as "knowledge policies," which include "the folkways of schooling, regulations of policymakers, and practices of teachers that influence student opportunity to learn" (p. 222).

Tracking is one example of a knowledge policy with the potential to affect students' opportunity to learn. In fact, multiple studies have indicated that students' learning opportunities are constrained by tracking practices. According to Darling-Hammond (1997), "students in lower tracks receive a much less challenging curriculum that accounts for more of the disparities between what they learn and other students learn than does their entering ability level" (p. 127). In mathematics, research indicates that students in high-track classes have access to "high-status" knowledge (ideas and concepts), whereas students in low-track classes repeat the same basic computational skills year after year (Oakes 1985, 1995). Students in low-track classes are significantly more likely than those in high-track classes to spend time each week doing worksheet problems, and students in low-track classes are less likely to be asked to write their reasoning about solving a mathematics problem (Weiss 1994). Thus, through a learning lens, students in low-track classes do not have the opportunities to learn "significant mathematics with depth and understanding" (NCTM 2000).

Viewed through an equity lens, the picture becomes even bleaker, as we observe that minority and low-income students are more likely to experience such inequities in opportunities to learn. For example, Oakes (1995) examined the tracking policies of two school districts and demonstrated that the track placements in both districts were racially skewed. African American and Latino students were much less likely than white and Asian students with the same test scores to be placed in high-track classes. She noted two mechanisms that appear to support those disparities: (1) differences in parents' knowledge of their power to intervene in the placement process and (2) teacher and counselor recommendations that included both "objective" data and more subjective judgments about behavior, personality, and attitudes. She concluded from those findings that "grouping practices have created a cycle of restricted opportunities and diminished outcomes, and exacerbated the differences between African American and Latino and White students" (p. 689). As a result of those practices, minority students within racially mixed schools continue to be disproportionately overrepresented in low-track mathematics courses (Oakes et al. 2000). Thus, although tracking is a knowledge policy with the potential to constrain the learning opportunities of all students placed into low-track classes, the impact is felt more profoundly by certain student groups.

Another knowledge policy described by Tate and Johnson (1999) is teacher quality. According to Darling-Hammond (2000), "substantial evidence from prior reform efforts indicates that changes in course taking, curriculum content, testing or textbooks make little difference if teachers do not know how to use these tools well and how to diagnose their students' learning needs" (p. 37). Without teacher quality, curriculum and instructional strategies have little chance of being effectively implemented to support student learning with understanding. Several studies of mathematics reform have demonstrated the importance of teacher knowledge of both content and pedagogy for the effective use of strategies and materials associated with student learning with understanding (Borko et al. 1992; Clarke 1995; Clarke 1993; Eisenhart et al. 1993; Johnson 1995; Manouchehri 1998; Manouchehri and Goodman 1998; Stein, Baxter, and Leinhardt 1990). Moreover, the importance of teacher quality for students' learning is further demonstrated through studies of student achievement. For example, in an analysis of data from the National Assessment of Educational Progress, Darling-Hammond (2000) found that teacher certification status and holding a degree in the field being taught were significantly and positively correlated with students' achievement. In fact, "the most consistent highly significant predictor of student achievement in reading and mathematics in each year tested is the proportion of well-qualified teachers in a state; those with full certification and a major in the field they teach" (p. 27). Similarly, Fetler (1999) analyzed the mathematics scores of high school students in California. He found that after controlling for student poverty, teacher experience and preparation were significantly related to student achievement. In particular, schools with larger numbers of teachers on emergency permits had lower average achievement scores. Thus, when viewed through a learning lens, the qualifications and knowledge of the teacher are clearly important.

Moreover, the impact of the teacher on students' learning must also be considered in conjunction with the inequitable distribution of teacher quality. In fact, Darling-Hammond (1997)

403

argued that "perhaps the single greatest source of inequity in education is [the] disparity in the availability and distribution of well-qualified teachers" (p. 273). For example, Sanders and Rivers (1996) found that although African American and white students made comparable progress when assigned to teachers of comparable effectiveness, African American students were nearly twice as likely to be assigned to the most ineffective teachers. Similarly, although more than one-fourth of high school mathematics students are taught by out-of-field teachers, those proportions are highest in high-poverty schools and high-minority classes (Darling-Hammond 1997). In a national survey, Weiss (1994) found that more than half of high school mathematics classes with populations that are at least 40 percent minority are taught by teachers without a degree in the field. Ingersoll (1999) found similar results in high-poverty schools (schools in which 50 percent or more of the students receive free or reduced-price lunch). Whereas 27 percent of secondary mathematics teachers in low-poverty schools (schools in which 10 percent or fewer of the students receive free or reduced-price lunch) have neither a major nor a minor in the field, approximately 43 percent of teachers in high-poverty schools do not have at least the equivalent of a minor in the field. Certainly, such measures as certification status and degree in the field are but proxies for the knowledge and depth of understanding necessary to help students learn. However, the predictive power of those indicators with respect to student achievement and the disparities in their distribution highlight the significance of teacher quality to any discussion of equity and opportunity to learn in mathematics education. In fact, Darling-Hammond (1997) asserted that "much of the difference in school achievement among students is due to the effects of substantially different school opportunities and, in particular, greatly disparate access to high-quality teachers and teaching" (p. 270). Those differences in opportunity to learn are highlighted when an equity lens is applied to teacher quality.

Students' opportunities to learn mathematics are also shaped by the assessment policies of the school district, since assessment policies often influence the nature of content and pedagogy in the classroom. In particular, the influence of high-stakes standardized tests can lead teachers to curtail the amount of time given to the types of curriculum and instruction that are associated with learning with understanding (NCTM 2000). For example, Gutstein (1998) found that pressure to cover the material to be included on district-mandated standardized tests prevented middle school mathematics teachers from implementing the reform curriculum Mathematics in Context (MiC). "The district emphasis on these tests has a chilling effect on teachers' willingness to let go of the old (even if it did not work) and embrace the new (e.g., MiC)" (p. 22). Similar examples of the impact of standardized tests can be found throughout the literature on mathematics education (Ball 1990; Clarke 1994; Eisenhart et al. 1993; Knapp and Peterson 1995; Manouchehri 1998; Manouchehri and Goodman 1998; Putnam et al. 1992; Romberg 1997; Schifter and Fosnot 1993; Stein and Brown 1997; Webb and Romberg 1994; Wilson 1990). In general, the foregoing examples demonstrate that the emphasis of many standardized tests on computational skills makes teachers less likely to attempt to change their practices to focus more on conceptual understanding. Thus, a learning lens applied to assessment policies highlights the negative impact of high-stakes tests on what happens in the classroom.

An equity lens, however, reveals that the negative impact is felt more acutely in certain classrooms. The influence of high-stakes standardized tests is often greater in high-minority classrooms (Darling-Hammond 1994; Madaus 1994; Shepard 1991; Weiss 1994). For example, in a nationwide survey, teachers in high-minority classrooms reported using test-specific instructional practices more often than teachers in low-minority classrooms (Madaus et al. 1992). In high-minority classrooms, about 60 percent of the teachers reported teaching test-taking skills, teaching topics known to be on the test, increasing emphasis on tested topics, and starting test preparation more than a month before the examination. Those practices were reported significantly less often by teachers in low-minority classrooms. Moreover, mathematics teachers in high-minority classes indicated more pressure from school district officials to improve test scores than teachers in low-minority classes. The example of the impact of the Texas Assessment of Academic Skills (TAAS) also illustrates the effects that high-stakes tests can have on students' opportunity to learn. The results of surveys of Texas teachers indicate that they devote a substantial amount of instructional time to preparing students specifically for the TAAS, with most teachers reporting that they begin test preparation more than a month before the test (Haney 2000). According to McNeil and Valenzuela (2000), teachers report spending several hours a week drilling students on practice examinations. In that effort, commercial test-preparation materials become the de facto curriculum in many schools, reducing mathematics to sets of isolated skills. Teachers report that the time devoted to instructional activities that engage students in higher-order problem solving is severely reduced (or disappears completely) in the press to prepare students for the TAAS. Instructional focus on test preparation is unevenly distributed and more likely to exist in schools attended by low-income and minority students. Thus, an equity lens reveals the disproportionate impact of testing policies on the learning of certain student groups.

Conclusion

In this article, I have attempted to build on the argument made by the NCTM Research Committee (2005) of the need to apply a "critical equity lens" to mathematics education. I have used their metaphor as a means to highlight some of the important considerations with respect to learning and equity. I argue that realizing the vision outlined in *Principles and Standards* will require ongoing examination of mathematics education through the dual lenses of equity and learning. Moreover, I suggest that this examination must occur at the level of the classroom and at broader policy levels.

The overlay of an equity lens on the classroom reveals that classroom processes intended to promote students' learning with understanding do not necessarily contribute to the deeper learning of all students. However, other examples demonstrate that such differences in participation and learning can be overcome with greater attention to the factors that contribute to those differences (e.g., unfamiliarity with the expectations for communication).

405

As Secada (1996) noted, unless we attend to the impact of those practices on all students, we risk the possibility that our efforts intended to promote student learning will, in fact, work against equity. We must be aware of the potential for differential impact. With such awareness and sensitivity, we are better positioned to create classrooms that promote the learning of all students. This goal requires that we simultaneously attend to issues of learning and equity in the classroom.

In addition to a focus on the classroom, the equity lens must also have the capacity to "zoom out" from the classroom to consider other factors that shape students' opportunities to learn mathematics with understanding. For mathematics education researchers, that perspective requires a blurring of the lines that we have traditionally imposed between classroom research and policy. To have a full understanding of equity concerns, we must be able to take in the wide-angle picture that includes the knowledge policies that affect students' learning. The same is true of classroom teachers and others involved in the day-to-day operation of schools. We must be careful not to fall into the trap outlined by Secada (1991) of attributing students' learning strictly to conditions within the individual or the individual's family. Rather, we must consider the big picture to include the student's prior mathematics learning experiences and the influences of the different knowledge policies on those experiences.

Finally, equity is not a stand-alone consideration that can be examined in isolation. If we seek to fulfill the vision outlined in *Principles and Standards for School Mathematics* (NCTM 2000), we must examine equity in conjunction with the other facets of school mathematics. We must consider learning, teaching, curriculum, technology, and assessment with the equity lens as an overlay. The members of the NCTM Research Committee (2005) note that "equity concerns exist in any educational endeavor, whether or not they are made explicit" (p. 95). If we treat the equity lens as something to be used only in isolation, only by a few people, or only with respect to certain learning situations, the vision of school mathematics outlined in *Principles and Standards* will remain nothing more than a mirage.

REFERENCES

Ball, Deborah. "Reflections and Deflections of Policy: The Case of Carol Turner." *Educational Evaluation and Policy Analysis* 12, no. 3 (1990): 263–75.

Boaler, Jo. "Learning from Teaching: Exploring the Relationship between Reform Curriculum and Equity." *Journal for Research in Mathematics Education* 33 (July 2002): 239–58.

Borko, Hilda, Margaret Eisenhart, Catherine A. Brown, Robert G. Underhill, Doug Jones, and Patricia C. Agard. "Learning to Teach Hard Mathematics: Do Novice Teachers and Their Instructors Give Up Too Easily?" *Journal for Research in Mathematics Education* 23 (May 1992): 194–222.

Carpenter, Thomas P., Elizabeth Fennema, Penelope L. Peterson, and Deborah Carey. "Teachers' Pedagogical Content Knowledge of Students' Problem Solving in Elementary Arithmetic." *Journal for Research in Mathematics Education* 19 (November 1988): 385–401.

Carpenter, Thomas, Elizabeth Fennema, Penelope L. Peterson, Chi-Pang Chiang, and Megan Loef. "Using Knowledge of Children's Mathematics Thinking in Classroom Teaching: An Experimental Study." *American Educational Research Journal* 26 (November 1989): 499–531.

Carpenter, Thomas, Cristina Gomez, Celia Rousseau, Olof Steinthorsdottir, Carrie Valentine, Lesley Wagner, and Peter Wiles. "An Analysis of Student Construction of Ratio and Proportion Understanding." Paper presented at the Annual Meeting of the American Educational Research Association, Montreal, Que., 1999.

Clarke, Barbara. "Expecting the Unexpected: Critical Incidents in the Mathematics Classroom." Unpublished doctoral dissertation, University of Wisconsin–Madison, 1995.

Clarke, David. "Influences on the Changing Role of the Mathematics Teacher." Unpublished doctoral dissertation, University of Wisconsin–Madison, 1993.

_____. "Ten Key Principles from Research for the Professional Development of Mathematics Teachers." In *Professional Development for Teachers of Mathematics*, 1994 Yearbook, edited by Douglas B. Aichele, pp. 37–48. Reston, Va.: National Council of Teachers of Mathematics, 1994.

Cobb, Paul. "Supporting the Improvement of Learning and Teaching in Social and Institutional Context." In *Cognition and Instruction: Twenty-five Years of Progress,* edited by Sharon M. Carver and David Klahr. Mahwah, N.J.: Lawrence Erlbaum Associates, 2001.

Cobb, Paul, and Erna Yackel. "Constructivist, Emergent, and Sociocultural Perspectives in the Context of Developmental Research." *Educational Psychologist* 31 (1996): 175–90.

Darling-Hammond, Linda. "Performance-Based Assessment and Educational Equity." *Harvard Educational Review* 64, no. 1 (1994): 5–30.

_____. *The Right to Learn: A Blueprint for Creating Schools That Work.* San Francisco: Jossey-Bass, 1997.

_____. "Teacher Quality and Student Achievement: A Review of State Policy Evidence." *Education Policy Analysis Archives* 8, no. 1 (2000): 1–48.

Eisenhart, Margaret, Hilda Borko, Robert Underhill, Catherine Brown, Doug Jones, and Patricia Agard. "Conceptual Knowledge Falls through the Cracks: Complexities of Learning to Teach Mathematics for Understanding." *Journal for Research in Mathematics Education* 24 (January 1993): 8–40.

Fennema, Elizabeth, Thomas Carpenter, Victoria Jacobs, Megan Franke, and Linda Levi. "New Perspectives on Gender Differences in Mathematics: A Reprise." *Educational Researcher* 27, no. 5 (1998): 19–21.

Fetler, Marie "High School Staff Characteristics and Mathematics Test Results." *Education Policy Analysis Archives* 79, no. 9 (1999): 1–22.

Gutierrez, Rochelle. "Beyond Essentialism: The Complexity of Language in Teaching Mathematics to Latina/o Students." *American Educational Research Journal* 39 (2002a): 1047–88.

_____. "Enabling the Practice of Mathematics Teachers in Context: Toward a New Equity Research Agenda." *Mathematical Thinking and Learning* 4, no. 2–3 (2002b): 145–87.

Gutstein, Eric. "Lessons from Adopting and Adapting Mathematics in Context, a Standards-Based Mathematics Curriculum, in an Urban, Latino, Bilingual Middle School." Paper presented at the Annual Meeting of the American Educational Research Association, San Diego, Calif., 1998.

Haney, Walt. "The Myth of the Texas Miracle in Education." *Education Policy Analysis Archives* 8, no. 41 (2000).

Ingersoll, Richard. "The Problem of Underqualified Teachers in American Secondary Schools." *Educational Researcher* 28, no. 2 (1999): 26–37.

Johnson, Loren. "Extending the National Council of Teachers of Mathematics' Recognizing and Recording Reform in Mathematics Education: Documentation Project through Cross-Case Analysis." Unpublished doctoral dissertation, University of New Hampshire, 1995.

Knapp, Nancy, and Penelope Peterson. "Teachers' Interpretations of 'CGI' after Four Years: Meanings and Practices." *Journal for Research in Mathematics Education* 26 (January 1995): 40–65.

Lubienski, Sarah. "Mathematics for All? Examining Issues of Class in Mathematics Teaching and Learning." Unpublished doctoral dissertation, Michigan State University, 1996.

Madaus, George. "A Technological and Historical Consideration of Equity Issues Associated with Proposals to Change the Nation's Testing Policy." *Harvard Educational Review* 64, no. 1 (1994): 76–95.

Madaus, George, Mary West, Maryellen Harmon, Richard Lomax, and Katherine Viator. *The Influence of Testing on Teaching Math and Science in Grades 4–12*. Boston: Boston College, Center for the Study of Teaching, Evaluation, and Educational Policy, 1992.

Manouchehri, Azita. "Mathematics Curriculum Reform and Teachers: What Are the Dilemmas?" *Journal of Teacher Education* 49 (1998): 276–86.

Manouchehri, Azita, and Terry Goodman. "Mathematics Curriculum Reform and Teachers: Understanding the Connections." *Journal of Educational Research* 92, no. 1 (1998): 27–41.

McNeil, Linda, and Angela Valenzuela. "The Harmful Impact of the TAAS System of Testing in Texas: Beneath the Accountability Rhetoric." Cambridge, Mass.: Harvard Civil Rights Project, 2000.

Murrell, Peter. "In Search of Responsive Teaching for African American Males: An Investigation of Students' Experiences of Middle School Mathematics Curriculum." *Journal of Negro Education* 63, no. 4 (1994): 556–69.

National Council of Teachers of Mathematics (NCTM). *Curriculum and Evaluation Standards for School Mathematics*. Reston, Va.: NCTM, 1989.

———. *Professional Standards for Teaching Mathematics*. Reston, Va.: NCTM, 1991.

———. *Principles and Standards for School Mathematics*. Reston, Va.: NCTM, 2000.

NCTM Research Committee. "Equity in School Mathematics Education: How Can Research Contribute?" *Journal for Research in Mathematics Education* 36 (March 2005): 92–100.

Oakes, Jeannie. *Keeping Track: How Schools Structure Inequality*. New Haven, Conn.: Yale University Press, 1985.

———. "Two Cities' Tracking and Within-School Segregation." *Teachers College Record* 96 (1995): 681–90.

Oakes, Jeannie, Kate Muir, and Rebecca Joseph. "Coursetaking and Achievement in Mathematics and Science: Inequalities That Endure and Change." Madison, Wis.: National Institute of Science Education, 2000.

Putnam, Ralph, Ruth Heaton, Richard Prawat, and Janine Remillard. "Teaching Mathematics for Understanding: Discussing Case Studies of Four Fifth-Grade Teachers." *Elementary School Journal* 93 (1992): 213–28.

Romberg, Thomas. "Mathematics in Context: Impact on Teachers." In *Mathematics Teachers in Transition*, edited by Elizabeth Fennema and Barbara Scott Nelson, pp. 357–80. Mahwah, N.J.: Lawrence Erlbaum Associates, 1997.

Sanders, William, and June Rivers. *Cumulative and Residual Effects of Teachers on Future Student Academic Achievement*. Knoxville, Tenn.: University of Tennessee Value-Added Research and Assessment Center, 1996.

Schifter, Deborah, and Catherine Fosnot. *Reconstructing Mathematics Education: Stories of Teachers Meeting the Challenge of Reform*. New York: Teachers College Press, 1993.

Secada, Walter. "Educational Equity versus Equality of Education: An Alternative Conception." In *Equity and Education*, edited by Walter Secada, pp. 68–88. New York: Falmer Press, 1989.

_____. "Diversity, Equity, and Cognitivist Research." In *Integrating Research on Teaching and Learning Mathematics*, edited by Elizabeth Fennema, Thomas Carpenter, and Susan Lamon, pp. 17–53. Albany, N.Y.: State University of New York Press, 1991.

_____. "Race, Ethnicity, Social Class, Language, and Achievement in Mathematics." In *Handbook for Research in Mathematics Teaching and Learning*, edited by Douglas Grouws, pp. 623–60. New York: Macmillan Publishing Co., 1992.

_____. "Equity and a Social Psychology of Mathematics Education." Paper presented at the Annual Meeting of the International Group for the Psychology of Mathematics Education, North American Chapter, San Jose, Calif., 1993.

_____. "Urban Students Acquiring English and Learning Mathematics in the Context of Reform." *Urban Education* 30 (1996): 422–48.

Secada, Walter, and Patricia Berman. "Equity as a Value-Added Dimension in Teaching for Understanding in School Mathematics." In *Mathematics Classrooms That Promote Understanding*, edited by Elizabeth Fennema and Thomas Romberg, pp. 33–42. Mahwah, N.J.: Lawrence Erlbaum Associates, 1999.

Shepard, Lorrie. "Will National Tests Improve Student Learning?" *Phi Delta Kappan* 73, no. 3 (1991): 232–38.

Stein, Margaret, Juliet Baxter, and Gaea Leinhardt. "Subject-Matter Knowledge and Elementary Instruction: A Case from Functions and Graphing." *American Educational Research Journal* 27 (1990): 639–63.

Stein, Margaret, and Catherine Brown. "Teacher Learning in a Social Context: Integrating Collaborative and Institutional Processes with the Study of Teacher Change." In *Mathematics Teachers in Transition*, edited by Elizabeth Fennema and Barbara Scott Nelson, pp. 155–92. Mahwah, N.J.: Lawrence Erlbaum Associates, 1997.

Tate, William F., "Race-Ethnicity, SES, Gender, and Language Proficiency Trends in Mathematics: An Update." *Journal for Research in Mathematics Education* 28 (December 1997): 652–79.

_____. "Access and Opportunities to Learn Are Not Accidents: Engineering Mathematics Progress in Your School." Southeast Eisenhower Regional Consortium for Mathematics and Science at SERVE (University of North Carolina at Greensboro), 2005. (Available at www.serve.org)

Tate, William, and Howard Johnson. "Mathematics Reasoning and Educational Policy: Moving beyond the Politics of Dead Language." In *Developing Mathematical Reasoning in Grades K–12,* 1999 Yearbook of the National Council of Teachers of Mathematics (NCTM), edited by Lee V. Stiff, pp. 221–33. Reston, Va.: NCTM, 1999.

Tate, William, and Celia Rousseau. "Access and Opportunity: The Political and Social Context of Mathematics Education." In *Handbook of International Research in Mathematics Education,* edited by Lyn English, pp. 271–99. Mahwah, N.J.: Lawrence Erlbaum Associates, 2002.

Webb, Norman, and Thomas Romberg. *Reforming Mathematics Education in America's Cities: The Urban Mathematics Collaborative Project.* New York: Teachers College Press, 1994.

Weiss, Iris. "A Profile of Science and Mathematics Education in the United States: 1993." Chapel Hill, N.C.: Horizon Research, 1994.

Wilson, Suzanne. "A Conflict of Interests: The Case of Mark Black." *Evaluation and Policy Analysis* 12 (1990): 309–26.

Zevenbergen, Robyn. "'Cracking the Code' of Mathematics Classrooms: School Success as a Function of Linguistic, Social, and Cultural Background." In *Multiple Perspectives on Mathematics Teaching and Learning,* edited by Jo Boaler, pp. 201–24. Westport, Conn.: Ablex Publishing Corp., 2000.